Coal and Coke in Pennsylvania

Carmen DiCiccio

Commonwealth of Pennsylvania
Pennsylvania Historical and Museum Commission
Harrisburg, 1996

**THE PENNSYLVANIA
HISTORICAL AND MUSEUM
COMMISSION**

Tom Ridge
GOVERNOR

Timothy Buchanan
CHAIRMAN

COMMISSIONERS

William A. Cornell, *Vice Chairman*
James M. Adovasio
Thomas C. Corrigan, *Representative*
John A. Herbst
Edwin G. Holl, *Senator*
Janet S. Klein
Nancy D. Kolb
John W. Lawrence, M.D.
Stephen R. Maitland, *Representative*
George A. Nichols
LeRoy Patrick
Allyson Y. Schwartz, *Senator*
Eugene W. Hickok, *ex officio*
Secretary of Education

Brent D. Glass
EXECUTIVE DIRECTOR

©1996
Commonwealth of Pennsylvania
ISBN 0-89271-072-1

Table of Contents

Introduction 1

The Emergence of Coal in the Age of Wood, 1740-1840 4

Transportation, Iron and Railroad Revolution as Impetus for
   Expansion of the Coal and Coke Industries, 1840-1880 31

The Golden Age of King Coal, Queen Coke, and Princess Steel,
   1880-1920 61

Retrenchment, Decline, and the Mechanized Mine, 1920-1945 .. 150

Bibliography 209

Index 218

This publication has been financed entirely with Federal funds from the National Park Service, Department of the Interior. However, the contents and opinions do not necessarily reflect the views or policies of the Department of the Interior, nor does the mention of trade names or commercial products constitute endorsement or recommendation by the Department of the Interior. Under Title VI of the Civil Rights Act of 1964 and Section 504 of the Rehabilitation Act of 1973, the U.S. Department of the Interior prohibits discrimination on the basis of race, color, national origin, or handicap in its federally assisted programs. If you believe you have been discriminated against, or if you desire further information, please write to: Office for Equal Opportunity, U.S. Department of the Interior, Washington, DC 20240.

Acknowledgments

In completing this book, I have incurred many debts which I am happy to acknowledge. Specific thanks are due to many people without whose help and encouragement this volume would not have been completed. They have helped, advised, prodded, and supported the author in the completion of this study of the coal and coke industry of Pennsylvania.

A special debt of gratitude is owed to the Pennsylvania Historical and Museum Commission (PHMC) and America's Industrial Heritage Project for financial support that enabled me to complete the study, and to the Commission for making this publication possible. I was fortunate to receive sound and useful advice in my research from the following members of the PHMC's Bureau for Historic Preservation: Dan Deibler, Gregory Ramsey, William Sisson and Ira Smith. All read the manuscript with great care and made valuable suggestions.

I am indebted to the many retired coal miners and their families of the former company towns that field-visited in the course of this study. These wonderful people were gracious and supportive, providing me invaluable maps, photographs and other written materials in their possession. Oral interviews conducted during these frequent field visits provided a clearer picture of daily life in these isolated coal communities and of work underground.

I am grateful to the staffs of the Historical Society of Western Pennsylvania and the State Archives of Pennsylvania, to the libraries of the University of Pittsburgh, and Indiana University of Pennsylvania, and to the Carnegie Library of Pittsburgh and the Cambria County Library of Johnstown for their competent and professional assistance in providing me with pertinent materials from their collections. Anthony Grazini, geologist with the USX Corporation, generously supplied materials and many invaluable photographs from the H. C. Frick Coke Company archives in Fayette County that were used to illustrate the text.

This book has benefited enormously from the interest, encouragement and criticism of all these fine people. To all, I extend my sincere gratitude and appreciation. Of course, I alone am responsible for any errors of fact and interpretation found in the text.

Carmen DiCiccio

Introduction

"A diamond is a chunk of coal made good under pressure." —Anonymous

This book began as a coal and coke context written as a working guide for the nomination of coal and coke extractive facilities of western Pennsylvania to the National Register of Historic Places. The survey, begun in September 1991 and completed in June 1993, was prepared for the Bureau for Historic Preservation of the Pennsylvania Historical and Museum Commission (PHMC), with funding from the America's Industry Heritage Project (AIHP). A nine-county regional project, the AIHP was funded by Congress in 1988 to preserve, promote, and interpret the heritage of the coal, iron and steel, and transportation industries of Bedford, Blair, Cambria, Fayette, Fulton, Huntingdon, Indiana, Somerset, and Westmoreland Counties as it relates to the industrial development of the region and the nation. Congress created and funded the Southwestern Pennsylvania Heritage Preservation Commission to direct the AIHP.

In the first phase of this multiple-property documentation, a comprehensive coal and coke bibliography, *Bituminous Coal and Coke Resources of Pennsylvania, 1740-1945,* was compiled for the writing of the historic context. Diverse sources were consulted, including state and federal documents and maps, AIHP reports, both published and unpublished, PHMC reports, dissertations and masters theses, and general and scholarly books on the industry. Information gained from on-site investigations and previous surveys was also incorporated into the context. Informants at individual sites provided written and oral history essential to understanding the community and the extractive facility. Retired miners identified the extant buildings and structures at the abandoned mines and archaeological remains. Long-time residents provided maps, photographs, and written materials which permitted an accurate picture of how the community and the extractive facilities were originally laid out, and their evolution over time. These sources were used to generate a list of extant coal and coke resources for on-site visits. The physical remains of these abandoned extractive sites varied from archaeological remains to pristine, though abandoned, facilities that included all the buildings and structures required to mine coal, maintain mining, and transport the fuel to market.

Two principal property types associated with the coal and coke industry were delineated in the context: extractive facility and mining community, including residential and nonresidential buildings. A list of potential resources for future nomination to the National Register of Historic Places was created by using the following criteria to select coal and coke resources, over fifty years old: (1) integrity of the site, (2) historical significance, (3) architectural significance, (4) technological significance, (5) association with prominent person(s), (6) significance in labor history, and (7) significance in ethnic and/or immigration history.

The North American Review in 1836 called Pennsylvania "the Key-Stone State not solely by reason of its geographical position and its magnitude but on account of its natural resources also." "The Keystone Arch" was the phrase used by Governor Bigler (1852-1855) when describing the state's abundant mineral wealth. Rich and extensive deposits of iron ore, limestone, natural gas, oil and coal underlay the Commonwealth. This immense storehouse of natural resources made Pennsylvania one of the principal mineral-producing states of the nineteenth and twentieth centuries. The exploitation of these natural resources was a driving force in the development of the state's extensive industrial economy.

The coal and coke industries played a critical role in the industrial development of the state and the nation. Vast coal deposits underlying much of Pennsylvania made it the pioneer state in the establishment of the coal industry of the United

States. Coal deposits were discovered in Pennsylvania by fur trappers, explorers, and settlers before the Revolutionary War. Pennsylvania became the nation's coal bin, producing more than half the nation's coal until the eve of the Civil War. Coal was the state's principal mineral resource, with reserves estimated at over 107 billion tons (84 billion tons of bituminous coal and 23 billion tons of anthracite and semianthracite coal). It was the leading commercial producer of anthracite (hard) coal; its output peaked at 100,445,299 net tons in 1917, employing a labor force of 156,148 men.

Bituminous (soft) coal deposits underlay some 14,200 square miles of Pennsylvania, approximately one-third the state's area. These vast coal deposits are found in a number of irregular-shaped fields in the western half of the state, and in a few scattered coal fields in the northern part of the state bordering New York. Coal has been mined continuously in the counties of southwestern Pennsylvania bordering the Monongahela River since its discovery during the second half of the eighteenth century. As demand increased after 1840, the industry developed in the neighboring counties of western Pennsylvania. At its peak, coal was mined in twenty-nine western counties, most of it concentrated in the six southwestern counties of Allegheny, Cambria, Fayette, Greene, Washington, and Westmoreland. Coke was manufactured in western Pennsylvania beginning in the 1830s.

The state's bituminous coal regions include the following principal coalfields: (1) Main Bituminous Field—Allegheny, Armstrong, Beaver, Blair, Butler, Cambria, Cameron, Center, Clarion, Clearfield, Clinton, Elk, Fayette, Greene, Indiana, Jefferson, Lawrence, McKean, Mercer, Somerset, Venango, Washington, and Westmoreland Counties; (2) Broad Top Field—Bedford, Fulton, and Huntingdon Counties; (3) North-Central Fields (five small fields)—Bradford, Lycoming and Tioga Counties; (4) Georges Creek Field—Somerset County. Bituminous coal production surpassed that of anthracite coal during the 1890s and became the Commonwealth's most important mineral resource. Pennsylvania, the leading bituminous coal-producing state in the union, produced more than 30 percent of all the coal mined in the United States from 1840 until 1930, when West Virginia permanently surpassed its annual coal production. Over 18 billion tons of bituminous coal and 10.75 billion tons of anthracite had been mined in the state by 1970.

Coal is Pennsylvania's greatest mineral resource not only for its fuel value, but because its production made possible the growth of other essential industries in the past and today. The abundance and accessibility of high-grade coal and, later, high-quality metallurgical coke spurred production of the iron and steel, zinc, salt, and glass industries. The industry provided jobs for tens of thousands. Foreign-born workers immigrated from Europe to Pennsylvania in search of employment in the booming coal industry, adding their diverse ethnic, linguistic, social, and religious customs to the social fabric of the Commonwealth. The economy of entire regions within the state was dominated by the coal industry, defining their economic development by providing employment and housing for a majority of their inhabitants. Some of the towns created to house miners and their families are still visible on today's landscape and represent the physical legacy of this once vibrant and significant extractive industry.

This book is a survey of the bituminous coal industry of Pennsylvania from its colonial origins to the conclusion of the Second World War. It is not intended as a definitive text; instead, it is an introductory guide to this significant and fascinating industry within an extended time frame. There is a plethora of studies on the industry, but most are specialized and topical works. There are histories of mining technology, histories of social life in the company-owned towns, institutional histories of miners' unions and related biographies, histories of particular coal companies and coal towns, and histories of the industry in a particular, well-defined, and brief time period. Surprisingly, there are few narrative or chronological histories of the state's bituminous coal industry tracing the development of the industry from its colonial origins in the Pittsburgh region through the following two centuries. The story of an industry as diverse and significant as the coal and coke industry of Pennsylvania is naturally a complex one, and would be an undertaking of enormous proportions to prepare and publish.

The growth of the state's bituminous coal and coke industry is divided into four distinct historical periods: The Emergence of Coal in the Age of Wood, 1740-1840; Transportation, Iron, and Railroad as Impetus for the Expansion of the Coal and Coke Industry, 1840-1880; The Golden Era of King Coal, Queen Coke, and Princess Steel, 1880-1920; and Retrenchment, Decline, and the Mechanized Mine, 1920-1945. I have chosen to highlight the development of the industry by examining the following themes in each historical period: technology employed in the mining process (technology is defined broadly to include mechanical, engineering, and managerial activities and changes in the bituminous coal industry), the commercial uses of coal and its valuable by-products (coke and chemical by-products of coal), the transportation system employed to distribute coal and coke to market, the expansion and development of new industries within the coal resource marketplace, and the changing social composition of the industry's labor force and the daily lives of miners and their families in a dangerous and often repressive industry. The exact dating of these individual periods is not definitive because the principal themes that characterize each period develop gradually and incrementally and often do not conform to any model. Nevertheless, each of the four chronological periods identified is distinctive and is based upon a series of recurring themes in the industry.

The Emergence of Coal in the Age of Wood, 1740-1840

Introduction

Bituminous coal was mined on a small scale in a number of locations throughout the counties of western and central Pennsylvania for nearly a century prior to 1840. A limited commercial coal market developed but the trade was confined to a few localities within the Commonwealth.

Before 1840 the use of bituminous coal was restricted to the blacksmith's forge, steam engines, home heating, the glass and salt industries, and rarely, the smelting of iron ore in the blast furnace. The expansion of the bituminous coal industry of Pennsylvania was restricted during its first century by a sparse population; a primitive transportation system, generally limited to navigable rivers usually running north and south; limited retail markets; and an abundant supply of wood. Since coal was a bulky commodity, it could only at this time be transported by water. Furthermore, its market share was restricted by the dominance of wood. There was no energy crisis in the nation during this first century. The country was sparsely populated and wood was abundant, inexpensive, and near at hand, so coal use was limited to a few locations. Wood had been the principal energy source of the United States since the colonial era and as late as 1850 it provided 90.7 percent of the nation's energy. A number of requisites had to be met if the coal industry was to develop a broader commercial trade; all were essential and a failure in any one of them would have been detrimental to the expansion of the coal trade: (1) a good quality of coal, (2) a sufficient quantity, (3) cheapness and regularity of production, (4) cheapness of transportation, (5) a sufficiency of transportation, and (6) a good market.[1]

The Geology and Chemical Composition of Coal

Coal, which has been called "Nature's Black Diamond," is a black or brownish-black, combustible sedimentary organic rock, containing more than 50 percent carbonaceous material by weight. Coal is simply a rock that burns, and compared with other rocks is relatively light, weighing about eighty pounds per cubic foot, about one-half the weight of most other rocks. Coal is formed by the slow alteration of decaying plant life, buried million of years under water and without air. Heat and pressure also contribute to this alteration. Geologists regard coal as a rock because rocks are defined as all-natural solid substances, organic or inorganic, that compose the earth's crust. It is considered a mineral rock by geologists, and is known as mineral coal in trade, industrial, and legal affairs. Coal is a mineralized vegetative material deposited over a long period of time, its chemical composition modified by the effects of time, heat, and pressure. Coal, like natural gas and oil, although formed exclusively from decaying plant vegetation, is a fossil fuel. Three factors affect the formation of coal deposits: (1) initiation, maintenance, and repetition of environments that favor large-scale accumulation and preservation of vegetal sediment; (2) conditions within this depositional environment that favor biological degradation and alteration of the vegetal sediment to peat; and (3) geochemical processes that induce chemical coalification of the peat to higher-rank coal.[2]

Ancient forests were the birthplace of coalfields of Pennsylvania.

The Carboniferous period, some 225 to 350 million years ago, was the great coal-making age. The earth was warmer and more humid than today, with ferns, mosses, and tropical plants growing profusely in large swamps. When these plants died they fell into mud and water and were preserved from complete decay by the water, a series of vegetative layers developed. Moisture and bacteria converted these deposits of partially decayed organic material into peat, which is a spongy substance. Peat bogs were buried under sediment and compressed by the heat and pressure of the earth. Peat is not coal, but represents the initial stage in the development of coal, and under favorable geological conditions may give rise to coal seams. Peat is a dark-brown or black residue produced by the partial decomposition and disintegration of mosses, sedges, trees, and other plants growing in marshes. It is composed principally of carbon, hydrogen, and oxygen in varying proportion. Because of its high carbon content, peat will ignite and burn freely when dry. It had long been burned for energy in Europe, but was never used as a commercial fuel in the United States; instead, it was used almost entirely for soil improvement and fertilizer. Harvesting peat and draining bogs for agriculture have carved up the world's 2 million square miles of peat bogs, including some 90 million acres left in the United States.

Coal is a complex material composed chiefly of carbon, hydrogen, and oxygen. There are smaller amounts of sulfur and nitrogen, and small traces of elements ranging from aluminum to zirconium. "Coalification" is the technical term used to describe the formation of coal. The process is advanced largely by pressure and heat from the earth's core. Mud, sand, and debris seal the peat from air, which prevents its further decomposition. Pressure begins as soon as this material is buried under layers of sand and mud, and as these layers of sediment grow deeper their weight compacts them to a fraction of their thickness and drives out existing moisture and gaseous compounds, including oxygen, nitrogen, and hydrogen. The effect of this pressure is greater on deposits buried deeper in the earth. Pressure and high temperature release volatile matter or combustible gases, such as carbon monoxide, carbon dioxide, and methane (CH_4 or marsh gas) in large quantities leaving behind carbonaceous deposits called coal. Coal occurs in seams, sometimes called "beds" or "veins," that are interlaced or sandwiched between several layers of shale, sandstone, clay, and sometimes limestone.

Coal is not a homogeneous mineral rock; instead, its chemical composition varies within the same seam and even from coal mined at neighboring mines. Some underground mines extract more than one seam of coal simultaneously. Coal seams vary greatly in their thickness, from less than an inch to over fifty or more feet; in area, from a few acres to thousands of square miles; and in depth, from a few feet below the surface to several hundred feet underground. Geologists estimate that about three hundred years is required to deposit sufficient organic vegetation to form one foot of bituminous coal from eight to twelve feet of compressed vegetable matter. Most coal seams are horizontal; some coal beds are inclined, folded, or faulted, the result of geological forces. Commercial coal seams vary from two to twelve feet in thickness, although a few seams are nearly a hundred feet thick.

Several methods of classifying coal have been developed by geologists, using "chemical analyses" and physical tests that measure the progressive response of coal to pressure and heat. Coal is graded according to its size, appearance, weight, structure, cleanliness, heat value, and burning characteristics. The rank of coal pertains to the degree of metamorphism or geological change through which it has passed from the time of its original deposit as peat to the present. All coal is divided into the following ranks: anthracite, semianthracite, low-volatile bituminous, medium-volatile bituminous, high-volatile bituminous, subbituminous, and lignite.[3]

The rank of coal describes its physical qualities. All coal contains varying percentages of combustible matter, divided into volatile matter and fixed carbon, ash, moisture, and sulfur. The chemical composition and percentages of these elements are the chief factors in determining both the economic value and the usage of coal.

The chemical analysis of coal involves the determination of four principal constituents found in all ranks of coal: water, called moisture; mineral impurity, called ash-inert material left when coal is completely burned; volatile matter, consisting of gases and vapor expelled when coal is heated; and fixed carbon, the solid or carbon residue that burns at a higher temperature after the volatile matter has been expelled as gases. Volatile matter and fixed carbon are the principal ingredients in coal that produce heat when it is burned, while moisture and ash are inert ingredients which hinder the process of heat emission when coal is burned.

The Ranks of Coal			
Ranks of Coal	Fixed Carbon	Volatile Matter	Moisture
Lignite	28.7	25.8	45.5
Subbituminous	42.4	34.2	23.4
Low-rank bituminous	47.0	41.4	11.6
Medium-rank bituminous	54.2	40.8	5
High-rank bituminous	64.6	32.2	3.2
Low-rank semibituminous	75.0	22.0	3.0
Semianthracite	85.8	11.7	2.5
Anthracite	95.6	1.2	3.2 [4]

Carbon is the solid combustible matter of coal and with ash is what remains of coal after its volatile or gaseous matter has been driven off as smoke. The fixed carbon content of Pennsylvania ranges from 55 percent in western Pennsylvania to as high as 98 percent in the anthracite region of northeastern Pennsylvania, found in what is called superanthracite.

Volatile matter is determined by chemists by heating coal in a platinum crucible in an electric furnace at 950 °F for six minutes without contact with air. The loss of weight less moisture gives the percentage of its volatile matter. Volatile matter does not represent a single compound but contains hydrogen-, nitrogen-, and oxygen-gas compounds and some noncombustible matter, including traces of ammonia which are driven off as gases, tars, and oils when coal is heated and deprived of oxygen. The energy value of coal when burned is measured in Btus (British thermal units) and depends in large measure upon the relative percentage of fixed carbon and the volatile matter it contains. A Btu is the quantity of heat necessary to raise the temperature of one pound of water by 1 °F and is a convenient measurement to compare the energy content of various fuels. The number of Btus in a ton of coal is a rough measure of the heating properties of that coal, and varies from 7,000 °F in lignite to 13,850 °F in high-volatile bituminous coal in the Pittsburgh District, and to 14,350 °F in low-volatile coal found in the Broad Top coalfield. The higher the Btu rating in coal, the more heat it will produce when burned. Low-volatile coal (under 20 percent volatile matter) yields more Btus per pound than medium-volatile coal (27.5 to 35 percent volatile matter), while high-volatile coal (35 to 42 percent volatile matter) yields the lowest Btus. Low-volatile coal with high Btus in Pennsylvania is found in the Broad Top field, in southeastern Somerset and southern Cambria Counties. High-volatile coal, with lower Btus, is found in Clarion, Butler, and Lawrence Counties.

Ash and sulfur are the chief impurities that reduce the quality of coal. Ash is the inorganic matter or incombustible residue remaining after the combustibles of coal have been burned. It consists of silica, alumina, lime, and bisulphide of iron, along with smaller quantities of magnesia and alkalis. Some of these substances are combined with sulfuric acid as sulfates. A high ash content is objectionable in coal because it inhibits burning. The ash found in coal after combustion is caused by some inferiority during the pre-coal-making process by infiltration of foreign matter. Ash content ranges from as little as 2 percent in anthracite coal to 5 to 10 percent in medium- and high-rank bituminous coal and as high as 80 percent in lignite. The average ash content of coal from the famous Pittsburgh coal seam runs from 3 to 5 percent.

Sulfur is the most important of all impurities in coal today because on combustion it is converted mostly into sulfur dioxide, a colorless, extremely irritating gas or liquid that is a dangerous water and air pollutant. Sulfur is found in several forms in coal, as iron sulfides (pyrites and marcasites), as sulfate of lime or alumina, in an organic form combined with carbon and hydrogen, and in rare cases as free organic sulfur chemically bonded to the coal-forming plant material. The fumes from burning high-sulfur coal are extremely corrosive and a principal source of air pollution. Sulfur combines in the air chemically with oxygen and water to form acid rain that can destroy lakes and vegetation. The percentage of sulfur in coal is usually expressed separately because it is present in volatile matter, in fixed carbon, and in ash. A low sulfur percentage in coal is desirable for illuminating gas, for smithing purposes, for making metallurgical coke, and in the manufacture of pottery.

Two kinds of moisture are found in coal. The first is surface moisture produced by coal dampened by mine water, or by rain, or by being washed to alter its chemical composition at a preparation plant. The second type of moisture is the inherent water found within the coal. The former disappears when coal is exposed to dry atmosphere, but the latter absorbs a definite part of the energy in the coal when it is burned. Coal with moisture content of 10 percent would indicate two hundred pounds of free water present in a ton of coal; to disperse it would involve burning thirty-three pounds of average coal, making a total loss of 233 pounds of coal. High moisture content of coal decreases the heating value of the coal. Indiana, Illinois, and western Kentucky coal has somewhat higher moisture content than Pennsylvania coal.

Location of the Coalfields of the United States

The United States has the largest deposits of coal in the world and because of its extensive coal reserves has been appropriately called the "Saudi Arabia of coal." Coal is present in thirty-eight states, underlying a total of 458,600 square miles or 13 percent of the land area of the United States. Coal is by far the nation's most abundant fossil fuel, with resources estimated at 1.7 trillion tons, including more than four hundred billion tons which can be mined with known methods and existing technology. Coal provides the nation with 80 percent of its known fuel reserves. The original coal reserves of the United States were made up of 29 percent lignite, 28 percent subbituminous, 42 percent bituminous and semibituminous, and less than 1 percent anthracite/semianthracite.[5] These extensive coal deposits have been divided into seven major regions: Anthracite Region, Appalachian or Eastern Region, Middle Western Region, Western Region, Southwestern Region, Rocky Mountain Region, and Pacific Coast Region. Each of the principal coal regions has been divided into numerous subdivisions reflecting local variations in the quality of the coal and the thickness of the seams. The Appalachian Region covers an area of about 55,076 square miles, is a little over nine hundred miles in length, and ranges in width from 30 to 180 miles. It is located in nine states, including "western Pennsylvania, and parts of the southeastern part of Ohio, the western part of Maryland, the southwestern corner of Virginia, nearly all of West Virginia, the eastern part of Kentucky, portions of eastern Tennessee, the northwestern corner of Georgia, and nearly all of northern Alabama." This coal region accounted for about three-fourths of total annual production as recently as 1970.

Coal is divided broadly into two principal groups, anthracite ("hard" coal) and bituminous ("soft" coal), which are further divided into four subtypes: semianthracite, semibituminous, subbituminous (black lignite), and lignite (brown or woody lignite). Bituminous coal is by far the most abundant and widely occurring rank of coal found in the United States. There were more than 228 billion tons of

Appalachian or Eastern Coal Region	
States	Square Miles
Pennsylvania	12,656
Ohio	7,100
Maryland	550
West Virginia	15,900
Kentucky	10,700
Tennessee	3,700
Alabama	4,300
Georgia	170

Coalfields of the United States, 1960. U.S. Department of Energy.

bituminous coal in the United States in 1976, located in the hills of Appalachia stretching from Pennsylvania and Ohio southward through Alabama with other fields located in the flatter midwestern states, and scattered fields throughout the West.

Anthracite generally lies deeper in the earth than bituminous coal, with deposits found in Scotland, Wales, the former Soviet Union, China, and the United States. It is found in the smallest quantity among major ranks of coal, with a worldwide reserve estimated at seven billion tons. The majority of United States anthracite coal is located in the mountainous counties of northeastern Pennsylvania, with smaller deposits in Alaska, Arkansas, Colorado, Massachusetts, Rhode Island, New Mexico, Utah, Virginia, Washington, and West Virginia. Anthracite has the highest percentage of fixed carbon and lowest percentage of volatile matter of all coal. It has a low sulfur content and burns slowly with a blue flame, producing little soot or smoke. The United States Bureau of Mines defined anthracite as "a hard, black, lustrous coal having 92 percent or more, but less than 98 percent, fixed carbon and 8 percent or less, but more than 2 percent, volatile matter, on a dry, mineral-matter-free basis."[6]

Subbituminous coal, frequently called black lignite, is dull black coal that contains about 15 to 30 percent moisture. The heat content of subbituminous coal ranges from sixteen to twenty-four million Btus per ton. Major deposits of subbituminous coal are found in Alaska, Colorado, Montana, and New Mexico. It burns easily and is used for household heat, in industrial plants, and for generating electricity.

Lignite is brownish-black coal in which the vegetal matter has been altered more than in peat, but not as much as in subbituminous coal. It is the lowest rank of coal mined in the United States. Deposits of lignite occur in North and South Dakota, Montana, Texas, and western Canada. The terms lignite and brown coal are used interchangeably in the United States. Lignite has a moisture content of 35 to 45 percent, and relatively low heat value of 6 to 7.5 million Btus per ton. This low heat value means it must be consumed locally, to limit transportation cost, which would otherwise make it uneconomical to use. The composition of lignite is ranked between coal and peat. Peat is distinguished from lignite by the presence of free cellulose and a high moisture content often exceeding 70 percent. The heat content of air-dried peat is about 50 percent moisture and produces about nine million Btus per ton. Lignite is mined in California, Louisiana, Montana, North Dakota and Texas. Subbituminous and lignite are used primarily for fueling steam-boilers for electric generation. Processes to transform lignite and subbituminous coals into gases are currently being developed. Low-volatile bituminous, also known as semi-bituminous and smokeless coal, is located in Pennsylvania, Maryland, West Virginia, Alabama, Arkansas, and Oklahoma. It is mined in the central Pennsylvania Georges Creek, Upper Potomac, Pocahontas, and New River fields. This coal is between bituminous coal and anthracite in rank, averaging 15 to 20 percent volatile matter. Medium-volatile is mined principally in West Virginia and Pennsylvania. High-volatile coal occurs in all coal-producing states except North and South Dakota. Montana, Wyoming, Colorado, New Mexico and Washington are the principal subbituminous-coal producing states.

Coalfields of Pennsylvania

Anthracite, semianthracite, bituminous, and semibituminous coal deposits all occur within Pennsylvania. The principal commercial coalfields of Pennsylvania are confined to rocks of the Pennsylvanian period of earth history, formed from 270 to 310 million years ago in Paleozoic times, long before the age of the dinosaurs. The coalfields in Pennsylvania were originally deposited as virtually flat-lying deposits and were physically connected. They were then separated by the process of erosion over tens of thousands of years. Subsequent periods of mountain building in the state folded and broke rocks in varying degrees. Mountain building was responsible for changing the chemical composition of the coal seams that experienced this process.

The Chemical Composition of Different Ranks of Coal by Counties in Pennsylvania:						
District	Grade	Moisture	Volatile Matter	Fixed Carbon	Ash	Sulfur
Anthracite Region	A	2.8	1.2	88.2	7.8	0.9
Sullivan	SeA	3.4	9.3	75.6	11.7	0.8
Tioga	SeB	2.3	20.9	66.9	9.9	1.3
Center	SeB	2.9	19.9	69.7	7.5	1.9
Clearfield	SeB	3.3	19.9	69.0	7.8	2.0
Clearfield	B	2.8	24.3	66.3	6.6	0.9
Indiana	B	1.0	26.1	63.8	9.1	2.7
Butler	B	4.6	33.0	54.4	8.0	1.3
Cambria	SeB	3.3	12.5	77.9	6.3	1.0
Cambria	B	3.1	26.0	64.4	6.5	1.4
Westmoreland	B	2.7	30.4	57.8	9.1	1.3
Allegheny	B	3.7	34.0	56.8	5.5	1.4
Broadtop Field	SeB	2.1	15.5	76.0	6.4	1.1
Somerset	SeB	2.5	12.5	78.8	6.2	1.1
Somerset	SeB	2.6	21.5	68.0	7.9	1.7
Fayette	B	2.8	30.0	59.8	7.4	1.2
Washington	B	1.4	34.6	57.8	6.2	0.8 [7]

Anthracite Coalfields

The major commercial anthracite fields in the United States are located exclusively in the northeastern counties of Pennsylvania within a fourteen hundred square mile area, of which only 472 square miles contain anthracite.[8] Over 99 percent of the anthracite production of the United States is concentrated in this small area of Pennsylvania. The balance is produced in Arkansas, Colorado, Virginia and New Mexico. This compact region has four irregular deposits or fields in counties located in northeastern Pennsylvania. Three of the fields, Northern, Western Middle, and Southern, occupy valleys or basins, whereas the Eastern Middle Field occupies a plateau-like tableland. The four anthracite fields are as follows: (1) The Northern Field (in the Wyoming and Lackawanna Valleys in Luzerne and Lackawanna Counties) extends from Forest City to Shickshinny, a distance of fifty miles, with a maximum width of six miles extending through Luzerne, Lackawanna, and small portions of Susquehanna and Wayne Counties. The largest cities in the Northern Field are Scranton and Wilkes-Barre. (2) The Western Middle Field, lies southwest of and adjoins the Eastern Middle Field, extending thirty-six miles in length and four to five miles in width in Northumberland, Columbia, and Schuykill Counties. The principal urban center is Shamokin. (3) The Eastern Middle Field, lying about fifteen miles south and southeast of the western end of the Northern Field, has a maximum length of twenty-six miles and a maximum width of ten miles. It is centered on Luzerne County with extensions in Schuykill, Carbon, and Columbia Counties. Hazleton is the principal urban center. (4) The Southern Field, or Schuylkill Field, extends from Mauch Chunk (present-day Jim Thorpe) to Dauphin, a distance of seventy miles, with a maximum width of eight miles. The Southern Field is the largest area in size, occupying 180 square miles. It extends northeast-southwest in Schuykill, Carbon, Dauphin, and Lebanon Counties. Pottsville is the largest city in the field.[9] A number of small, detached anthracite areas, including the Bernice Basin of Sullivan County, are located north of the Northern Field. The Bernice Field consists of three basins about fifty miles northwest of the western end of the Northern Field.

More than two hundred seams of coal have been mined in the anthracite fields. The original anthracite reserves of Pennsylvania were 24.4 billion tons, of which 8 billion tons were extracted between 1830 and 1959, representing 99 percent of all

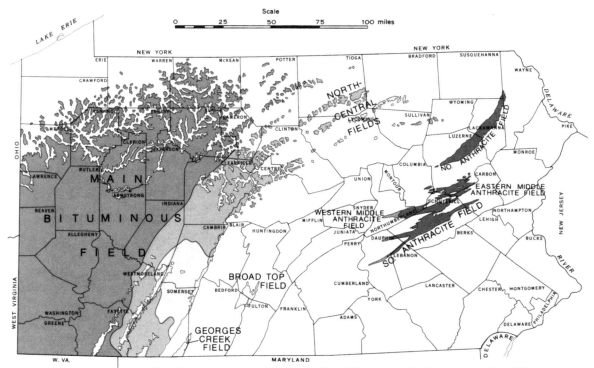

Bituminous and anthracite coalfields of Pennsylvania. Commonwealth of Pennsylvania.

anthracite coal mined in the United States. About 8 billion tons of the remaining 16.4 billion tons can be extracted employing contemporary mining methods. The Pennsylvania anthracite fields have been subjected to great cracks or faults in the earth surface. This process created intense heat and pressure associated with intense folding and faulting of the surface. Rocks were compressed, fractured, and heated, and this geological action drove out part of the volatile matter from the marsh deposits and pressed the carboniferous matter into coal. This action was more pronounced in the eastern part of the state and decreased westward in the state. Nearly all the volatile matter in anthracite coal was driven off, leaving a jet black coal with a high luster and a high carbon content.

A typical analysis of Pennsylvania anthracite is fixed carbon 87 percent, volatile matter 3.50 percent, sulfur .65 percent, ash (white) 5.90 percent, moisture (water) 2.95 percent.[10] True anthracite coal is 91 to 98 percent carbon while semianthracite, located in Sullivan and Wyoming Counties, is 85 to 90 percent carbon. Semianthracite is softer and less lustrous than anthracite and is mined in Arkansas, Virginia, and Alaska on a limited commercial scale.

Anthracite coal, also known as "hard coal" and "black diamond," is characterized by its black luster and hardness. Anthracite has no equal in the world as a home fuel. Its high carbon content, low percentage of volatile material, and low sulfur content make anthracite a slow-burning, clean, high-Btu-value fuel. It is difficult to ignite, since its ignition temperature is approximately 925 to 970 °F, but when lit it burns cleaner and longer than bituminous coal.[11] Anthracite was first discovered in Rhode Island and Massachusetts about 1760 and has since been discovered in other states, including Virginia and Pennsylvania, and in the state of Sonora, Mexico. It was discovered in eastern Pennsylvania, first appearing in two locations as "stone coal" on a map prepared by John Jenkins Sr. in 1762. Parshall Terry and a company of Connecticut pioneers found coal at the mouth of Mill Creek on the banks of the Susquehanna near present-day Wilkes-Barre. Obodiah Gore, a blacksmith from Wilkes-Barre, used anthracite for heating iron at his forge in 1769. Anthracite was shipped from the Wyoming Valley down the Susquehanna River to Harrisburg in 1775 and from there was transported to Carlisle for use in a forge for the manufacture of firearms. Philadelphia businessmen investigated the feasibility of shipping anthracite coal along the Lehigh River in 1793, and in 1803 the Lehigh Coal Mining Co. was formed. Two arkloads containing two hundred tons of anthracite were floated down the Lehigh and the Delaware River to Philadelphia. The coal was sold in Philadelphia, about one hundred miles south, but the coal would not burn and was broken up and used as gravel in the city's walkways. Anthracite was generally ignored as a viable fuel, because it was very difficult to ignite and to keep burning efficiently. Its near carbon purity made it the purest coal in the world, but this chemical property made it difficult to ignite. It was nearly impossible for early consumers to burn it in conventional wood stoves. Frederick Glass and Oliver Evans of Philadelphia were successful in burning anthracite in a grate in 1802. In 1808 Judge Jesse Fell of Wilkes-Barre designed a grate that could withstand the intense heat produced by burning anthracite. Josiah White and Erskine Hazard, owners of the Fairmount Pennsylvania Nail and Wire Works at the Falls of the Schuykill River near Philadelphia, overcame these ignition problems during the War of 1812. The British naval blockade during the war had prevented fuel from reaching their factory. The owners were determined to find an alternative for their usual charcoal fuel, and they discovered that anthracite would become red hot if it was not constantly stirred. They used it to fire their iron furnaces.

There was no systematic attempt to mine anthracite in any quantity until about 1820; the anthracite coalfields were too far from the port cities of Philadelphia and New York. Transportation cost made anthracite more expensive than bituminous coal imported from England and Virginia. Ultimately canals, and later railroads, were constructed by coal companies to transport anthracite to these markets. White and Hazard built the Lehigh Canal in 1818 to transport their coal to market, it being seventy-two miles from White Haven (upper Lehigh River valley) to Easton on the Delaware River. The first delivery of 365 tons of anthracite to Philadelphia via Mauch Chunk was completed in 1820 by the Lehigh Coal Company. The commercial anthracite industry began with this shipment. The Lehigh Coal Company, organized in 1818 with capitalization of $55,000, merged with the Lehigh Navigation Company under the new business name of the Lehigh Coal & Navigation Company, with capitalization exceeding $100,000. The output of the anthracite trade increased from 365 tons shipped to Philadelphia via the Lehigh Canal in 1820 to some 8 million tons in 1860. Anthracite was originally used for domestic heat, and in factories, railroad locomotives, and steamboats during this pioneer period. It was introduced as an alternative to charcoal in the smelting of iron ore about 1840, when it was burned in iron blast furnaces of eastern Pennsylvania.

Bituminous Coalfields of Pennsylvania

The bituminous coalfields of Pennsylvania are located entirely within the northern extremity of the Appalachian Field. About 90 percent of the nation's coal production now comes from this field, which geologists call "the greatest storehouse of high-rank coal in the United States if not in the world."[12] The region is subdivided into three areas—Northern (Pennsylvania, Ohio, Maryland), Southern (West Virginia, Virginia, eastern Kentucky, and northern Tennessee), and Alabama (including Georgia and southern Tennessee). The Pittsburgh district, for example, is a subdivision of this coalfield and is itself divided into smaller fields. The western field is known as the Panhandle District; east of this is the Gas coalfield, with the Westmoreland coalfield to the north, and the Klondike Coke District to the east with the Connellsville Coke District forming the eastern border.[13]

Pennsylvania's bituminous coal deposits are located in a number of irregularly shaped fields in the western half of the state and in a few coalfields in the north bordering New York state. These coalfields belong to the Upper Carboniferous Formation, the lowest lying within the Pottsville Series and the highest being contained in the Dunkard Series. While anthracite coal is located in a concentrated region, bituminous coal underlies 14,200 square miles of Pennsylvania, or approximately one-third of the area of the state. Coal seams are found in nearly every county west of the Allegheny Mountains and are located in several outlying coalfields east of the mountains. Bituminous coal has been mined commercially in more than thirty western and central counties since its discovery during the second half of the eighteenth century by fur trappers, pioneers, and explorers. The expansive bituminous coal regions of Pennsylvania are divided into the following four principal fields:

> (1) Main Bituminous Field: Allegheny, Armstrong, Beaver, Blair, Butler, Cambria, Cameron, Center, Clarion, Clearfield, Clinton, Elk, Fayette, Greene, Indiana, Jefferson, Lawrence, McKean, Mercer, Somerset, Venango, Washington, and Westmoreland Counties; (2) Broad Top Field: Bedford, Fulton, and Huntingdon Counties; (3) North-Central Fields (five small fields): Bradford, Lycoming, and Tioga Counties; (4) Georges Creek Field: Somerset County. A majority of the Georges Creek Field is located in the western part of Maryland, although a small portion extends north into southern Somerset County. The Georges Creek Field, the most important coalfield in Maryland, is a canoe-shaped basin about twenty-five miles long and six miles wide. The region is located almost entirely in the western part of Allegany County, Maryland, in the valley between the Dans and Savage Mountains. The field is composed mostly of high-grade, low-volatile bituminous coal called "smokeless coal." This coal is similar to that located in the Windber region, Somerset County, Pennsylvania.[14]

The coal seams of Pennsylvania developed during the Carboniferous Age and usually occur in succession as follows: sandstone, limestone, clay, coal, shale, and sandstone, in ascending order. This succession is not always present but is found more often than any other order. There are at least forty-two coal seams in the state, although not all are commercial seams. Washington, Waynesburg, Sewickley, Redstone, Pittsburgh, Upper Freeport, Lower Freeport, Upper Kittanning, Middle Kittanning, Lower Kittanning, Clarion, Brookville, and Mercer are the principal bituminous seams. Western Pennsylvania coal was extracted largely from the Pittsburgh seam (79.9 percent), Thick Freeport (12.8 percent), and Upper Freeport (4.5 percent), with smaller production from the Lower Freeport, Kittanning, Redstone, Sharon, and Sewickley (2.8 percent) by 1939.

In central Pennsylvania coal was extracted from the Lower Kittanning (49.8 percent), Upper Freeport (20.3 percent), Lower Freeport (15.4 percent), Upper Kittanning (11.3 percent), and other miscellaneous seams (3.3 percent). Most of these seams average less than five feet in thickness, although some attain a local thickness of seven to eleven feet.[15] Coal quality varies, both locally and regionally, throughout the expansive bituminous coalfields of Pennsylvania, from a few inches up to ten to twelve feet in thickness. The Pittsburgh seam, underlying four states, is the best known and most valuable coal seam in the state, averaging from five to eight feet in thickness and covering 5,729 square miles in area.

The bituminous coal mining industry of the state has been dominated from its inception by this seam. It is the most famous seam of high-volatile gas coal and coking coal in existence. Geologists have called this coal bed "the world's most valuable single mineral deposit" because it has yielded more mineral value than any single mineral deposit in the world. Production from this seam began in western Pennsylvania in 1759, in Ohio about 1795, in the western counties of what became

West Virginia about 1800, and in Maryland about 1804.[16] It contained 54 billion tons of coal after the Civil War, and its economic value was appraised then as greater than would be the total output of the California gold mines for one thousand years. As late as 1939, nearly 80 percent of the entire coal production of western Pennsylvania was extracted from this seam. The chemical composition of this seam in Allegheny County contains 57 to 65 percent fixed carbon, 30 to 35 percent volatile matter, 4 to 14 percent ash, and sulfur usually under 1 percent. This was the principal seam of American coal production until the 1970s. The compact Connellsville coke district, the premier coking region in the nation, is located entirely within the Pittsburgh seam. As late as 1922, an estimated 98 percent of all coal mined in Fayette County and 96 percent of coal mined in Westmoreland County came from this seam.[17]

Bituminous coal, unlike anthracite, was less affected by the process of mountain building although its chemical composition is not uniform throughout the state. Pennsylvania's soft coal ranges from high-volatile coal to low-volatile (semibituminous) coal that exhibits regional variation in percentage of fixed carbon, volatile matter, moisture content, and calorific value.

The composition of coal changes by counties from east to west, from low-volatile bituminous coal (fixed carbon from 78 and 86 percent) through medium-volatile bituminous coal (fixed carbon between 69 to 78 percent) to high-volatile coal (fixed carbon less than 69 percent). Generally, the fixed carbon percentage of coal beds decreases from east to west, while the percentage of volatile matter increases from east to west throughout the coalfields. Coal in Cambria, Clearfield, and Clinton Counties and in the Broad Top coalfield, which is known as low volatile semibituminous or "smokeless" coal, contains 18 to 20 percent volatile matter. Farther west the volatile matter increases to 36 percent in the Pittsburgh District, and rises as high as 40 percent in Mercer, Butler, and Lawrence Counties.

Volatile and Carbon Content of Bituminous Coal in Pennsylvania

		Volatile Matter	Carbon
Bituminous (high volatile)	Pittsburgh, Allegheny Co.	39.3 percent	60.7 percent
Bituminous (high volatile)	Irwin Basin, Westmoreland Co.	37.1 percent	62.9 percent
Bituminous (high volatile)	Greensburg, Westmoreland Co.	34.4 percent	65.6 percent
Bituminous (medium volatile)	Punxsutawney, Jefferson Co.	28.5 percent	71.5 percent
Semibituminous (low volatile)	Huntingdon, Bedford Co.	16.1 percent	83.9 percent
Semibituminous (low volatile)	Moshannon, Clearfield Co.	26.6 percent	73.4 percent
Semibituminous (low volatile)	Snowshoe, Clearfield Co.	26.4 percent	73.6 percent
Semibituminous (low volatile)	Blossburg, Tioga Co.	24.5 percent	75.5 percent[18]

Pioneer Development of the American Bituminous Coal Industry

The economic development of the bituminous coal industry in the United States was slower than the anthracite industry, although its initial discovery and development began nearly a century earlier. The first recorded discovery of bituminous coal in North America was made on Cape Breton Island, Canada, during the 1670s. Nicolas Denys, governor of French Acadia, in an account published in Paris in 1637, described "mines of coal within the limits of my concessions and upon the border of the sea."[19] French settlers mined coal on a small scale near Louisburg, Cape Breton, during this period. Richard Cowling Taylor published statistics in 1848 showing that the discovery of coal in the United States was made by Father Louis Hennepin, a French Jesuit missionary. He was part of a French exploration party under the command of Robert de La Salle which paddled down the Illinois

River in hope of reaching the Mississippi. Coal was discovered by them on December 1679 above Fort Creve Coeur, near Ottawa, Illinois, about eighty miles from present-day Chicago. Father Hennepin deemed the discovery on the Illinois River of sufficient importance to mark its location on his map as "charbon de terra."

Coal was used commercially in Europe as early as circa 1300 A.D., although it did not become a major energy source until the advent of the Industrial Revolution in the mid-eighteenth century. The word "coal" derives from the Anglo-Saxon "col," which originally referred to charcoal. The spelling "cole" was used until about three hundred years ago, at which time the present spelling of the word was adopted. European immigrants to the United States, although aware of its existence, turned to this mineral fuel only after the abundant supply of wood began to dwindle. Coal was found outcropping at the surface throughout the thirteen colonies following Father Hennepin's discovery in Illinois.

By the end of the 1750s the discovery of coal had been reported in Pennsylvania, Ohio, Kentucky, and West Virginia, and about 1762 anthracite was found in Pennsylvania. Bituminous coal was mined at two principal locations in the United States until about 1800—along the James River of Virginia a short distance north of present-day Richmond, and in the area around present-day Pittsburgh and the Monongahela Valley. Coal was mined at the James River by French Huguenots, who settled on the river in 1699 at a place now called Manakin about fourteen miles north of Richmond in Chesterfield County. Small amounts of "fossil coal" or "stone coal" were originally used locally for home heating and at the blacksmith's forges as a supplement to charcoal. Furnaces at Massaponax, Virginia, used coal as early as 1732. The earliest written record of commercial mining in the colony dates from 1750 when African American slaves were used to extract coal at a mine owned by an English company on the James River near Richmond. Virginia colonial records of 1758 note that six hundred bushels of Virginia coal were shipped to England. Samuel Davis advertised coal for sale at Richmond for twelve pence per bushel in the *Virginia Gazette* of July 1776. Thomas Wharton Jr. and Owen Biddle of Philadelphia were authorized by the General Assembly of Pennsylvania in 1776 to purchase coal from Virginia. An iron furnace, located at Westham on the James River, used coal in the manufacture of shot and shell during the Revolutionary War until the facility was destroyed by General Benedict Arnold in 1783. Coal was exported to Philadelphia, New York, and as far north as Boston before the American Revolution from the Richmond district. Coal sold in Philadelphia for a shilling and six pence a bushel in 1789.

The existence of rich coal deposits in southwestern Pennsylvania had been known to trappers, hunters, and settlers since the 1740s when coal was observed outcropping at the surface or in river beds. An early reference to bituminous coal in the region appears on a map made by John Pattin, an Indian trader, about 1752. Pattin's map indicates a "sea coal" outcropping at a site along the Kiskiminetas River, a few miles below Saltsburg on the Indiana-Westmoreland County line. Lewis Evans, a Philadelphia cartographer, in *Analysis of Travels and Exploration in the British Colonies*, located coal at Licking Creek a few miles below Venango. An early reference to the rich Pittsburgh seam is found in a letter written by Captain Adam Stephen, George Washington's second in command during a military expedition in the Ohio Valley in May 1754. He wrote about his experience along the Monongahela River, near Redstone (present-day Brownsville), to a friend in London: "Most of the hills on both sides of the Ohio are filled with excellent coal and a coal mine was in the year 1760 opened opposite Fort Pitt on the River Monongahela for the use of the Garrison."[20]

Swiss-born Colonel Henry Bouquet instructed James Burd to construct a road at a point near Christopher Gist's plantation at Mount Braddock, Fayette County, to Fort Redstone at Brownsville, Fayette County, in 1759. Burd discovered coal two miles from the Monongahela River near Brownsville, and an entry in Burd's diary of September 22 notes that "the camp moved two miles to Coal Run. This run is

Drift mine entry with loading dock on the river for loading coal on flat boats. Carnegie Library of Pittsburgh.

entirely paved in the bottom with fine stone coal, and the hill on the south is a rock of the finest coal I ever saw. I burned about a bushel of it on my fire."[21] This is one of the earliest authentic written descriptions of coal in southwestern Pennsylvania.

The first map showing coal near Pittsburgh, according to noted coal historian Howard N. Eavenson, was by B. Raber on *A Plan of Fort Pitt and Points Adjacent* in 1761. Captain Thomas Hutchins, who visited Fort Pitt in 1760, observed a coal mine on the opposite side of the Monongahela River. The mine was located on Coal Hill, later renamed Mount Washington, opposite Fort Pitt at the Point. Major Edward Ward of the Fort Pitt garrison was the superintendent of the Coal Hill mine. Coal was dropped down the hill in a wooden chute and transported across the Monongahela River by flatboat to the fort. Coal was used to heat the garrison's kettles and heat the fort during the winter. The Coal Hill mine was the site of the first recorded underground mine fire in the nation. Charles Beatty and Reverend Duffield, two Presbyterian ministers, visited Pittsburgh to preach at Fort Pitt in 1766. Beatty described coal mining at Coal Hill and wrote, "A fire being made by workmen not far from where they dug the coal, and left burning when they went away, by the small dust communicated itself to the body of the coals and has set it on fire, and has been burning almost a twelve month entirely underground."[22] This drift-entry mine was still burning, shooting sulfurous fumes through crevices in 1820, when the *Pittsburgh Gazette* reported, "it was like a volcano." From this period coal production grew, and Allegheny County became for many years one of the largest producers of bituminous coal in Pennsylvania.

The Penn family purchased the entire bituminous coalfields of Pennsylvania from the chiefs of the Six Nations in November 1768, except for that portion north of Kittanning which they acquired in 1784.[23] They purchased this land for $10,000—less than a cent per acre. On January 6, 1769, John Penn, lieutenant governor of Pennsylvania, acting on orders from the colony's proprietors, Thomas and

Richard Penn, instructed John Lukens, surveyor general, to survey some five thousand acres of known coal properties, including Fort Pitt and the "Cole Mine" on Coal Hill. After the American Revolution, in 1784, the Penns sold rights to mine coal on the "Great Seam" for £30 per lot in the hills around Pittsburgh. Bountiful coal reserves were discovered outcropping at the side of hills and in several stream valley basins in the early frontier communities of southwestern Pennsylvania. The founders of these pioneer communities, including Connellsville, Brownsville, and Canonsburg, offered to each purchaser of land "the right to take coal for their use forever gratis from adjacent coal banks."[24] Most residents of these communities saw no real economic future for these vast coal deposits underlying their properties. Cornelius Woodruff, a tavern owner in Connellsville, was a visionary who clearly foresaw the potential economic value of coal. He wrote on the flyleaf of a book about 1800:

> For those who will come after us, will find vast and undeveloped mines of material for men to work upon, treasure of untold wealth that now are hid from us . . . It will give employment to millions, not only for war, but peaceful occupation and the wants of life . . . Some great invention will be made to carry on commerce and communication in this to-be-great country.[25]

The bituminous coal industry of Pennsylvania developed first in the Pittsburgh seam in the southwestern counties of Allegheny, Westmoreland, Fayette, and Washington bordering the Monongahela River. Small-scale coal-mining activity also developed during this pioneer period in a number of coal-producing counties throughout the state. Clearfield County, along the West Branch of the Susquehanna River, was the next mining area chronologically. Samuel Boyd patented the first tract of coal land near Oldtown, Clearfield County, in November 1785. The first commercial shipment of coal from the county occurred in 1804 when his son, William Boyd, shipped an ark full of coal from the village of Clearfield down the Susquehanna River to Columbia, Lancaster County, a distance of 260 miles. P. A. Karthaus opened a mine at the mouth of the Little Moshannon Creek, Clearfield County, in 1813. Karthaus shipped semibituminous coal on arks, each holding eight hundred to one thousand bushels, down the Susquehanna River to Port Deposit, Maryland, near the head of the Chesapeake Bay, on the way to Philadelphia, in 1828. The coal sold for about thirty-three cents a bushel or $8.75 to $9.30 a ton. Other coal arks followed and all the towns along the Susquehanna River were familiar with semibituminous coal from Clearfield County. Coal was discovered near Snowshoe, Centre County, in 1819 by a party of hunters who saw coal outcropping near a spring. It was used by the blacksmiths of Bellefonte, about twenty-five miles from Snowshoe. Shipments outside Centre County were small until the construction of the Lock Haven and Tyrone Railroad, and its connection to the Bellefonte and Snow Shoe Railroad, in 1857-1859.[26]

Coal mining developed north of the Susquehanna River in Tioga, Bradford, and Lycoming Counties in the North-Central Fields, a series of five small coalfields. The Patterson brothers discovered coal at Peters Camp (present-day Blossburg Borough), Tioga County, in 1792 while clearing a road for a party of settlers moving from Williamsport, Pennsylvania, to Seneca Lake, New York. David Clemons opened a drift-entry mine on Bear Creek about 1815. R. C. Taylor undertook the first detailed geologic survey of the region in 1832. Taylor also surveyed a railroad line from Blossburg to Tioga. Area blacksmiths used coal as fuel in their forges. Coal was discovered in Bradford County by Abner Carr while hunting on Towanda Mountain, in 1812. Semibituminous coal was used by blacksmiths and for home heating locally shortly after its discovery.

The Broad Top Mountain coalfield is an isolated field comprising about fifty square miles, and underlies parts of Bedford, Fulton, and Huntingdon Counties. This coal region is an independent and isolated region located between the anthracite district of northeastern Pennsylvania and the eastern boundary of the bituminous coal basin. It is "in the form of a high tableland mesa known as the Broadtop Mountain, which lies between two mountain ranges, Sidling Hill and Tussey Mountain, in the Appalachian Valley."[27] The northern end of the Broad Top field covers a relatively small area in the southern part of Huntingdon County, located primarily in Carbon, Wood, and part of Todd Townships. Coal was concentrated in the northeast corner of Bedford County, while a very small part of the Broad Top coalfield extends into the northwest corner of Fulton County. This isolated semibituminous coalfield has been described as "freak (coal) veins . . . [that] shouldn't be there."[28] There were thirteen coal seams, although the principal commercial seams mined in ascending order of value were Fulton, Barnett, and Kelly. Coal was discovered by Nathan Port Horton, a local blacksmith at Shreeves Run, near Coaldale village, Bedford County, in 1765. Horton used local coal that outcropped at the surface in his forge. Samuel Riddle, a Bedford lawyer and pioneer coal operator in the region, had first shipped coal by boats by about 1800 from the area down the Juniata River to eastern markets. J. Peter Lesley, geologist of the First Geological Survey of Pennsylvania, first surveyed the coal resources of the region in 1855.

Semibituminous coal is lower in volatile matter and sulfur than bituminous coal found in the Pittsburgh seam. These chemical properties made it a superb smithing and steamer coal because it produces more heat and less ash, and burns more uniformly than bituminous coal, which has higher percentages of water and volatile matter. The principal deposits of semibituminous coal, outside of Pennsylvania, are found in the Pocahontas and New River Fields of West Virginia in the Georges Creek Field, Allegany County of western Maryland, and in western Arkansas.

Transportation and Early Markets

Most coal was transported by wagon to the surrounding region in the Monongahela Valley and was consumed locally until about 1800. Coal was later shipped on flatboats on the Monongahela River and floated down the Ohio and the Mississippi to river towns after 1800. The first recorded shipment of coal from Pittsburgh was made in 1803. A company of French merchants shipped 350 tons of coal on a ship called the *Louisiana*. The coal, acting as ballast on the ship, was transported to New Orleans and then to Philadelphia, where it was sold for 37.5 cents a bushel (between nine and ten dollars a ton). Coal was usually bartered for molasses and sugar or exchanged for French and Spanish gold at New Orleans. A small river traffic in coal had developed with western markets on the Ohio River from the Pittsburgh district by the time of the War of 1812. Regular coal shipments began to be made in 1817 down the Ohio River to Mayville, Cincinnati, and Louisville, which were principal early coal markets. Coal was mined in the winter, and these small commercial operators waited until spring to ship their coal out on flatboats to cities and towns on the banks of the Ohio River.

The Monongahela and Youghiogheny Rivers served as the principal route for the shipment of bituminous coal from the Monongahela Valley to these western commercial markets. A variety of crafts—flatboats, arks, rafts, keelboats and steamboats—were all used to carry coal on the rivers. The larger mines shipped coal in the spring on flatboats called French-Creek boats. These boats, constructed at French Creek, a tributary of the Allegheny River, were used to transport pig iron, walnut lumber, salt, and agricultural products to Pittsburgh. Local coal operators obtained these boats and enlarged their sides to hold more coal. Each boat averaged from 68 to 79 feet long, 16 feet wide, and from 4 1/2 to 5 feet deep, and held between four and six thousand bushels of coal. Pairs of these boats were sometimes lashed together with ropes and fitted with steering oars. The boats were sold for lumber or broken up when they reached their down-river destinations. Keelboats and barges were also being used in 1810 to transport coal to market. Coal was shipped to market in the spring and fall. There were few shipments in the summer because of low water, and in the winter because of ice on the river. River transportation was slow because of low water and the numerous rapids. The boats moved with the current and were guided by long sweeps or oars by a crew of five to eight men. A trip from Pittsburgh to Cincinnati, a distance of 457 miles, usually took five days to complete.

Mining Technology

The abundant outcropping of coal from the Pittsburgh seam along the counties bordering the Monongahela and Youghiogheny Rivers made possible the development of the small commercial mining industry during the last quarter of the eighteenth century. Coal cropped out of the hills fronting the Monongahela River from Pittsburgh to Brownsville. An outcrop of coal is a seam or end of a seam of coal appearing at the earth's surface that may be visible or may be covered with a thin layer of earth. An estimated thirty thousand tons were being produced annually in and around Pittsburgh by 1790. Before 1840 coal output was an educated guess because no government agency in the Commonwealth kept records of coal operators or coal production until after this date. The earliest bituminous coal miners and owners of western Pennsylvania were farmers who mined the coal they found outcropping in their fields or on a hillside fronting the river. These operations were small scale and seasonal. Coal was quarried like stone from coal seams that were exposed at the surface. Farmers mined coal during the winter with pickaxes and crowbars until it was necessary to follow the seam underground. Coal was tied up in rawhides and rolled down the hill to the riverbank, where it was emptied into wagons or boats and the hides returned to the pit mouth to be reloaded. Coal was loaded by hand onto wagons or carts and sold by the bushel for home fuel or the blacksmith's forge. Coal was also sold locally, in 1800, for home heat for $1.25 a ton. The borough government of Pittsburgh passed an ordinance on January 1, 1802, requiring coal carts and wagons to be clearly marked with their capacities measured in bushels. This ordinance is believed to be one of the first examples of government regulation of the weight and measure of coal. A variety of measurements were used by coal operators to weigh coal during the nineteenth century. A bushel of coal is equal to 80 pounds and 25 bushels equal a ton. A barrel of coal is equal to 200 pounds or 2 1/2 bushels. A long ton contains 2,400 pounds while a short ton contains 2,000 pounds. Coal was weighed at the mines using the long-ton measurement while the short-ton measurement was used when it was sold to consumers.

Coal production ceased when spring arrived and farmers began planting their crops. After the outcropped coal was mined out locally, some farmers who were fearful of going into the darkness of an underground mine sold or leased the mineral rights on their properties. Some farmers dug small drift-entry underground mines into the side of a hill, variously called "country banks," "wagon mines," "dig holes," "gopher holes," or "father-and-son mines." It was common practice for workers to work underground from four or five o'clock in the morning until six o'clock in the evening. Underground work was not steady, however, and workers were idle for several months of the year. Then they worked on coal barges and made regular trips as far south as New Orleans.

The original coal operators were farmers who doubled as part-time miners since they owned the property where the coal was located. As coal became more marketable, local business and professional men became coal operators by purchase or lease of coal properties, and hired farmers or day laborers to work in their mines. Among the early producers and shippers in the Monongahela Valley were James and Robert Watson, who were credited with regular coal shipments to New Orleans around 1817. George Ledlie of Birmingham Borough, Herron & Peterson, Colonel William L. Miller, D. Bushnell, Fawcett & Brothers, John Gill, and William H. Brown were pioneer coal owners in the Monongahela Valley. There are very few written accounts of the activities of these pioneer mine operators.

Thomas Hulme, a transplanted English miner, observed the higher wages and improved standard of living at a coal mine at Wheeling in 1819:

> [Coal] cost 3 cents per bushel to be got out of the mines. This price, as nearly as I can calculate enables the American collier [miner] to earn upon an average, double the number of cents for the same labor that the collier in England, can earn; so that as the American collier can, upon an average buy his flour for one third the price that the English collier pays for his flour, he receives six times the quantity of flour for the same labour.[29]

Glowing accounts of the American coal industry by Hulme and other British travelers enticed British miners to emigrate to the United States. A small number of British miners emigrated after the American Revolution and found work in the nascent coal industry. Mass immigration of skilled miners from Great Britain did not occur until the period between 1840 and 1880. English-speaking miners were attracted by higher wages and safer mining conditions. These skilled British miners worked underground with farmers and field hands during this period. Farmers and farm hands represented a valuable labor reserve needed by coal operators during periods of brisk business or when full-time workers threatened to strike. These full-time skilled British miners resented the part-time and seasonal miners, calling them "winter diggers," "wheats," "corn crackers," "hay johns," "pumpkin rollers," "greenies," "scissorbills," and "sagers."

Eastwick Evans, a visitor to Pittsburgh in 1818, observed a number of coal operations in the Pittsburgh area. He described the quality of the coal: "[The hills] on the west of the Monongahela constitute a horizontal strata six inches [feet?] thick and apparently unlimited in its direction through the mountain. This coal is superior to that of England."[30] There were about forty or fifty coal pits carved out of the hills flanking the Monongahela River and operating on the bank of the river south of Pittsburgh by 1814. These mines produced about one million bushels of coal annually.[31] There were ten mines alone overlooking Coal Hill producing five million bushels of coal annually by 1837. Zadok Cramer (1773-1813), a transplanted Quaker from New Jersey, arrived in the borough of Pittsburgh in 1811. He was the publisher of *Cramer's Almanac* and a business and travel guide almanac, *The Navigator*. Cramer wrote an early account of mining activity in southwestern Pennsylvania in *The Navigator*. The passage, written in 1814, describes the cost of coal, the location of mine sites, and contemporary mining techniques employed to extract and transport coal in western Pennsylvania during the first two decades of the nineteenth century:

> Little short of a million bushels [of coal] are now consumed annually; the price formerly six cents, has now risen to twelve cents keeping pace with the increased price of provisions, labor etc. . . . There are forty or fifty pit openings [on Coal Hill] including those on both sides of the river. They are worked into the hill horizontally, the coal is wheeled to the mouth of the pit in a wheelbarrow, thrown upon a platform and from there thrown into wagons. After digging in for some distance, rooms are formed upon each side, pillars being left at intervals to support the roof. The coal is in the first instance, separated into solid masses, and is afterward broken into small pieces for the purpose of transportation. A laborer is able to dig upward of 100 bushels per day.[32]

Coal deposits are worked by surface mining and/or by deep mining underground. Removal of coal from the surface is known as surface or "strip" mining and is employed when the coal seam lies close to the surface of the earth; otherwise, mining takes place underground. Surface mining or quarrying of coal is probably the oldest form of coal production. When the coal seam outcrops at the surface it is possible to remove it without tunneling underground. Early miners simply used picks and shovels to remove the surface coal. Large-scale surface mining was delayed until the last quarter of the nineteenth century and the introduction of large-scale mechanical earth-moving equipment.[33]

Once the surface coal was removed it was essential to follow the seam underground. When the outcropping seam ran into the ground the coal-picker who followed it underground became a miner. Three methods of underground mining were employed to reach coal seams that were too deep to make surface-mining techniques feasible. The type of underground opening was determined by topography and depth of cover. Drift, slope, and shaft are the principal types of underground mine entry. The drift entry was the easiest and earliest method of opening an underground mine during this era. Shaft and slope-entry mining were rarely employed by bituminous coal companies of Pennsylvania during this period. A slope is an inclined opening at almost any angle from the horizontal. Coal cars in most slope mines are pulled by a rope operated by a hoisting engine, or in some cases the coal is carried up the slope by conveyors. A shaft is a vertical opening in which coal cars and miners enter and depart in a cage powered by a steam engine. Drift and slope types of openings are usually driven in coal or rock, but in most cases the entries are driven directly into the coal seam. A shaft mine entry must always be sunk through rock to reach the coal seam. Shaft and slope entries were rare during this period because these openings were costly to sink, and hoisting machines expensive to install. The hauling technology was extremely primitive.

A drift mine is a tunnel driven into the coal seam at the point where the seam emerges from the side of the hill or mountain. The tunnel gradually slopes usually southward following the seam. Opening a drift-entry mine was known as "driving to the rise." It did not require much technical skill or capital on the part of the operator since no excavation through rock was required. This opening was essentially a large tunnel dug where the coal seam outcropped or was exposed on the hillside or a gully. The opening was driven into the coal at a slight upward inclination to permit natural drainage of water from the mine. This entry was used only when the coal seam was nearly level and above water level. Extracted coal was usually run out directly onto railroad tracks in wooden carts in the drift entry, hauled at first by

Early small "country bank" mine of western Pennsylvania. Historical Society of Western Pennsylvania.

men and dogs and later by mules or horses to the pit-mouth entry, which was an opening about six feet high and eight feet wide. Coal was then dumped down a wooden incline to the river tipple where it was weighed and loaded onto a variety of riverboats.

Most coal was extracted from the drift-entry mines using a system of mining called room-and-pillar. The new underground mine was opened by creating passages known as "main entries" or "main headings" in bituminous mines. The entries were driven to the coal seam as straight as possible to avoid turns in railroad tracks or sudden changes in the direction of the air current.

From these main entries, tunnels or entries were driven at right angles creating a series of "rooms," "chambers," or "breasts." Most mining activity occurred in these rooms. Each room was about eight yards wide and between these rooms were solid blocks of coal called "pillars" or "columns," measuring twenty to one hundred feet wide, left standing between the rooms to support the roof of the mine. The rooftop of the mine must be supported or it will eventually cave in. Timber was used as supplementary roof props. This method of mining coal was generally inefficient because there was no possibility of extracting coal pillars entirely. Some coal from the "pillars" can be "robbed" or removed after they are no longer needed to support the room but "pillar drawing" is dangerous work and often causes the surface to subside. Pillars were frequently wider than the rooms themselves, and if they were not drawn approximately half the coal seam was left underground. A mining engineer, in the 1870s, estimated about one-third to one-half of all coal was abandoned in the mines, mostly in the form of pillars left standing. This was the most popular system of extracting coal in the United States until widespread longwall mining and surface mining were introduced after World War II.

Commercial Uses of Bituminous Coal

Bituminous coal was used for a variety of commercial and residential purposes from 1740 to 1840. Pittsburgh, Allegheny City, and thirteen towns within a radius of five miles, with an estimated population of fifty-five thousand, had established a modest commercial coal trade by the end of this period. The coal trade contributed nearly $1 million to the economy out of a total annual business of $3 million. It was used in stoves to heat houses, and in the glass and salt industry. The frontier communities of Pittsburgh, Connellsville, Brownsville, and Washington were principal consumers of coal mined in the region. Coal had largely replaced wood as a household fuel in the Monongahela Valley by the 1790s. Pittsburgh was known as the "smoky city" by 1810 because coal, not wood, was the primary fuel used in the heating of private homes. A mechanic from Salem, Massachusetts, who visited the city on October 19, 1817, commented that "coal is used for domestic purposes, as well as in their factories, and the city being hemmed in by the surrounding mountains, the air is always smoky. Coal is about six cents a bushel. It makes the best fire I ever saw, equal to the best walnut wood."[34] Still, wood was preferred for cooking stoves because it could be lit with ease and because coal burned indoors gave off offensive odors.

The steam engine was another early user of coal. Plentiful coal provided a cheap fuel for the production of steam power. Oliver Evans (1755-1819), of Philadelphia, introduced the first steam engine in Pittsburgh in 1809. Evans was the also the inventor of the automatic flour mill. Its operation is described in his book, *The Young Millwright and Miller's Guide,* first edition, 1795. Steam engines using bituminous coal were employed to drive machinery at flouring mills, gristmills, rolling mills, breweries, glass manufactories, and nail factories. Coal furnishing power in a gristmill was consumed at a rate of about twenty bushels per day, at a total cost of a dollar a day by 1810. Zadok Cramer reported three manufacturing establishments using Pittsburgh coal in 1814 and eight establishments using coal steam power in 1817. By 1833, thirty-two ironworks operated in Pittsburgh alone, including ten engine and machine shops, three foundries, eleven rolling and nail mills, and eight miscellaneous shops. These facilities employed 1,080 men and consumed 97,497 bushels of coal in 1833. Pittsburgh was consuming four hundred tons of bituminous coal per day (ten thousand bushels) for domestic and light industrial uses by 1830.

Coal was the principal fuel used in the expanding glass and saltmaking industries of southwestern Pennsylvania. As these industries grew, so did local coal production in the area. Albert Gallatin established a glass factory on Georges Creek, near New Geneva, Fayette County, on September 20, 1797. General James O'Hara and his partner, Major Isaac Craig, opened the first glass factory in Pittsburgh on the south side of the Monongahela River, nearly opposite the Point and Fort Pitt. It was a frame factory with an eight-pot furnace for the production of window glass and bottles. This site was selected so the factory would be near the coal deposits located on Coal Hill. The south side of the Monongahela was the center of the glass industry of western Pennsylvania for most of the nineteenth century. Coal was used to fire the glass kilns. Glass manufacturers in the region consumed 1,480,000 bushels (59,200 net tons) of coal in the production of window glass, vial glass, bottles, and flint glass in 1837.[35]

Saltmaking was an important early industry in western Pennsylvania. Coal was used to separate salt from the brine. William Johnston bored a number of wells near the Conemaugh River, Armstrong County, in 1812 and found enough brine to furnish salt for one million inhabitants.

Before Johnston's discovery salt had been imported from the east by packhorses over the Allegheny Mountains. This discovery initiated the local saltmaking industry with salt wells sunk along the Conemaugh, Kiskiminetas, and Mahoning Creeks, as well as along the Monongahela and Allegheny Rivers. Over two hundred thousand tons of coal per year were being used by the salt industry to produce salt for domestic consumption by 1825. Salt production had become a major industry in Armstrong and Indiana Counties in western Pennsylvania by 1830. There were twenty-four salt wells producing sixty-five thousand tons of salt in 1830. Initially horse power was used to pump brine from the wells, but horses were replaced by coal-powered steam engines. Salt brine pumped from the wells was boiled in large pans to evaporate the water. Coal supplied the heat to separate the salt from the brine. Each boiler furnace used almost two hundred bushels of bituminous coal daily. A contemporary account written in 1828 notes that "coal is the only fuel used in the evaporation. This fuel costs nothing but the quarrying, about 3/4 of a cent per bushel, and is run out, and down to the furnaces upon railways without any hauling. All the furnaces lean against the coal hills, in which are two strata of this fuel 4 or 5 feet thick and inexhaustible."[36] Some one hundred thousand tons of bituminous coal were consumed annually by the saltmaking industry between 1815 and 1870.

Coal, Coke, and the Iron Industry

The iron furnace originated in the Rhine provinces at the beginning of the fourteenth century and spread to other European nations. Charcoal was the principal fuel in iron production. The first great improvement in the iron production process was the substitution of coke for charcoal fuel. The history of coke-making dates back to the late sixteenth century in England, where the process was first called "charking" coal. Coal was piled in a mound or placed in a pit and was burned slowly. Charcoal was used in the iron-smelting furnaces and forges of both the United States and Great Britain until the eighteenth century. A voracious appetite for wood in iron production quickly depleted forests and forced British ironmakers to find a substitute fuel for charcoal by the beginning of the eighteenth century. Several patents were issued in the sixteenth and seventeenth centuries for various processes using coal and coke for iron smelting, although all these processes generally proved unsatisfactory.

Historians believe Abraham Darby, an English Quaker, first successfully smelted iron using coke in 1735. He leased an iron plant at Colebrookdale near the Severn River in Shropshire, England, intending to smelt iron with charcoal, but the rising cost of charcoal made the enterprise unprofitable. Darby turned then to low-sulfur coal and made coke that was a suitable fuel for iron smelting. He produced "pit coal pigs," which were used as bar-iron stock for rolling. Darby sold cast-iron pots, kettles, and other small goods manufactured from iron produced by coke at a reduced cost. Darby's coke-making process was not immediately embraced, but improvement in the manufacture of coke and newer designs of the coke blast furnace made by his son improved the process. Coke smelting became more efficient and economical than charcoal, and coke-produced iron was being widely used in England after 1760. The Darby family managed the ironworks at Colebrookdale for five generations.

The production of coke is essentially a distillation process in which coal is burned under controlled conditions, "cooking" out volatile gaseous matters, including tars, oils, and gases, at a temperature between 900 and 1150 °C, so that the fixed carbon and ash are fused together. Coke, known as "the bones of coal," is a hard and extremely porous residue which has a dull to submetallic luster, and is dark gray to silvery gray in color. Coke was spelled "coak" by British ironmakers and the introduction revitalized the once-sagging British iron industry. The high carbon percentage of coke gives it the capacity to generate heating power superior to coal or charcoal. Coke is strong enough to support a load of iron ore in a blast furnace. It has a heat value of about 25 million Btus per ton. The quantity of coke derived from the controlled burning of coal varies from about 50 to 85 percent. Coal to become profitable in coke production should yield at least 65 to 70 percent coke. In theory, coke functions like charcoal once in the blast furnace. Both fuels are placed at the bottom of the furnace and iron ore is dumped on top. The carbon in coke or in charcoal combines during combustion with the oxygen in the ore, freeing the iron to exist in pure form. Coke burns less easily than charcoal because it requires a much hotter fire. Raising the temperature in the furnace demanded years of experimentation and was accomplished after the steam engine replaced the water wheel in working the bellows that blew the air (the furnace was kept "in blast" day and night) into the iron furnace. British ironmakers used low-sulfur coal to make iron, and the new iron process called puddling removed impurities from brittle coke-smelted iron. These technological advances improved both the quality and the quantity of iron production in Great Britain. The charcoal furnaces had all but disappeared from the landscape by 1796. The British iron industry more than quadrupled as iron output increased from seventeen thousand tons in 1740 to sixty-eight tousand tons in 1788. All but eleven of Great Britain's 173 iron blast furnaces were using coke for smelting iron in 1806, with an annual output of 273,113 tons.[37]

American ironmakers did not adopt ironmaking technology using mineral fuel (coal or coke) as rapidly as Great Britain. The American industrial revolution based on mineral fuel was delayed until the 1830s. Charcoal was the principal fuel in American ironmaking for more than a century because there was no wood shortage. American ironmaking was centered in Pennsylvania after 1716, when the first successful ironmaking venture was undertaken on Manatawny Creek, a few miles above Pottstown, Berks County. Pennsylvania was the foremost iron-producing colony in America after 1750, with the construction of new iron plantations in the Schuylkill, Delaware, and Susquehanna River valleys. At least twenty-one blast furnaces, forty-five forges, four bloomeries, six steel furnaces, three slitting mills, two plate mills, and one wire mill were constructed in Pennsylvania between 1717 and 1776. Ironmasters had located their blast furnaces in rural areas to be near a ready supply of raw materials. Three principal ingredients are required to transform iron ore into iron and they were widely distributed in Pennsylvania—iron oxides such as magnetite, limonite, red and brown hematite, and carbonate; fuel (charcoal); flux (limestone or oyster shells were used to remove the waste matter, called slag); and fire. About thirty to forty pounds of limestone were fed into a typical furnace for every four to five hundred pounds of iron ore, depending on the purity of the latter. A blast furnace of this era produced two tons of iron daily and consumed charcoal made from one acre of woodland. A wooded area of 240 acres yielded five to six thousand cords of wood annually to feed a typical rural blast furnace. Large quantities of charcoal were also used in forges and blacksmith shops, in addition to blast furnaces. The Hopewell Village Iron Plantation, Berks County, Pennsylvania, consumed fifteen thousand cords of wood annually to produce charcoal, which is an almost pure carbon fuel that burns with an intense heat. Charcoal was produced by the slow combustion of wood under carefully controlled conditions. The use of charcoal was a rational fuel choice given the abundance of wood and the isolation before the 1830 of known coal deposits in the state.

A variety of groups and organizations in Pennsylvania attempted to find an alternative fuel for charcoal in the production of pig iron after 1820. Ironmasters, state government officials, and members of the business and intellectual communities all promoted experimentation to find an alternative mineral fuel for charcoal. The increased scarcity of wood and subsequent rise in charcoal prices, as well as the increasing competition from inexpensive British iron goods, prompted this frantic search. There was a demand for cheaper iron by consumers, and all these impulses inspired a search for a new fuel to replace charcoal and create a new method to manufacture iron. Experiments with anthracite, raw bituminous coal, and coke as a mineral fuel to smelt iron ore were undertaken. Hazard's *Register of Pennsylvania,* in 1835, suggested that the General Assembly award premiums for the smelting of ore with anthracite. The General Assembly of Pennsylvania passed an act on June 16, 1836 (P.L. 799), "to encourage the manufacture of iron with coke or mineral coal," and authorized the formation of companies for the manufacture, transportation, and sale of iron made with coal or coke.[38] Governor Joseph Ritner (he was a member of a family in the iron business), in his annual address to the General Assembly in 1838, stated that "the successful union of stone coal [anthracite] and iron ore, in the arts, is an event of decidedly greater moment of our state than any has occurred since the application of steam in aid of human labor."[39] The general assembly authorized the formation of the Geological Survey of Pennsylvania in 1836 to begin an extensive five-year geological survey to map and record the state's natural resources. The survey spawned three subsequent geologic and topographic surveys funded by the state during the next century. Pennsylvania was the first state to undertake a systematic geological survey of its coal resources by locating deposits and determining the quantity and the quality of coal that underlay the state. Coal operators were dependent on prospectors to locate coal seams. The typical coal prospector used a pick, a hammer, a clinometer for measuring angles of dip and strike in formation, and an aneroid barometer for determining altitude. Coal prospectors studied rock formations in order to determine the geological age of the earth strata. The Geological Commission was granted a budget of $6,400 to pay the salaries of a geologist, a chemist, and two assistants. Henry D. Rodgers established the First Geological Survey of Pennsylvania. The survey stopped its research temporarily in 1842 because of a lack of funds, although Rodgers published the first survey in two large volumes in 1858. The finding of this original survey was augmented by later surveys funded and conducted by the state.

The Franklin Institute of Philadelphia, founded in 1824 for the promotion of applied science and the mechanical arts, offered gold medals to "the person who shall manufacture in the United States the greatest quantity of iron from the ore during the year, using no other fuel than bituminous coal or coke, the quantity to be not less than 20 tons."[40] F. H. Oliphant produced iron from coke at the Fairchance Furnace near Uniontown, Fayette County, and forwarded samples of pig iron to the Franklin Institute in 1836. Nicholas Biddle, president of the Bank of Philadelphia (successor to the Second Bank of the United States), and some business associates offered a $5,000 reward to the first person who could keep an iron furnace in blast continuously for three months using a mineral fuel. Bostonian William Lyman was the first ironmaker to claim a first prize offered by Biddle and associates. He successfully burned anthracite at his Pioneer Furnace in Pottsville, Schuylkill County, in 1839.

Anthracite has such high carbon, compared to bituminous coal, that it was believed it could be used alone in the iron furnace. However, all experiments with anthracite, or anthracite mixed with coke or charcoal, failed. Anthracite, as a metallurgical fuel, required an extremely high fire in the furnace, one that could not be easily attained by American ironmasters using a cold blast to heat their furnace by blowing cold air from outside the furnace under pressure directly onto the surface of the charcoal. The temperature of the cold blast used in these iron furnaces ranged from 100 °F in summer to 10 °F in winter.[41]

James B. Neilson, George Crane, and David Thomas carried the process to its successful culmination.[42] Neilson, superintendent of the Glasgow Gas Works, with his associates Mr. Mackintosh and Mr. Wilson, achieved a technological breakthrough in 1828 when they designed and patented a new furnace that could maintain high and consistent temperature using anthracite. They designed bellows that blew preheated air at 300 °F, rather than cold air, into the bottom of the furnace. Doing this, they discovered that the ore in the furnace was reduced more quickly, and with a smaller consumption of fuel. Neilson patented the principle of the hot-blast furnace in both England and France and continued to improve the furnace and the process at the Clyde Iron Works. In 1837 George Crane, an iron manufacturer from the anthracite coal region of Yniscedwin, Wales, and his manager, David Thomas, used Neilson's new hot-blast furnace process and successfully made iron using anthracite as a metallurgical fuel.[43] Weekly iron production in England rose from thirty-five tons to four hundred tons in the forty years following the first use of the hot blast.

The Rev. Dr. Frederick W. Geissenhainer, a German-born Lutheran minister, had secured a United States patent on December 19, 1833, for smelting pig iron with anthracite using a strong hot blast, which he attempted successfully, in an experimental furnace that he constructed in New York City in 1831. In 1836 he built the Lucy Furnace near Pottsville, Schuylkill County, and made a few tons of pig iron with anthracite. Geissenhainer's experiment was the first successful attempt to use anthracite as an alternative to charcoal to smelt iron in the United States. He died at Lebanon, Pennsylvania, shortly after this successful experiment and was unable to improve on his furnace's design.

Josiah White and Erskine Hazzard, owners of the Fairmount Nail and Iron Works, had been trying to make pig iron using anthracite as fuel in their furnaces in eastern Pennsylvania since the 1820s. Hazzard, White, and a group of Philadelphia investors formed the Lehigh Crane Iron Company and tried unsuccessfully to use anthracite in their iron blast furnaces. They traveled to Wales and tried to persuade George Crane to return with them to Pennsylvania and construct a hot-blast furnace. David Thomas, Crane's iron-plant manager and the actual designer of the hot-blast furnace in Wales, came to the Lehigh Valley and constructed a number of anthracite hot-blast furnaces for the iron company. Thomas fired the first anthracite-fueled hot-blast furnace at Catasauqua, near Allentown, on July 4, 1840. The furnace employed the hot-blast process perfected by James Neilson in Scotland in 1828. Thomas adapted Neilson's concept of heating air in the chamber before it was blasted into the furnace. The furnace produced fifty to sixty tons of foundry iron a week, and anthracite iron was made at this facility until it closed in 1879. This early success made the Lehigh Crane Iron Company the leading anthracite iron firm in the state.

Anthracite played the same pivotal role in reviving the iron industry of Pennsylvania as coke had played in Great Britain. Anthracite was abundant and its low cost made possible the production of the cheapest iron ever made in America. Anthracite was being used instead of charcoal fuel in a majority of iron blast furnaces by 1859. The hot-blast process designed by Thomas, known as "The Father of the Anthracite Industry," required less fuel to raise the temperature to the melting point of the ore and reduced fuel consumption in the furnace. The new furnace, unlike contemporary charcoal iron blast furnaces, was distinguished "by its greater height and overall size, and the large dimensions and great power of its blowing engines."[44] The advantages of the hot-blast furnace over the conventional charcoal blast furnace were numerous: "[I]t permitted the use of anthracite which resulted in substantial savings in smelting costs, its inherent mechanical simplicity and relatively cheap installation cost, and its adaptability to existing cold blast furnaces."[45]

The new method of making iron with anthracite caused an explosive growth of iron manufacturing in Pennsylvania, as hot-blast furnaces were constructed along the Susquehanna River, in the Wyoming-Lackawanna region, and in the Schuylkill and Lehigh Valleys. The use of anthracite made it possible for iron manufacturers to shift iron manufacturing from a rural to an urban environment. Pig-iron production which amounted to slightly less than fifty thousand gross tons in 1810, had risen to almost six hundred thousand tons by 1850, produced at some 370 iron establishments.[46]

Development of the Coke Industry of Western Pennsylvania

Bituminous coal had shown itself at an early date to be a useful fuel for the heating and puddling of pig iron, since there was no direct contact between metal and fuel in the process; therefore chemical impurities in the coal or coke did not matter. There was little success achieved in the use of either coal or coke in the iron blast furnaces of western Pennsylvania before the 1830s. Several reasons can account for this tardiness: there was an ample supply of timber near the furnace to produce charcoal, iron manufacturers preferred pig iron made from charcoal, technological problems remained in the coking process, the proper blast for the iron furnace was hard to control, and there was the need for better grades of coking coal. Coke was used originally in Pennsylvania by blacksmiths and foundrymen. However, the principal value of coke is its ability to reduce iron ore in the blast furnace. Coal was first coked to remove part of the sulfur and arsenic and to reduce the smoke. Coke is an effective blast furnace fuel because it is strong enough to support the ore and limestone, and yet sufficiently porous to be affected quickly by the air blast. The production of coke was perfected in eighteenth century England, but the process was not understood generally or applied in the United States until the 1830s. A number of English and Welsh ironworkers, familiar with the manufacture and use of coke in smelting iron in blast furnaces, had immigrated to the United States in the decades following the American Revolution. John Beal was one of these transplanted skilled artisans who offered to share his coke-making expertise with local ironmakers. Beal placed an advertisement in the *Pittsburgh Mercury* on May 27, 1813, volunteering to "convert stone coal into 'coak'":

> To proprietors of blast furnaces:
> John Beal, lately from England, being informed that all the blast furnaces are in the habit of smelting iron ore with charcoal, and knowing the great disadvantage it is to proprietors, is induced to offer his service to instruct them in the method of converting stone coal into coak. The advantage of using coak will be so great that it cannot fail becoming general if put to practice. He flatters himself that he has had all the experience that is necessary in the above branch to give satisfaction to those who feel inclined to alter their mode of melting their ore.
> John Beal, Iron Founder.[47]

It is not known if any local ironmaster answered this advertisement. He opened Beal and Company, an extensive foundry on the Monongahela River bank. Coke as a substitute for charcoal was first made in Fayette County, which was the principal iron-producing county in southwestern Pennsylvania between 1790 and 1830. The Alliance Iron Furnace, which opened on November 1, 1790, on Jacobs Creek, two and a half miles west of the Youghiogheny River, was the first blast furnace and forge located west of the Allegheny Mountains. Sixteen ironworks developed in the county between 1790 and 1800. There were ten furnaces, one air furnace, eight forges, three rolling mills and slitting mills, one steel furnace, and five triphammers in the county by 1811. There were three iron furnaces in southwestern Pennsylvania in the eighteenth century: Greene Furnace in Greene County; John Probst's Westmoreland Furnace near Ligonier, Westmoreland County; and George Anschutz's, William Amberson's, and Francis Beelen's Shadyside Furnace near Pittsburgh at Two Mile Creek.

The history of the early coke industry in the United States, as in England, is rather confused, as authorities dispute events and dates in chronicling its development. Some historians believe the first manufacture and use of coke in this country began at the Allegheny Furnace in Blair County in 1811, while others credit Colonel Isaac Meason, in 1818. Meason's Plumsock Iron Works was the first rolling mill west of the Alleghenies, founded in 1815 on Redstone Creek near Upper Middletown, nine miles east of Brownsville, Fayette County. Thomas C. Lewis, a transplanted Welsh-born ironworker and the superintendent of Meason's iron furnace, was responsible for producing coke used in the rolling mill to heat and puddle iron and roll iron bars. Lewis employed three men to mine coal from the Redstone seam, probably the earliest captive coal mine in the county. Coal was first made into coke in turf-covered "mounds" or "ricks" in a manner similar to the making of charcoal from wood. Meason placed an advertisement in the *Pittsburgh Gazette* on June 15, 1818, advertising that his ironworks had "an inexhaustible pit of stone coal within one hundred yards of the forge. Three men with a horse and cart are sufficient to raise coke, and haul to the forge all the coal necessary for keeping the works in full operation."[48] The resulting coke was of unsatisfactory quality because coal extracted from the Redstone seam was hard coal with a high sulfur content. Meason returned to the use of charcoal within a few years.

The rising cost of charcoal fuel for blast furnaces and the success of eastern iron manufacturers using anthracite prompted iron manufacturers of western Pennsylvania to find a way to use the abundant bituminous coal, either as raw coal or as coke, to make high-quality iron. A number of iron manufacturers throughout western and central Pennsylvania began to experiment with the production of pig iron using coke and raw bituminous coal in their blast furnaces. These experiments in the 1830s and 1840s encountered technical difficulties similar to those that had plagued English ironmasters nearly a century earlier. The weakness of the blast in the furnace was a major technical problem. The first coke iron furnace built in the United States was at Bear Creek Furnace, south of Parker on Bear Creek, Armstrong County, in 1819. It was believed to be the largest furnace in the United States in 1832. Coke was used at the furnace to make iron, but after a few attempts the owners found that the (cold) air blast of five pounds to the inch was insufficient to raise the furnace temperature for the successful use of coke. The operators abandoned the use of coke after producing two or three tons of coke iron. The furnace changed back to the use of charcoal and went out of blast prior to 1850. Peter Ritter and John Say rebuilt the Karthaus Furnace at Karthaus, Clearfield County, on the west branch of the Susquehanna River in 1836 under the name of the Clearfield Coke and Iron Company. The furnace made pig iron from coke in place of charcoal. The iron furnace was subsequently acquired by a group of investors led by Henry C. Carey, John White, and Burd Patterson. They hired William Firmstone to convert the furnace to a hot-blast furnace, but poor transportation and an inferior grade of iron ore put an end to coke production the next year. Boston entrepreneurs attempted to smelt iron ore with coke at Farrandsville, Clinton County, but their attempts failed after they had made several hundred tons of coke pig iron between 1837 and 1839. A furnace at Frozen Run, Lycoming County, made pig iron from coke in 1838, but the furnace closed because of the inferior quality of the pig iron, and the furnace owners discontinued coke iron production and returned to the use of charcoal the next year.

Furnace men complained that coal miners were sending them dirty coal, while the latter retorted that furnace men did not know how to coke the coal properly, or failed in its proper application in the furnace. Coke was first successfully used for the production of pig iron by ironmaster William Firmstone at the Mary Ann Furnace in Trough Creek Valley, Huntingdon County, in 1835. He used semibituminous coal shipped to his iron plantation from the Broad Top Mountain in Huntingdon County. The furnace made gray forge iron from coke but the furnace used coke for only a month. Fideleo H. Oliphant made one hundred tons of coke iron at the Fairchance Iron Furnace near Uniontown, Fayette County, in 1836. The Georges Creek Coal Company built the Lonaconing Furnace, eight miles northwest of Frostburg, Maryland. The furnace was 50 feet high by 14 1/2 feet in the bosh and made seventy tons per week of good foundry iron using coke instead of charcoal, beginning in 1837. Five iron furnaces were in blast in Pennsylvania producing iron using coke by the 1840s.

Coke Making Technology

Coke is the residue of the destructive distillation of bituminous coal. It is not a fixed, uniform product, chemically or physically, but varies according to the type of coal or coal mixture used in its production and the methods used in its production. Coal was unsuitable for commercial use in the iron industry of this period because of the high phosphorous and sulfur content found in greater or lesser quantities in all bituminous coal. The process of coke making consists of baking the coal so that these elements are burned off. Coke was first produced in Pennsylvania employing a method similar to that used to make charcoal. Coal was first converted to coke in "mounds" or "ricks." Coal, which must be in lumps, was piled in the open air in a circular mound called a coking pit, with the lumps set at sharp angles so that airspaces remained. The coal pile was covered with nonflammable material and then ignited. It was burned slowly in a moist, smoldering heat that drove out impurities. The exterior oxygen was cut off by sealing all vents when the pile was ablaze. Coal was converted into coke in five to eight days, although this simple method of production was very wasteful, yielding less than 50 percent coke. It was also a slow process, but it was still used after the introduction of the dome-shaped brick beehive ovens, especially in areas where the demand for coke was small or its manufacture had just begun.

The first production of coke in a beehive coke oven took place within what are now the city limits of Connellsville, Fayette County, during the 1830s. These new ovens were called beehive ovens because the interior of the original ovens was shaped like a beehive. Mr. Nichols, a transplanted Englishman, is generally credited with building the first beehive coke oven in the U.S. Nichols is reputed to have made coke at Durham, England, and was responsible for a series of unsuccessful coke-making experiments at Meason's Union Furnace (Dunbar Furnace), south of Connellsville in the 1790s. He supervised the construction of a small oven on the property of Lester LeRoy Norton, who operated a small textile mill near Connellsville. Norton subsequently added a small foundry next to his textile mill in 1831. John Taylor, a local farmer and stonemason, was contracted by Nichols to construct a single twelve-foot-square oven of stone, with a hive of bricks, at a site near the foundry on Connell Run in 1833. Coal was obtained from Taylor's nearby Plummer Mine on a wedge of land formed by the junction of Mounts Creek with the Youghiogheny River, just north of Connellsville. Nichols made coke in the single beehive oven and also made it in ricks on the ground. Coal from the Pittsburgh seam made excellent coke that was in demand by local iron manufacturers in Fayette County. It was transported to Meason's foundry at Plumsock, down the Youghiogheny River to McKeesport, then up the Monongahela to Brownsville, then hauled in wagons to the foundry. The first beehive coke ovens were constructed during the 1830s but their widespread use for coke production was delayed until the 1850s.[1]

Notes

[1] Henry F. Walling and O. W. Gray, *New Topographic Atlas of the State of Pennsylvania* (Philadelphia: Steadman & Lyon, 1873), p. 27.

[2] Shyamal Majumdar and E. Willard Miller, eds., *Pennsylvania Coal: Resources, Technology, and Utilization* (University Park: Pennsylvania State University Press, 1983), p. 27.

[3] Priscilla Long, *Where the Sun Never Shines* (New York: Paragon Press, 1989), p. 4; Harold Bargar and Sam H. Schurr, *The Mining Industries, 1899-1939: A Study of Output Employment and Productivity* (New York: National Bureau of Economic Research, Inc., 1944), p. 162; Elwood S. Moore, *Coal: Its Properties, Analysis, Classification, Geology, Extractions, Uses, and Distribution* (London: John Wiley and Company, 1940), pp. 42-43.

[4] Moore, *Coal*, p. 116.

[5] Edward T. Devine, *Coal: Economic Problems of the Mining, Marketing and Consumption of Anthracite and Soft Coal in the United States* (Bloomington, Illinois: American Review Service Press, 1925), p. 14.

[6] Majumdar and Miller, *Pennsylvania Coal*, p. 56.

[7] Adam T. Shurick, *The Coal Industry* (Boston: Little, Brown and Company, 1924), p. 365; J. S. Burrows, "Geology and Location of the Coal Fields of Pennsylvania," *Coal Age*, vol. 6 (1914).

[8] James M. Swank, *Introduction to the History of Ironmaking and Coal Mining in Pennsylvania* (Philadelphia: published by author, 1878), p. 113.

[9] Moore, *Coal*, p. 380; E. Willard Miller, *Pennsylvania: Keystone to Progress* (New York: Windsor Publication, 1986) p. 68; Edward Eyre, Hunt Tyron, Fred G. Tyron, and J. H. Willits, eds, *What the Coal Commission Found* (Baltimore: Williams and Wilkins Company, 1925), p. 283.

[10] William Nichols, *The Story of American Coals* (Philadelphia: Lippincott Company, 1897), p. 31.

[11] David L. Salay, *Hard Coal, Hard Times: Ethnicity and Labor in the Anthracite Region* (Scranton: The Anthracite Museum Press, 1984), p. x; Robert D. Billinger, *Pennsylvania's Coal Industry* (Gettysburg: Pennsylvania Historical Association, 1954), p. 5.

[12] Shurick, *The Coal Industry*, p. 16; Devine, *Coal: Economic Problems*, p. 174.

[13] Shurick, *The Coal Industry*, p. 15; Harold M. Watkins, *Coal and Men: An Economic and Social Study of the British and American Coal Fields* (London: George Allen & Unwin Ltd., 1934), pp. 92-96; Moore, *Coal*, pp. 371-375.

[14] Shurick, *The Coal Industry*, p. 12; Charles E. Beachley, *The History of the Consolidation Coal Company, 1864-1934* (New York: Consolidation Coal Company, 1934). The Maryland Mining Company was the first company organized in the Georges Creek Field, established in the same year as ground was broken for the construction of the Chesapeake & Ohio Canal and B&O Railroad in 1828. The firm failed financially and was absorbed, along with the Georges Creek Coal and Iron Company (1835) and the Maryland and New York Company (1838), by the Consolidation Coal Company after it was organized in 1860. This firm was later controlled by the Rockefeller interests. Consolidation Coal Company became one of Pennsylvania's largest coal companies after it began acquiring mining properties of the Somerset Mining Company, Somerset County, at the beginning of the twentieth century.

[15] *History of Pennsylvania Coal* (Harrisburg: Pennsylvania Department of Mines, 1955), pp. 50-51; *Pennsylvania's Mineral Heritage* (Harrisburg: Pennsylvania State College School of Mineral Studies, 1944), p. 208.

[16] Howard N. Eavenson, "The Early History of the Pittsburgh Coal Bed," *Western Pennsylvania Historical Magazine*, vol. 22 (1939).

[17] Richard T. Wiley, *Monongahela: The River and Its Region* (Butler: Ziegler Press, 1937), p. 148; *Pennsylvania's Mineral Heritage*, p. 208; James Sisler, *Bituminous Coal Fields of Pennsylvania Part 2* (Harrisburg: Fourth Geological Survey, 1926), p. 19; Frederic Quivik, *Connellsville Coal and Coke* (Washington, D.C.: National Park Service unpublished, 1991), p. 3.

[18] Majumdar and Miller, *Pennsylvania Coal*, p. 32.

[19] Tommy Ehraber, "King Coal," *Pitt Magazine*, vol. 5, no. 1 (February 1990).

[20] Howard N. Eavenson, *Coal Through the Ages* (New York: New York American Institute of Mining and Metallurgical Engineers, 1942), p. 50.

[21] Howard N. Eavenson, *The First Century and a Quarter of American Coal Industry* (Pittsburgh: privately printed, 1942), p. 23.

[22] William E. Edmunds and Edwin F. Koppe, *Coal in Pennsylvania* (Harrisburg: Department of Environmental Resources, 1970), p. 2, 15; Lewis Clark Walkinshaw, *Annals of Southwestern Pennsylvania* (New York: Lewis Publishing Company, 1939), p. 202.

[23] Eavenson, *Coal Through the Ages*, p. 52; Billinger, *Coal Industry*, p. 29.

[24] Majumdar and Miller, *Pennsylvania Coal*, p. 33.

[25] George Swetnam, *Pittsylvania Country* (New York: Duell, Sloan, & Pierce, 1951), p. 228.

[26] Eavenson, *Coal Through the Ages*, pp. 55-57; Nichols, *The Story of American Coals*, p. 66.

[27] Jon D. Baugham and Ronald L. Morgan, *Tales of the Broad Top*, vol. 11 (n. p., 1979), p. 44.

[28] Ibid.

[29] Billinger, *Pennsylvania's Coal Industry*, pp. 34-35.

[30] Wiley, *Monongahela*, p. 177.

[31] George H. Ashley, *Bituminous Coal Fields of Pennsylvania* (Harrisburg: Department of Forests and Water, 1928), p. 179.

[32] Howard N. Eavenson, *The Pittsburgh Coal Bed: Its Early History and Development* (New York: American Institute of Mining and Metallurgical Engineers, 1938), p. 21.

[33] G. F. Deasy and P. R. Griess, *Atlas of Pennsylvania Coal and Coal Mining* (University Park: Pennsylvania College of Mineral Resources, 1959), p. 14; Frank H. Kneeland, "Large Stripping Operation," *Coal Age,* vol. 8, no. 4 (September 1915); P. J. McAuliffe, "Stripping a Mine by Hydraulic Methods," *Coal Age,* vol. 5, no. 14 (July 1913); *U.S. Thirteenth Census: Mines and Quarries* (Washington, D.C., 1909). Strip mining began near Danville, Illinois, in 1866 when horse-drawn plows and scrapers were used to remove the overburden so that the coal could be hauled away in wheelbarrows and cars. A steam-powered shovel was used to excavate some ten feet of overburden from a three-foot seam near Pittsburg, Kansas, in 1877.

[34] Eavenson, *The First Century,* p. 177.

[35] Ibid., p. 189; William Bining, "The Glass Industry of Western Pennsylvania, 1797-1857," *Western Pennsylvania Historical Magazine,* vol. 19 (1936).

[36] Eavenson, *The First Century,* p. 184.

[37] Douglas Fisher, *Epic of Steel* (New York: Harper and Row, 1963), p. 47, 52.

[38] Swank, *Introduction,* p. 71.

[39] Ibid.

[40] Ibid., p. 24; Kershaw Burbank, "Noble Ambitions: The Founding of the Franklin Institute," *Pennsylvania Heritage,* vol. 18 (September 1992).

[41] Anthony F. C. Wallace, *St. Clair: A Nineteenth Century Town's Experience with a Disaster-Prone Industry* (Ithaca: Cornell University Press, 1987), p. 85.

[42] Donald L. Miller and Richard E. Sharpless, *The Kingdom of Coal: Work, Enterprise, and Ethnic Communities* (Philadelphia: University of Pennsylvania Press, 1985), pp. 60-67.

[43] Ibid.

[44] William Sisson, Bruce Bomberger, and Diane Reed, "Iron and Steel Resources of Pennsylvania, 1716-1945" (Pennsylvania Historical and Museum Commission, Harrisburg, 1991), p. 32.

[45] Ibid., p. 30.

[46] John W. Oliver, *History of American Technology* (New York: Ronald Press, 1956), p. 168.

[47] Eavenson, *The First Century,* p. 172.

[48] Billinger, *Pennsylvania's Coal Industry,* p. 35; Evelyn Abraham, "Isaac Meason: The First Iron Master West of the Alleghenies," *Western Pennsylvania Historical Magazine,* vol. 20 (1937).

Transportation, Iron and Railroad Revolution as Impetus for the Expansion of the Coal and Coke Industries, 1840-1880

Introduction

The future growth and prosperity of the bituminous coal industry of Pennsylvania was dependent upon successfully breaking the energy monopoly of wood, and upon the creation of a viable transportation system. Coal, despite its limited early use and markets, was slowly becoming a new, revolutionary energy source in the United States, as production and consumption increased with each passing decade between 1850 and 1880. The demand for coal grew with the industrial expansion of "smokestack" America during this period. Anthracite (14.3 percent) and bituminous (26.7 percent) coal was producing 41 percent of the nation's energy supply by 1880, in contrast to 57 percent from wood. The sooty age of "King Coal" would commence by the end of this period, a successful challenge and eclipse of wood's long-term energy hegemony.

Population growth and rapid industrial development created severe wood shortages in parts of the country during this period. The shortage of wood and charcoal for use in the iron industry caused the price of wood to soar. Coal was the sole alternative fuel to wood, and served this energy need until the introduction of natural gas and petroleum in the 1880s.

Pennsylvania was the pioneer state in the establishment of the commercial coal industry in the United States. It had extensive reserves of high-quality anthracite and bituminous coal within its boundaries. Howard N. Eavensen, a former professor of mining engineering at the University of Pittsburgh and a past president of the American Institute of Mining and Metallurgical Engineering, observed in 1942: "[U]nlike the bituminous part of the coal industry the production of anthracite has been fairly well publicized; in fact until about 1845 whenever the coal industry of Pennsylvania was mentioned in papers, magazines or books, anthracite only was meant."[3] The anthracite production of Pennsylvania was greater than all bituminous coal production nationally until 1870. This was the principal coal as measured in production, monetary value of coal production, and employment in Pennsylvania until the 1890s. About 25,000 of the nation's 36,500 miners worked in the anthracite coal region producing 73 percent of the total value of American coal in 1860.[4] Between 1840 and 1860 the bituminous coal industry in Pennsylvania had greatly expanded, but its production was dwarfed by the state's anthracite industry. Fewer than five thousand miners were employed at the early bituminous drift-entry mines of Pennsylvania by 1860.

The bituminous coal industry had established a secure but still modest place for itself in the nation's economy. The combined annual production of anthracite and bituminous coal rose from 2 million tons in 1840 to 8.3 million tons in 1850, and 20 million tons nationwide in 1860. More than three-quarters of this total production was concentrated in Pennsylvania. Ohio was second with 600,000 tons annually; Virginia (including the western counties that became West Virginia in 1863) produced 350,000; and Illinois and Maryland mined about 250,000 tons each in 1860.

Energy Sources as Percentage of Aggregate Energy Consumption in the U.S.

	Bituminous Coal	Anthracite Coal	Wood
1850	4.7 percent	4.6 percent	90.7 percent
1855	7.3 percent	7.7 percent	85.0 percent
1860	7.7 percent	8.7 percent	83.5 percent
1870	13.8 percent	12.7 percent	73.2 percent
1880	26.7 percent	14.3 percent	57.0 percent [1]

Coal Production in Millions of Tons

Years	Bituminous	Anthracite	Total	Percent Increase
1850	4,029	4,327	8,356	
1855	7.543	8,607	16,150	
1860	9,057	10,984	20,041	1850-1860, 139.8 percent
1865	12,349	12,077	24,426	
1870	20,471	19,958	40,429	1860-1870, 101.7 percent
1875	32,657	23,121	55,778	
1880	50,757	28,650	79,407	1870-1880, 96 percent [2]

Anthracite and Bituminous Industry of Pennsylvania in 1860

	Number of Establishments	Capitalization	No. of Employees	Value of Product
Anthracite	176	$13,888,250	25,126	$11,869,574
Bituminous	134	$ 3,721,780	4,651	$ 2,876,579[5]

31

Transportation

Transportation innovations in Pennsylvania broadened the commercial markets of the bituminous coal industry in terms of both production and employment. The state's bituminous production was concentrated in the Monongahela Valley between Pittsburgh and Brownsville, a distance of some fifty miles. Production in this compact region was 357,140 tons in 1840, of which 70 percent was consumed locally.[6] The western coal trade had been restricted by poor navigation on the Monongahela to the river towns on the Ohio and Mississippi Rivers. Only limited quantities of coal could be shipped during the summer because of low water levels on the river. The lack of an adequate transportation system had restricted coal export from the coal-producing counties of western Pennsylvania. The needs of the eastern seaboard markets had been filled by anthracite producers of Pennsylvania and Virginia and coal importation from England and Wales.

Colonel Robert W. Milnor, chief engineer of the Monongahela Navigation Company, in a report submitted to the General Assembly of Pennsylvania on December 24, 1839, defined the need for river improvements and their economic advantages for the local coal trade:

> During the year 1837, a large number of flat boats were loaded at various points along the Monongahela, but, at that period when the owners wished to carry it to market, there was not sufficient depth of water on the ripples to enable them to float to the Ohio river . . . in October of last year there were 150 flat boats at the coal landing up the Monongahela River, which had been waiting upwards to three months for the rise of water, in order to get to market.[7]

The Monongahela Navigation Company, a private corporation, was formed in 1836 to begin slack water improvements on the Monongahela River from Pittsburgh to the West, organized primarily to aid the local coal trade. The company constructed a series of locks and dams on the river. Locks 1 and 2 were completed by 1841. Each lock consisted of a single chamber 190 feet long and 50 feet wide, and their completion by 1844 provided 55 1/2 miles of continuous slack water transportation between Pittsburgh and Brownsville. Connellsville coke manufacturers on the Youghiogheny River, the principal tributary of the Monongahela, petitioned the General Assembly of Pennsylvania to establish a navigation company to build locks and dams on the Youghiogheny. The Connellsville and West Newton Navigation Company was incorporated in 1841 to make the river navigable between these two river towns but the company never completed the task. The Youghiogheny Navigation Company, incorporated in 1843, completed construction of a series of locks and dams on the river in 1850. The new locks and dams on the rivers controlled the depth of water and permitted year-round river transportation, except when the river was obstructed by ice.

The western coal trade was not created by the Monongahela Navigation Company, but slack water river improvements were responsible for greatly increasing the volume of the coal trade. Coal was the principal product of the Monongahela Valley, although agricultural products, especially fruits and grains, were important goods shipped downstream. These river improvements enabled an expansion of mining from Pittsburgh to New Geneva above Brownsville in the 1840s. Dozens of new mines were opened as a consequence of these improvements and demands for coal by the river communities along the Ohio and Mississippi Rivers. There were about sixty small mining operators in Allegheny, Fayette, Washington, and Westmoreland Counties clustered around the Monongahela and Youghiogheny Rivers by 1859.[8] The volume of coal shipments transported from the region rose significantly with each passing decade. Some 750,000 bushels of coal were shipped in 1844, the year the internal improvements were completed. Coal, and later coke, shipments increased to 12.2 million bushels in 1850, 22.2 million bushels in 1855, and 37.9

million bushels in 1860. The region mined and shipped 1,076,820,914 bushels of coal in 1877.[9]

"French Creek" flatboats, barges, and arks had been the principal modes of river transportation for coal until this period. The construction of the *New Orleans*, built at Sukes Run near Pittsburgh on the Monongahela River, began the steamboat construction industry of western Pennsylvania. Nicholas Roosevelt, Robert Fulton, and New York business associates spent nearly forty thousand dollars and one year in its construction. Roosevelt, with his wife, piloted the *New Orleans*, powered by local coal, from Pittsburgh to New Orleans in 1811. Steamboats and steam engines were manufactured in Pittsburgh, and later Cincinnati, after this successful voyage. Pittsburgh built forty-five steamboats and Cincinnati thirty-two in 1847 alone. Steamboats were not immediately used to transport coal down the river to market. Captain David Bushnell towed three small coal barges, each loaded with two thousand bushels of coal, using his small stern-wheeler steamboat, the *Walter Forward*, to Cincinnati in 1845. Bushnell's successful trip revolutionized coal transportation by ushering in the era of the steam towboat. Steamboat transportation of coal was in vogue within five years on the western rivers. Barges and flatboats were built in large numbers to carry cargoes of coal, attached to steamboats for trips down the Monongahela River to the Ohio River and then south to the markets of Ohio and the lower Mississippi Valley. A steamboat could tow or push a dozen barges, with each barge holding about 12,000 bushels of coal. Some 184,200 tons of coal were shipped on the Monongahela to the Ohio River and Cincinnati in 1845, and more than 1.5 million tons were being transported by 1860. Fewer than 4,200 barrels of western Pennsylvania coal had been shipped via the Monongahela River to New Orleans in 1816. Some 1,510,000 barrels of coal had been shipped to New Orleans by 1861.

The potentially lucrative eastern seaboard coal trade remained untapped. Its demands for coal were satisfied chiefly by the bituminous coalfields of Maryland and Virginia, anthracite coal shipped from eastern Pennsylvania, and imported English and Welsh coal. The coal industry was responsible, to a great extent, for the initial development of canals and railroads in the United States. In northeastern Pennsylvania they were built to haul anthracite to the principal cities on the eastern seaboard. The Mauch Chunk Railroad was the first steam railroad in the state and the second in the United States. The construction of this nine-mile-long railroad was begun in 1818 and completed nine years later by the Lehigh Coal and Navigation Company. At least a dozen railroads were constructed in the anthracite region in the next two decades to carry coal from the mines to the river and canals and then to market.

The first attempt to tap the potentially lucrative eastern coal market by coal operators in western Pennsylvania came, during the age of the canal, with the successful completion of Governor DeWitt Clinton's Erie Canal. The success after 1825 of the New York canal, which connected New York City to Buffalo and Lake Erie, spawned an era of canal building throughout the nation. Pennsylvania's own canal era began with the completion of the Pennsylvania Main Line Canal during the 1830s. This 394-mile system cost about $38 million to construct, and was a combination railroad, canal, and incline railroad system. The canal started at Columbia on the Susquehanna River and traveled west through the Juniata River valley to Hollidaysburg, then over the Allegheny Mountains via the Allegheny Portage Railroad and inclined planes to Johnstown. The canal traveled from Johnstown through the valleys of the Conemaugh, Kiskiminetas, and Allegheny Rivers to Pittsburgh. The canal, with its thirty-seven-mile Portage Railroad with ten inclines, took eight years to complete. The first boat arrived from Philadelphia at Allegheny City, north of Pittsburgh, in 1834. Coal operators of the Monongahela Valley saw the development of the Main Line Canal and the Allegheny Portage Railroad as a potentially viable transportation route to sell their coal in the eastern seaboard markets.

Moncure Robinson, chief engineer for the Allegheny Portage Railroad, concurred with their sentiment and urged the installation of more-powerful stationary

The Allegheny Portage Railroad's inclined plane utilized stationary steam engines to haul canal boats over the Allegheny Mountains. Pennsylvania State Archives.

steam engines for the five inclined planes located west of the summit, to haul the heavy coal-laden barges over the Allegheny Mountains from Johnstown to Hollidaysburg. Neither Robinson's nor the coal operators' high expectations that the canal and the Portage Railroad would open these potentially profitable eastern markets came to fruition. In fact, very few coal-laden barges traveled east on the Pennsylvania Canal System over the Portage Railroad from the mines of the Monongahela Valley. A mere five tons of coal were transported from Pittsburgh over the canal in 1832. The maximum annual tonnage of coal shipped east from the Pittsburgh region never reached thirty-thousand tons between 1832 and 1880. Some 29,234 tons of coal were shipped to eastern markets in 1854, the highest annual tonnage.[10]

The inclined railroad and the canal provided an outlet for coal mined on the eastern edge of the Great Allegheny coalfield. The Broad Top Mountain coal region is an isolated semibituminous coal region forming the eastern boundary of the bituminous coal basin. Parts of the region are situated in Bedford, Huntingdon and Fulton Counties. Semibituminous coal, iron, and grain had been shipped down the Juniata River to the canal from these counties and from Blair and Centre Counties since its completion in 1832. Some 973 tons of semibituminous coal from the Broad Top coal region and Centre County had been shipped on the Western Division of the Pennsylvania Canal by 1843.

Coal samples from mines located on the Monongahela River were shipped east in 1835 and tested for making "gas-coal" for illumination purposes at the gas works of Philadelphia and New York. The report found this coal suitable for such use, but little was actually shipped east until the advent of the railroad several decades later. The completion of the Pennsylvania Canal did not provide a viable solution to the transportation problem for the coal-producing counties bordering the Monongahela River. Excessive freight rates, narrow locks, and competition from Virginia and Maryland coal producers, anthracite producers from eastern Pennsylvania, and coal imported from Great Britain made the western bituminous coal trade with the east unprofitable at this time.

The opening of the few miles of the Baltimore and Ohio Railroad and the South Carolina railroads during the 1830s marked the beginning of the railroad era in the United States, and although these early railroads were limited, they proved the economic value of steam railway transportation. Eleven state legislatures had granted more than two hundred railroad charters by 1835. The Baltimore and Ohio Railroad was organized in 1827 by a group of progressive businessmen in Baltimore who believed the construction of a railroad "from Baltimore to some eligible point on the Ohio river would tap the growing western trade and make Baltimore second only to New Orleans as the great outlet to the west."[11]

The locomotives of the 1830s and 1840s were powered primarily by wood, but the increasing scarcity and the subsequent higher cost of wood prompted railroad companies to search for an alternative fuel that was both inexpensive and plentiful. The B&O was an early user of anthracite coal, which cost the company seven dollars per ton in the 1830s. The firm began experimenting with the use of bituminous coal in its locomotives in 1837, when the railroad reached as far west as Cumberland, Maryland. Bituminous was obtained from the nearby Georges Creek coalfield of Maryland and Pennsylvania. Bituminous coal began to replace anthracite by 1840-1841 because it was a more economical fuel. A number of technical problems in pre-Civil War locomotives, however delayed the widespread use of bituminous.

Coal, unlike wood, generated hotter fires that burned through the locomotive's fire box, created voluminous smoke, and generated sparks that flew from the smokestack and often set nearby farm buildings on fire. Railroad operators expended large sums of money and time overcoming these problems, which persisted for nearly a quarter century, but on the eve of the Civil War the principal railroads—the Reading, the Pennsylvania, and the Baltimore and Ohio—were all operating coal-burning locomotives. The B&O Railroad consumed seventy-five thousand tons of coal in its locomotives in 1856. All freight locomotives operated by the Pennsylvania Railroad were burning coal, not wood, by 1862. Railroads became a principal market for bituminous coal during this period and remained so until the widespread usage of diesel locomotives began during the 1930s.

The American railroad era began in earnest during the 1850s. The inability of the Pennsylvania canal system to provide a viable system to transport coal to eastern markets was one of the reasons for the formation of the Pennsylvania Railroad Company (PRR). The railroad was incorporated by the General Assembly of Pennsylvania on April 13, 1846, and the construction of the 249-mile railroad linking Harrisburg to Pittsburgh was soon begun. The railroad entered Pittsburgh on December 10, 1852, and its completion began a railroad-building boom throughout the state. By 1880 railroads had opened up several isolated but rich coalfields in western and central Pennsylvania.

Meanwhile, the Westmoreland Coal Company of Philadelphia pleaded with the directors of the Pennsylvania Railroad to reduce transportation rates so that they could ship their coal to the gas works of Philadelphia and New York. General William Larimer Jr., one of the principal owners of this large antebellum coal company, argued before the directors of the Pennsylvania Railroad:

> Gentlemen: If you will grant us a freight rate that will permit us to enter into a contract with the city gas works to supply it with coal your company will enjoy the benefits of earning on that tonnage and we shall be able to reap benefits for our miners and our shareholders. All of us would be benefitted and none injured in the slightest degree. Other sales would assuredly follow.[12]

Larimer convinced the directors of the PRR in 1855 to reduce rates to less than six dollars a ton to haul coal east. The company acquired twenty-four railroad cars in 1856 and began to ship coal by rail to the eastern market; this was the first use of railroads to transport coal to this market.

The Westmoreland Coal Company had been incorporated on July 5, 1854, by the General Assembly of Pennsylvania. The firm had its corporate office at the Fidelity-Philadelphia Trust Building, Philadelphia, with a regional office at Irwin, Westmoreland County. The original board of directors were General Larimer of Pittsburgh, John Covode of Lockport, James Magee and John Scott of Jeannette, and Herman Haupt of Philadelphia. Stockholders were prominent businessmen from Philadelphia, Baltimore, and Pittsburgh. Covode was elected the first president of the Westmoreland Coal Company while William F. Caruther served as its first mine superintendent until 1872, at which time he became the paymaster for five years. The Oak Grove mine (later renamed the Old North Side mine) near Irwin was the company's first mine. From these humble beginnings this coal company acquired more coal properties and opened mines principally located in the Irwin Coal Basin in Westmoreland County. It was one of the earliest and largest mining companies in Pennsylvania or the nation. Today the Westmoreland Coal Company is the oldest incorporated bituminous coal company in the United States retaining its original name.[13]

The introduction of coal gas for illumination purposes was a major factor in the expansion of the coal mining industry of western Pennsylvania. Mine operators located in the counties bordering the Monongahela Valley had found a viable new market for their coal to supplement their western river trade on the Ohio and Mississippi Rivers. They were soon shipping coal by rail to the fifty-eight gas works in

Philadelphia and New York. The Philadelphia Gas Works, one of the largest and earliest gas works operated by the city government, became a principal consumer of western Pennsylvania coal, replacing coal from Virginia and Wales by the end of the 1850s. Coal converted into coal gas heated homes and illuminated churches, auditoriums and street lamps. Some 50,904 tons of gas-coal from the Irwin Gas Basin, Westmoreland County, were shipped in 1855 by the P.R.R. Westmoreland County shipped some 206,636 tons in 1860, 866,498 tons in 1870, and 943,117 tons in 1880 to eastern markets.

Like bituminous coal, semibituminous coal of Pennsylvania, known locally as "smokeless" coal, was shipped by rail to distant commercial markets before the Civil War. "Smokeless" coal was an excellent steam and smithing coal mined in the North Central Fields and the Broad Top Field. The North Central Fields, composed of five small coalfields located in Tioga, Bradford, and Lycoming Counties, and the Broad Top Field shipped semibituminous coal to eastern markets after 1840. Wood was still an abundant and low-cost fuel in most parts of the nation, although some communities saw their supply of timber dwindle. The southern counties of New York bordering the northern counties of Tioga and Bradford, Pennsylvania, faced a critical wood shortage during the 1830s. The General Assembly of New York formed a committee to find a solution to this energy crisis. The committee reported that "the bituminous coal has been brought to us either from Great Britain or from that part of our country which borders on, and lies west and south of the Susquehanna." It was known by this time that coal deposits were located in the mountains at the head of the Tioga River. The General Assembly promoted the construction of the the Corning and Blossburg Railroad to run between Blossburg and Corning to connect with the Erie Railroad, which reached north to Corning in 1838. The railroad, constructed by the Tioga Navigation Company, connected the Chemung Canal and the Erie Railroad with the semibituminous coalfields of Tioga County, permitting coal shipment to the southern counties of New York. Some 4,325 tons were initially sent by rail to New York State in 1840 and 25,966 tons in the following year. The Arbon Coal Company, founded by William Magee, was the first coal company in Tioga County to ship coal by rail, supplying three rolling mills in Troy, New York. The firm continued in the soft coal business until succeeded in 1842 by W. M. Mallory and Company, a firm that failed during the Panic of 1857. Imported coal was still used at salt works and by blacksmiths throughout western New York. Regional mines had shipped about a half million tons of coal to New York State by 1860. The Blossburg Coal Company developed a large tract of coal land in the Blossburg Field four miles west of Blossburg Borough in 1866.

A railroad was constructed in neighboring Bradford County from Towanda to Barclay to open up the Barclay Mining District in the county. Bradford County, located east of Tioga County, contains the eastern extension of bituminous coal in Pennsylvania. The Barclay Coal Company was formed in September 1856, and shipped coal north to New York State. Towanda, Montezuma, Syracuse, Oswego, Utica, Troy, Rochester, and Buffalo were the principal coal markets. Coal was used for blacksmithing and rolling mills (44 percent), steamboats (20 percent), stationary engines (18 percent), saltworks at Syracuse (8 percent), and glassworks (10 percent). Semibituminous coal was also located in Lycoming County in an isolated field north of Williamsport. Coal was mined east of Lycoming Creek on Lick Run, but its production was restricted to local consumption until the construction of the Williamsport and Elmira Railroad. This railroad connected these isolated coal mines and provided eastern markets with coal for "locomotives, stationary engines, in the manufacturing of bar-iron and for house-hold purposes."

Two railroads served the Broad Top coal district. The Huntingdon and Broadtop Railroad and Coal Company (H&BT) tried to obtain a charter from the state legislature in 1846 to construct a railroad and engage in commercial coal production in the western side of Broad Top Mountain for eastern urban markets—especially Philadelphia. The Pennsylvania Railroad was extended from Philadelphia through

Huntingdon County to Altoona in 1850, reaching Pittsburgh by 1852.

The bill was passed in 1851 by the General Assembly, but was vetoed by the governor. The H&BT Railroad finally received a state charter in 1852, and a railroad line was completed linking the western Broad Top coalfield southward at Mount Dallas with the branch of the Pennsylvania Railroad which southward joins the Baltimore and Ohio at Hyndman, and northward joins the main line of the Pennsylvania Railroad at Altoona. The other end of the Huntingdon and Broad Top Railroad leads northward to the main line of the Pennsylvania Railroad at Huntingdon. Coal was first shipped from this railroad to the Pennsylvania Railroad in February 1856. The *Huntingdon Journal* reported in its August 13, 1856, edition on the opening of the railroad that "there is no end to the demand for Broad Top coal. Although hundreds of tons are brought in daily, yet the demand is greater than can be supplied." The railroad was transporting 267,720 tons of coal annually to eastern markets by the 1860s.[14]

East Broad Top Railroad was established in 1872-3, and is currently running as an excursion line.
Photo by Gordon R. Roth, Pennsylvania Historical and Museum Commission.

The East Broad Top Railroad (EBT), built in 1872-1873, was a three-foot-wide, thirty-mile narrow-gauge railroad constructed to transport semibituminous coal from the Broad Top mines of Huntingdon County northeastward to the main line of the Pennsylvania Railroad at Mount Union, where the coal was transferred to standard-gauge railroad cars for shipment to eastern markets. This narrow-gauge railroad also shipped sand, rock, lumber, and general freight to eastern markets. Broad Top coal was a high-grade, low-volatile semibituminous coal, which was principally shipped by rail east for use in general industries, railroad locomotives, coking, and domestic uses.

Bituminous coal was being mined in twenty-nine counties of Pennsylvania by 1860, but within a generation coal production in Bradford, Cameron, McKean, Venanago, Lycoming, and Warren Counties had become negligible. River improvements and the growth of the railroads were responsible for increasing bituminous coal production and markets. By 1880 the principal coal producing regions in Pennsylvania were the counties bordering on the Monongahela River and the semibituminous Broad Top coalfields, and the North Central. The Pittsburgh district, including the counties of Allegheny, Fayette, Washington, and Westmoreland, was the major bituminous coal region in Pennsylvania, producing about 5.5 million tons in 1870. Owing to its small hydrogen content, semibituminous coal was in great demand in eastern markets as a steam coal. The two semibituminous Broad Top and North Central Fields were producing 2.1 million tons of coal annually by 1870.

Railroads became a new and major user of coal and consumed nearly one-quarter of all mined coal as late as the 1930s. The transportation of coal became a principal source of revenue for many railroads. After the Civil War some railroads were constructed with the single purpose of hauling coal to market. More than thirty railroad lines were operating in Pennsylvania by 1859. Branch lines of the Pennsylvania Railroad Main Line began connecting the new mining communities scattered throughout western Pennsylvania. The Baltimore and Ohio Railroad followed the Youghiogheny River across

Coal Companies and Output in Semibitumious Coal Regions of Pennsylvania

1872-Semibituminous	Gross Tons
Fall Brook Coal Company, Blossburg	312,466
Morris Run Coal Coal, Blossburg	357,384
Blossburg Coal Company, Blossburg	321,207
Mcintyre Coal Company, Ralston	212,462
Towanda Coal Company, Towanda	252,329
Fall Creek Coal Company, Towanda	85,315
Snow Shoe, Centre County	95,257
Clearfield County	612,036
Broad Top, Huntingdon County	350,246[15]

the southwest corner of Westmoreland County with a branch line to Everson and Mount Pleasant. The Pittsburgh and Lake Erie Railroad followed the west bank of the Youghiogheny River and crossed into Rostraver Township. The Ligonier Valley Railroad, a narrow-gauge railroad, operated between Latrobe, Ligonier, and the new mining town of Wilpen. The Pittsburgh, Westmoreland and Somerset Railroad operated from Ligonier across the eastern county line into Somerset County.

Railroads, Iron and Coke

The character of the iron industry changed during this period. The railroads displaced farmers as the principal consumer of iron. Iron manufacturers of Pennsylvania had been chiefly producers of agricultural tools, grates, and stoves, but the increased demand for iron rails led to the formation of a number of iron manufacturers specializing in the production of rails. The construction of the Pennsylvania Railroad created new demands for iron rails. The percentage of the total American pig-iron output used in making iron rails reflects this demand for rails: 1849, 4.8 percent; 1855, 25.3 percent; 1860, 31.8 percent; 1865, 54.6 percent; 1870, 47.5 percent, 1875, 49.9 percent; 1880, 48.6 percent.[16] The construction of new railroads opened markets for coke and anthracite, the principal fuel in the blast furnace during this period. With this boom, the need for miles of iron rails for the expanding railroad industry placed demands on iron manufacturers for increased iron production at reduced costs. The early railroad companies ran on a strap rail fastened to wood. These rails were susceptible to mishaps, and the advent of heavier trains created a demand for improved and stronger rails. Wrought-iron rails, imported from England, began to replace the strap rails. American rail mileage increased from thirty thousand in 1850 to fifty-three thousand in 1870 and ninety-four thousand in 1880. Iron-rail production rose from less than 25,000 net tons in 1849 to 365,923 net tons in 1865, although the railroad industry was still heavily dependent on English imports.

The Mount Savage Rolling Mill, Allegany County, Maryland, became the first manufacturer to produce heavy rails in 1844 when it began to make the inverted "U" rail. The Montour Rolling Mill at Danville, Pennsylvania, which opened in 1845, also manufactured iron rails and made the first "T" rails in the country. A number of iron companies were established in Pennsylvania during the 1850s, including the Cambria Iron Works (Johnstown), Bethlehem Iron Company (Bethlehem) and Jones and Laughlin (Pittsburgh). All these iron companies were manufacturers of iron rails. The Cambria Iron Works, founded at Johnstown in 1852, constructed four coke blast furnaces to manufacture pig iron for iron rails. The Cambria Iron Works and the Great Western Iron Company (Bradys Bend, Armstrong County) turned out about one-seventh of the 150,000 tons of iron rails manufactured in the United States in the decade prior to the Civil War.[17]

Iron production in blast furnaces using anthracite had exceeded charcoal iron production by the mid-1850s. Charcoal pig-iron production declined from 100 percent in 1840 to 45 percent around 1855, and to only 25 percent in 1866. Hot-blast furnaces using anthracite produced nearly one-third of all the pig iron made in Pennsylvania by 1860. About a half-million tons of anthracite iron were produced in the United States in 1860, chiefly in eastern Pennsylvania. There were 28 anthracite iron furnaces in 1845, 60 in 1849, and 121 in 1853 on established coal routes. By the 1850s anthracite furnaces had been constructed, near or within the limits of eastern towns, including Danville, Pottstown, Steelton, Scranton, Allentown, Reading, Norristown, Phoenixville, and Columbia. Anthracite pig iron cost twelve dollars a ton in Pennsylvania compared with sixteen dollars a ton for charcoal pig iron. Anthracite was the principal fuel being used in iron smelting by the mid-nineteenth century and continued so until it was surpassed by coke during the 1880s. In contrast, the development and use of bituminous coal and coke for iron smelting in western Pennsylvania came about more slowly.

The development of the coke industry in Pennsylvania was closely intertwined with iron production, which became the principal market for coke during this period. Coke production in western Pennsylvania began in the first quarter of the nineteenth century, but its acceptance as a metallurgical fuel by iron manufacturers was slow. Coke usage in iron furnaces in western and central Pennsylvania began during the 1830s; however, its widespread use as a blast furnace fuel developed very slowly during the next two decades. Pig-iron production from coke or bituminous coal was a modest 54,485 tons in 1854. Pennsylvania produced 29,941 tons, Ohio 15,000 tons, and all other states produced 9,544 tons.[18] There were twenty-one blast furnaces in Pennsylvania and three blast furnaces in Maryland using coke as fuel by 1856. Their total production was 44,481 gross tons of pig iron.[19] The blast furnaces of eastern Pennsylvania used anthracite during the transition era between charcoal and coke. The annual production in1871 of pig iron smelted by anthracite was 957,000 tons, iron using bituminous coal and coke 570,000 tons, and iron smelted with charcoal 385,000 tons.[20] Less than 8 percent of all iron manufactured in Pennsylvania was using either coke or raw bituminous coal by 1855, but at the end of this period usage had soared to 45 percent. By 1870 more pig iron was being made with coke than with charcoal; coke use surpassed anthracite fuel in the manufacture of iron rails after 1875.

The coke industry, like the iron industry of western Pennsylvania, remained small and localized until after 1859. The Clinton Furnace, erected by Graff, Bennett and Company in 1859 near Pittsburgh, ushered in coke as a reliable blast furnace fuel. The Clinton Furnace was the first successful iron blast furnace in Allegheny County after the demise of the ill-fated Shadyside Furnace, built in 1793 by George Anshutz, an Alsatian, and his partners, William Amberson and Francis Beelen, on Two Mile Creek east of Pittsburgh. The furnace was abandoned in 1794 because of the high cost of importing iron ore from Armstrong County. The Clinton Furnace, located on West Carson Street in Monongahela Borough, a short distance below the Pittsburgh and Lake Erie Terminal, was a small furnace with a single stack, forty-five feet high, a twelve-foot bosh, and an annual capacity of twelve thousand tons. The furnace was located next to the Clinton Iron Works and was first blown in the fall of 1859. It made pig iron using coke manufactured from coal deposits near Mount Washington. Coke had been made in kilns and by the "mound" method at this location since 1833. The Clinton furnace operated for about three months until its owners, John Graff and James I. Bennett, shut it down because local coke had proven unsatisfactory in producing high-quality iron.

A trial run was made at the Clinton Furnace in 1860 with Connellsville coke imported by river barges from Fayette County. Management noted that "the result was so satisfactory that . . . arrangements [were] made to secure a continuous supply."[22] Joseph Weeks noted in 1885 that "it was not, however, until the development of the Connellsville Region, Pennsylvania, that the use of coke as a blast furnace fuel for the manufacture of iron itself in the country assumed any importance."[23] The successful experiment with Connellsville coke at the Clinton Furnace in 1860 acted as a powerful catalyst for a coal and coke boom in the Connellsville district.

The Connellsville coke district, which is located entirely within the Pittsburgh seam and lies at the northern end of the Appalachian coalfields, became the principal coke-producing region in Pennsylvania and subsequently the nation during the 1860s. This compact district straddling Fayette and Westmoreland Counties in southwestern Pennsylvania extends "from a point near Latrobe on the Pennsylvania Railroad, in a southwesterly direction through Westmoreland and Fayette counties, a distance of 42 miles, almost to the [West] Virginia line, with an average width of 3.5 miles, covering an area of 147 square miles, and excluding barren measures, originally contained 88,000 acres."[24] This area was originally called theYoughio-

Percentage of Pig Iron Production with Various Fuels

	Bituminous Coal and Coke	Anthracite	Charcoal
1855	8	46	47
1860	13	57	30
1865	20	52	28
1870	31	50	20
1875	41	40	18
1880	45	42	12 [21]

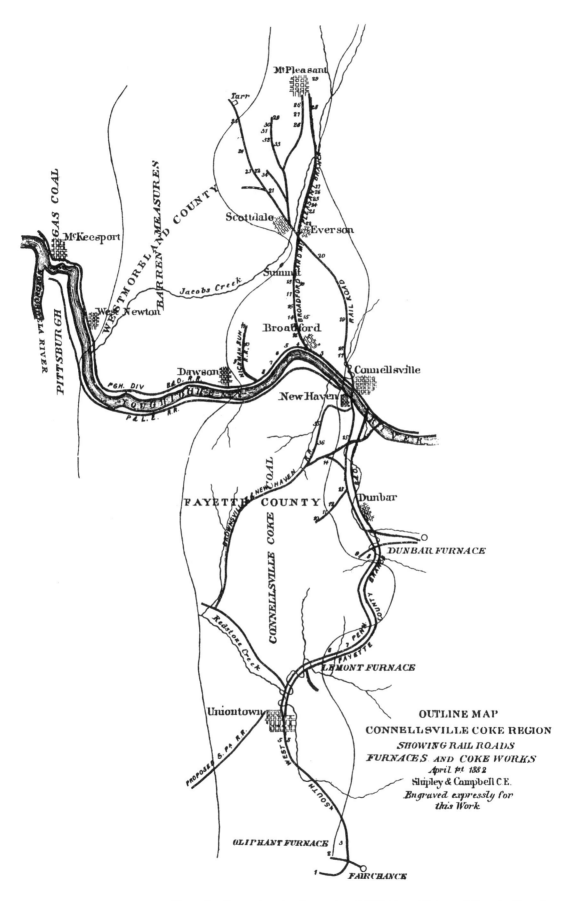

Connellsville coal and coke region, 1882. From History of Fayette County, Pennsylvania, *1982.*

gheny coke district because of the concentration of coke plants on or near the Youghiogheny River at Connellsville and at McKeesport, Allegheny County. The designation, Connellsville coke district, was first used to describe the district in an article in the *Keystone Courier* newspaper in 1879.

Coal from this part of the Pittsburgh seam was recognized as being physically and chemically suited for the production of high-quality metalurgical coke for the iron blast furnace, and soon acquired a national reputation. A contemporary observer noted that bituminous coal in this part of the Pittsburgh seam was "significant because there is no other seam that can compete with it in cheapness of production. There is no other coal so regular in form; so uniform in quality; of so convenient a thickness (8 to 9 feet); or so easily mined."[25] The carbon content of Pittsburgh coal ranged from 57 to 60 percent, volatile matter from 28 to 40 percent, ash content from 6 to 10 percent, moisture content under 2 percent, and most important its sulfur content was less than 1 percent.[26] Connellsville coal was used as run-of-the-mine coal in the blast furnaces, without the need to wash it to remove impurities often present in inferior grades of coal. Connellsville coke is of a silvery lustre and cellular, with a metallic ring, and its purity and strength made it the nation's best blast furnace coke.

In the 1860s, the market for coke shifted from foundries and forges to blast furnaces. Five more blast furnaces were constructed in or near Pittsburgh during the 1860s to use Connellsville coke. Each furnace had an annual capacity of eight to fifteen thousand tons of iron, several times that of the typical rural charcoal furnace. About twenty blast furnaces were erected in the greater Pittsburgh region between 1861 and 1887 and all used imported Connellsville coke. Ironmaking in the rural areas of western Pennsylvania began a period of rapid decline after 1860. Allegheny County alone produced 80 percent of the pig iron in the Pittsburgh region (Allegheny, Armstrong, Beaver, Butler, Washington, Westmoreland, and Fayette Counties) in 1880.

The iron and later the steel barons of Pittsburgh and the surrounding mill towns up and down the Monongahela River took advantage of the proximity of high-quality coke and excellent rail and river transportation to ship Connellsville coke to their facilities. Coke was shipped directly by the Baltimore and Ohio Main Line, the Mount Pleasant Branch of the Baltimore and Ohio, the Southwestern Railroad, and numerous branch lines recently constructed, or on river barges pulled by steamboats on the Monongahela River. Connellsville coke was soon shipped by rail to blast furnaces throughout Pennsylvania, Illinois, Michigan, and Ohio. It was also transported to California and the Rocky Mountain states where it was used in the smelting of gold and silver ore.

Coke was manufactured in Pennsylvania (92 percent) and Ohio (8 percent) in 1870. Alabama, Colorado, Georgia, Illinois, Indiana, Ohio, Pennsylvania, Tennessee, and West Virginia were the nine coke-producing states in 1880, although Pennsylvania, Ohio, and West Virginia produced a majority of American coke.

Pennsylvania produced 84.2 percent of the 2,752,475 tons of coke produced in the nation in 1880. The Connellsville district had seven thousand beehive coke ovens in 1880 and produced over two-thirds of the nation's coke. Fayette County produced 45.8 percent of all coke, while Westmoreland County was second with 27.4 percent. Allegheny County, the third most productive coke county in Pennsylvania, produced only about 3.5 percent.[28] The excellent quality of this coal, its accessibility, and large reserves were important factors leading to the concentration of the coke industry in this region. About 75 percent of all wage earners in this compact region, located less than fifty miles south of Pittsburgh, were employed in the coal-mining and coke industries by 1880. The beehive coke ovens of Pennsylvania manufactured over 65 percent of United States coke, with a majority of this production centered in the Connellsville district at the turn of the twentieth century.

The American Coke Industry in 1880			
(1880)	Establishments	Capital	Employees
Pennsylvania	104	$4,262,525	2,444
Ohio	15	$144,012	153
West Virginia	12	$330,000	163
U.S. Total	149	$5,545,058	3,068[27]

Coke Making Technology

Two principal methods were employed to produce coke during this period. Coal was originally converted in a manner similar to the production of charcoal from wood. Iron manufacturers made charcoal from wood by stacking logs in a conical pile with an opening at the top, covering the logs with earth, and building a smouldering fire inside the pile. The heat of the fire caused the logs to release their fluids in the form of steam and heavy oils, leaving behind charcoal. The earliest coke-production process was known by a variety of names—"coking in coke-fires," on "coke-hearths," "in ricks," "racks," and "on the ground." This was not a revolutionary coke-making process, but it was a popular method of coke production because it required little capital and no permanent investment in equipment, and coke could be made anywhere coal was available. This simple and inexpensive process, however, had a number of deficiencies which, over time, outreached the advantages. It was a wasteful and unpredictable method of coke production. It required a specific type of coal; weather influenced the process; and under adverse conditions as many as eight days were required for the conversion of coal to coke. A principal weakness of this inexpensive coke-making process was its inability to produce a consistently high-quality coke in sufficient quantities. Coke made by the mound process yielded from 50 to 55 percent carbon in contrast to an average carbon content of about 67 percent using the beehive coke oven. This primitive and inefficient method of coke production had largely disappeared as a viable process in

The 5,000 coke ovens of the H. C. Frick Coke Company, Connellsville region. Carnegie Library of Pittsburgh.

western Pennsylvania by the 1860s. Subsequently it was used occasionally in areas where the demand for coke was small and its manufacture had just begun.

Provance McCormick, James Campbell, and William Turner Sr. hired John Taylor, a local farmer and stonemason, to construct two beehive coke ovens on his farm at Hickman Run below Broad Ford on the Youghiogheny River in 1841-1842. Taylor constructed ovens with a fourteen-inch rise and a flat crown, each with a capacity of between sixty and seventy bushels of coal. Taylor provided the coal and the business partners manufactured sixteen hundred bushels of coke that was loaded onto a ninety-foot flatboat. They sailed to Cincinnati in the spring of 1842 in search of a buyer.[29] This river voyage was the first attempt to "export" coke from the Connellsville coke district, but the trip down the Ohio River was a dismal business failure. They were unable to find buyers because the furnace operators from southern Ohio were unfamilar with "the silvery porous cinders." Mordecai Cochran and his two nephews, James "Little Jim" and Sample Cochran, purchased the ovens. The new owners produced thirteen thousand bushels of coke and transported two coke-filled flatboats down the Ohio River to Cincinnati on April 1, 1843. The coke was sold to Miles Greenwood, a Cincinnati foundryman, for seven cents a bushel, the sellers receiving one-half in cash and the balance in old mill irons.[30] This transaction was the first commercial sale of coke produced in a beehive oven outside the Connellsville district. The ovens the Cochrans used near Connellsville subsequently became the universal method of coke production during this period.

The dome-shaped refractory, brick beehive coke oven developed near Connellsville subsequently became the universal method of coke production during this period. The design of the beehive oven evolved by trial and error during the nineteenth century as coke operators attempted to find an adequate oven design that would yield high-quality metallurgical coke. The beehive oven was an arched-roof circular brick room constructed of masonry firebrick and tile with an opening at the top. The space, between the lining and the outside walls, was filled with waste bricks and other material to prevent, as far as possible, the loss of heat to the exterior. On the top of each oven was a twelve-inch circular opening called the trunnel head or "eye," which could be covered with a metal lid. Coal was emptied into each oven through these openings. The interior of the oven was shaped like a beehive, measuring about twelve feet across the base and seven feet high, and was lined with heat-resistant refractory firebricks. There were no industry-wide standards for the size of the oven's interior, although larger coke operators, including the H. C. Frick Coke Company, did standardize the interior size of their ovens. The exact configuration of a coke plant was determined by local topography and operating costs. Operators constructed their ovens in batteries arranged in single rows called "banks," or in double rows called "blocks." Block ovens were usually constructed in conjunction with single-row bank ovens. The number of ovens at a coke plant during this period ranged from two ovens to continuous banks of as many as three hundred ovens. The manual coking process required five distinct steps: charging, leveling, quenching, drawing, and loading. Coal was delivered from the nearby mine to the wood tipple by horse-drawn wagons, then transported to the ovens. Coal was transported on larry or lorry cars (also known as "dinkey" cars) mounted on tracks and pulled by draft animals along the top of the ovens. Workers called "chargers" filled the cars with coal at the mine tipple, then positioned them over the oven to discharge the load into the trunnel opening. A charge of five to seven tons of coal filled the oven to a depth of about two feet when leveled. Some larry cars were constructed to charge two block ovens at the same time. Coal gases escaped through this opening as flames and noxious smoke. A worker called a "leveler" then leveled off the dumped coal in the oven by reaching in through the oven's front opening with a hoe, of at least twelve feet in length, leveling the coal within each oven. The two-by-three-foot entrance was sealed almost shut with firebricks and mud by a "dauber," to restrict air from entering the oven. The heat stored in the oven from the

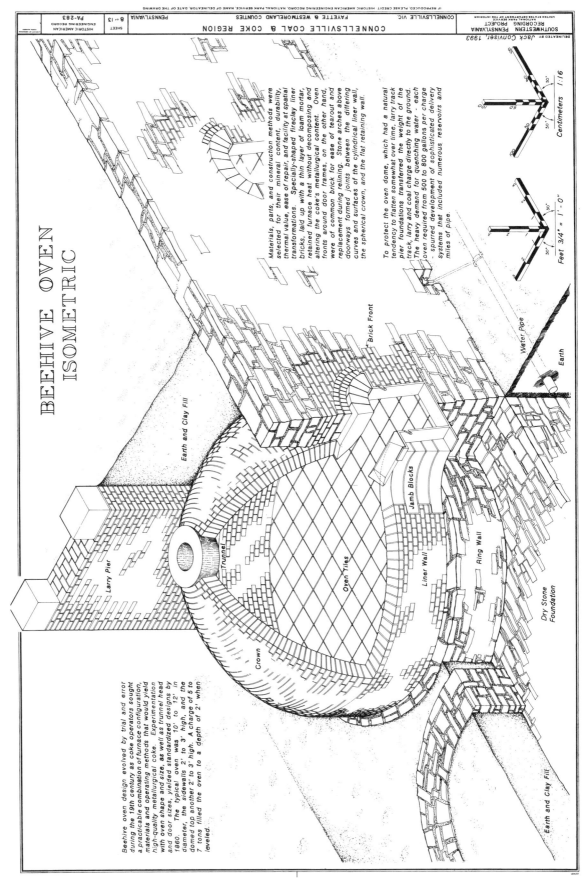

Cross section of a beehive coke oven in the Connellsville coke region. Southwestern Pennsylvania Recording Project, Historic American Engineering Record, 1993.

previous charge ignited the coal and the coke-making process began again. After the coal was converted into coke the firebricks were removed and the "puller" broke the coke loose and drew it from the oven. Pulling coke was the most arduous task at the ovens, requiring great physical strength to break up the hot coke and draw it from the oven. The coke was quenched by sprinkling it with water to prevent it from further baking. Water lines for this function ran the length of the ovens. Coke was then loaded into waiting railroad cars by hand, using a coke fork or large wheelbarrows. Every third oven in larger coke operations was fired at the same time so that the shifts of men were kept working uniformly. A contemporary description of the coke-making process using the beehive oven follows:

Coke ovens at Ellsworth, Washington County. From Coal Age.

> The charge of coal is dumped in at the crown of the oven and spread on the floor to an average depth of two feet to burn what is called 48-hour coke and 2 1/2 feet for what is called 72-hour coke. The front opening of the oven from which the coke is taken when finished having been nearly closed with brick and luted with loam. The heat of the oven from its previous coking fires the charge. As the coking progresses the air is gradually shut off by closing all openings. When the coke is thoroughly burned the door is opened, and the coke cooled (quenched) by a stream of water thrown upon it with a hose, after which it is taken out, and the yield of coke is from 63 to 65 percent of the coal charge.[31]

Pulling hot coke by hand from beehive coke ovens. Carnegie Library of Pittsburgh.

A standard charge per oven of this period was about five tons of coal, which produced approximately 3.1 tons of coke from Pittsburgh-seam coal. A well-built beehive oven converted as much as 70 percent of each ton of coal into coke, while approximately 30 percent of the coal was discharged into the atmosphere as noxious smoke.[32] The gaseous and liquid hydrocarbons were burned off in the beehive coking process, so that only carbon and ash remained in the oven. A ton of coal was reduced to twelve to thirteen hundred pounds of coke, and this reduced weight and bulk lowered the cost of transporting coke to the distant blast furnaces. Coking time in the ovens varied from forty-eight to seventy-two hours depending on the character of the coke desired and the kind of coal used.[33] The coking time affects the quality of coke: forty-eight-hour coke was generally sold as "furnace" coke and seventy-two-hour coke sold as "foundry" coke. This is a special coke used in furnaces to produce cast and ductile iron products. It is a source of heat, and also helps maintain the required carbon content of the metal product. Foundry-coke production requires lower temperatures and longer times than blast furnace coke. It took about three hours to draw and recharge each oven manually. Individual pieces of coke ranged in

size from a golf ball to a basketball. Metallurgical coke must be made with a coal having a low ash and sulfur content. Coke has a heat value of about 25 million Btus per ton, and high-quality coke is capable of burning in a steel mill blast furnace at a temperature ranging from 2,000 to 2,200 °F.

The name Henry Clay Frick is synonymous with the coke industry of western Pennsylvania. Frick was born in the heart of the Connellsville coke district at his maternal grandfather's farm near the village of West Overton, Westmoreland County, on December 19, 1849. He was the second child and first son of John W. and Elizabeth (nee Overholt) Frick. Frick was named after Henry Clay, a prominent antebellum Kentucky Whig politician, three-time presidential candidate, and statesman. The Overholts came to the American colonies, landing in Philadelphia in 1732, from Lower Palatinate of Germany. The Fricks came to America from Switzerland, landing in Philadelphia in 1767. Abraham Overholt, his maternal grandfather, was a prominent local land owner, miller, whiskey distiller, and the wealthiest resident in the county. The A. Overholt & Company distillery was constructed on the banks of the Youghiogheny River at Broad Ford near Connellsville before 1834. Frick attended school at West Overton, Alverton, and Mount Pleasant, and after completing his formal education worked as a clerk for Macrum and Carlisle, a Pittsburgh department store. He contracted typhoid fever in Pittsburgh and returned home to West Overton in 1868. After his recovery, Frick eventually became chief bookkeeper in his grandfather's distillery at an annual wage of one thousand dollars.

Coal in southwestern Pennsylvania was unsuitable for use in the iron industry because of its high phosphorous and sulfur content; it burned poorly in the primitive iron furnaces of the period. Coal baked in beehive ovens diminished these elements and was usable in the furnaces. There were only twenty-five small coke plants in the entire Connellsville region in 1870, and coke production was not a principal industry in the area's economy. Frick became involved in the coke industry with A. O. Tinstman, his cousin, and Joseph Rist, a friend, when they established Frick and Company in 1871 to furnish coke to iron foundries and blast furnaces in western Pensylvania. The partners purchased 123 acres of coal lands at Broad Ford at a cost of $52,995. Frick and Company had two coke plants in operation with two hundred beehive ovens by 1872, located near Broad Ford, Fayette County.

Beehive coke ovens in Connellsville region. Sign posted in six languages warning of potential danger. Pennsylvania State Archives.

The Panic of 1873 reduced the price of coke to less than ninety cents a ton and forced many coke operators to the brink of bankruptcy. Money was scarce and the depression made it nearly impossible for coke producers to obtain loans. Frick borrowed $10,000 from Judge Thomas Mellon, president and founder of Mellon and Sons on Smithfield Street in Pittsburgh, and used these funds to build fifty more ovens and to acquire smaller rivals who were anxious to sell their unprofitable coke plants. Frick was an astute businessman and in 1876 bought out his partners. He was prepared to monopolize the coke industry following the end of the depression, as the price of coke rose from about ninety cents to four and five dollars per ton, to meet rising demands from the expanding steel industry in the Pittsburgh district. Frick insisted to friends and associates that "coke's the thing they can't make steel without." The Bessemer steelmaking process permitted the rapid shift from iron to steel production and increased steel production. In the 1880s, Frick's coke plants in Fayette and Westmoreland Counties shipped high-grade metallurgical coke in nearly one hundred railroad cars daily to the foundries and blast furnaces of Pittsburgh mills, much of it to the Edgar Thomson Steel Works at Braddock, situated twelve miles down the Monongahela River from the Point. This Carnegie-constructed steel plant was the first Bessemer steel plant in the region, opening for steel rail production in 1875.

Bituminous Coal as a Coke and Charcoal Substitute

Raw bituminous coal known as splint or block coal was used directly in the blast furnaces by iron companies located in Mercer and Lawrence Counties, bordering Ohio. Block coal found in the Hocking Valley of Ohio and in Indiana was also dumped directly in the furnace under nearly the same conditions as Pennsylvania block coal. This peculiar kind of hard free-burning bituminous coal, because of its geological formation, splits along the seam like slate. Like anthracite, splint or block coal is a low-volatile coal that is slow to ignite and burn. Its chemical structure is of sufficient hardness to bear the charge and high temperature required in the blast furnace of this time. This coal was used directly in the blast furnace in the smelting of iron or the manufacture of pig or cast iron. Raw splint coal was first used successfully in iron production at Himrod & Vincent Company's Clay Furnace, Mercer County, in July 1845.[34] The Mahoning Iron Works of Wilkinson, Wilkes and Company of Lowell, Ohio, produced pig iron with splint coal in its blast furnace the following year. Raw block coal was used in the furnace, although ironmasters of larger furnaces used a small amount of coke to keep them open and enable them to take the blast with greater freedom. These pioneer successes led to an increased use of splint coal as a blast furnace fuel in the Mahoning and Shenango Valleys. There were eleven iron blast furnaces in the immediate region using splint coal by 1850—four furnaces in the Mahoning Valley, Ohio, and seven in Mercer County, Pennsylvania.[35] There were six blast furnaces in western Pennsylvania and thirteen in Ohio making iron using splint coal by 1856. Coal output in Mercer County exceeded twenty-five hundred tons per day during the Civil War; this was a considerable quantity for the period. By 1870, Mercer County was producing about five hundred thousand tons of splint coal per year for use in twenty-three blast furnaces in the county and for sale to furnace operators in neighboring western Ohio. Splint coal production in 1870 was 267,257 tons. Coal was transported from Mercer and Lawrence Counties by the Erie and Beaver Canal and the Erie & Pittsburgh (Mercer County) and Shenango & Allegheny (Lawrence County) Railroads to distant commercial markets.

The Arrival of Skilled British Miners and Mining Technology

Before 1840, the majority of mine workers was drawn from neighboring farms and villages. These farmers and day laborers worked underground in drift-entry mines in the winter when there was little demand for farm labor. The new demand for coal between 1840 and 1880 increased production and transformed the industry from a seasonal to a continuous year-round industry. This increased production placed a new demand for labor that was filled by English-speaking miners from England, Wales, Scotland, and Ireland. Some two thousand arrived during the 1840s and an additional thirty-seven thousand miners arrived from Great Britain during the 1850s.[36] A contemporary writer on the American coal industry observed that "the mining population of our Coal Regions is almost exclusively composed of foreigners, principally from England, Scotland, and Wales, with a few Irish and Scotsmen."[37] These skilled British miners toiled underground in both the anthracite and bituminous coalfields of Pennsylvania. English, Welsh, Scotch, Irish, German, Canadian, and native-born American were the principal nationalities of mine workers before 1880. These skilled English-speaking miners would dominate coal mining until waves of immigrant workers from eastern and southern Europe supplanted them as the principal coal-diggers in the nation after the 1880s.

The early immigrant miners gave the American coal industry a distinctly British flavor. They had served long and arduous apprenticeships in British mines and were skilled in all facets of mining. The coal-mining industry in Great Britain, Germany, Belgium, and France was well established by 1600. The coal output of Great Britain in 1660 was estimated at about 2.2 million tons, in 1700 at about 2.5 million tons, and in 1800 at over 10 million tons. The city of London alone consumed nearly a half-million tons in 1600, and some 850,000 tons annually at the beginning of the American Revolution. Speaking of the miners, historian Keith Dix noted that "these independent craftsmen learned their job during an extensive

Young "trapper-boys" with sunshine lamps. Carnegie Library of Pittsburgh.

apprenticeship period and once having mastered these skills worked largely without immediate supervision and at their own pace. The early pick miner was not only in control of his job but also in control of his own time in the job."[38] These British miners took virtual control of the American coal industry as "their mining methods, tools, and much of their terminology became part of the American technique. They also initiated the miners' labor movement."[39]

J. Peter Lesley Jr. (1819-1903), the state geologist of Pennsylvania, held an unfavorable opinion of these skilled British miners. He was chosen in 1874 to become the state geologist of the Second Geological Survey of Pennsylvania (1874-1889). Lesley castigated their mining skills and their inability to adapt to mining conditions found in the United States. He wrote in 1876 that

the mines in the State were (with some honorable exceptions) bossed by the commonest miners from foreign and quite different geological regions, who had suddenly exchanged the character and position of the hewers of coal and-pumpers of water at home, for the character and position of mining engineers in America. Ignorant, undisciplined, obstinate, narrow-minded, and superstitious by nature and habit, and rendered presumptuous by their strange advancement, they were unwilling to accept as they were unable to acquire a correct knowledge of our geology, so different from their own, and hated professional geologists because they had never lived in childhood, pick in hand underground,—because they taught new things hard to comprehend . . .[40]

Lesley's castigation aside, life and work deep underground in the premechanized nineteenth-century mine was defined, organized, and controlled by these immigrant miners. These British miners, known as "practical" miners, were proud that "no damned foreman can look down my shirt collar." Each skilled miner was employed as an independent petty contractor who entered into a contract with the individual mine owner. Each contract stipulated rates of pay per bushel, ton, or car of mined coal delivered in the entry, where it was collected by the "mule skinners," or "trip ridders." As piece-rate workers, they determined both the hours of their daily labor and the tempo of their work. Because rates were low, however, they were forced to work long hours underground, averaging from ten to fourteen hours a day. The miner's work was hard, dirty, and carried on at an average temperature of about 45 °F, although underground temperature can vary slightly with the extremes of the seasons. The ante-bellum coal miner worked alone or with a companion called his "buddy" or "butty." These skilled workers usually worked two in a room and exchanged mining duties. They hired and paid one or more day laborers to load the coal onto wooden wagons and transport it to the surface tipple.

The miners and their assistants entered the drift mouth, or down the mine's slope entry, by foot to their "room" before sunrise, carrying with them a variety of hand mining tools, their lunches, and their lamps. In a shaft-entry mine, miners were transported to the underground working face in a cage or elevator powered by a steam engine located in the hoist house. The principal method of opening a mine was the drift entry, although a number of slope and shaft mines also were opened by coal companies in Pennsylvania during this period. A slope entry, the second most common type, is employed when the coal seam outcrops on the surface, but is inclined downward at angles varying sixty or more degrees. The entry is really an inclined passageway cut into the earth to reach the seam. It is driven from an outcrop generally southward along the dip of the seam. It is more costly to maintain than the drift entry because mechanical equipment is often required to haul coal out of the mine and because of the need to expel accumulations of ground water. The angle of the tunnel's inclination is often too great for working mules to haul the coal cars up to the mine's entry. Mules hauled coal in small wooden wagons on tracks to the foot of the slope, and then the loaded coal cars were pulled up by cables or ropes operated by a coal-powered steam engine at the mine's entrance.

A shaft mine is a vertical entrance that is sunk when the coal deposit is located some distance from the surface, or when the coal seam is located below water level. A vertical tunnel is dug through sandstone and shale deposits until the coal seam is located. Miners then dig horizontal entries through the coal seam. Miners, equipment, coal, and waste are transported between the coal seam and the surface by an elevator system. This entry is capital intensive and requires the installation of elaborate hauling equipment to move coal, waste, miners, draft animals, and equipment vertically. Some slope mines and most shaft mines presented difficult problems of water drainage and ventilation. The entry was expensive to maintain because the coal operator had to provide mechanical ventilation, which meant digging separate air shafts and maintaining a variety of equipment for haulage purposes. A shaft mine located below the water table required the installation of water pumps because these mines were often wet or flooded by water seeping into the mine from underground springs. Water in the mine was a great barrier to coal mining and the constant pumping of water from the mine added to the cost. The folding of the deeper coal seams often interfered with natural drainage; therefore, water had to be pumped out constantly or the mine would soon fill up.

The general operation of mines in this period was still by the room-and-pillar system. The lengthy inventory of equipment needed to mine coal included a broad shovel, picks, dynamite or black powder, auger (six to eight feet in length), tampering box or stick, sledge hammer, wedge, and some source of illumination. Miners never saw sunlight except on those days during the long winter months when they did not work. They entered the dark mine before the sun rose and left it after the sun set. Miners brought a variety of light sources underground to illuminate their work area. An old Belgian miners' song expressed the importance of the miner's lamp: "[M]y lamp is my sun—And all my days are nights." Originally, miners used large candles to produce sufficient light underground to mine coal. An individual miner used six to eight candles during a typical shift. "Lard" oil lamps were introduced in the mines during the 1850s. They had a small conical font in the front of the lamp that held their fuel in a long spout that extended outward from the font. The fuel was a mixture of lard and oil—cottonseed oil, kerosene, or crude oil. The lard lamp was soon replaced by sunshine fuel, which was a mixture of paraffin wax and 3 percent mineral oil, which unlike the lard lamp burned without smoking. The "sunshine" lamp resembled a miniature pitcher with a wick extending from the spout and was attached to a miner's canvas cap. This lamp was hazardous to miners since the paraffin fumes eventually solidified in their lungs.

The Anton Brothers of Monongahela City, Washington County, manufactured sunshine lamps around 1874. Brothers George, John, Fred, and Christopher came from Bavaria in 1849 and settled in Monongahela City, Washington County. They operated a number of lamp factories in the Monongahela Valley between 1874 and 1918. The "STAR" brand lamp made by the Anton brothers was the best-selling wick lamp in the nation owing to its quality construction, although the lamp was primitive in function. The brothers held several patents on their designs, which made them competitive in international markets. They made a variety of miners' wick lamps for different uses, and sold them through jobbers in all the mining regions of North America. They called their lamps "Anton Lamps," although most

The Anton brothers of Washington County were the leading manufacturers of miners' oil lamps.
Mining Artifact Collector.

miners simply called them sunshine lamps. The introduction of the carbide lamp during the 1890s and the demise of the wick lamp destroyed their market. The brothers were unable to make the conversion to carbide lamps and closed their company in 1918.[41] The carbide lamp with a reflector was a superior lamp because it provided more light and was less expensive to operate.

Miners checked for air quality and explosive gases that might have accumulated during their absence. Mine air was seldom pure; instead the underground mine atmosphere carried a variety of poisonous or highly flammable gases. Pure air contains 20.9 percent oxygen, 78 percent nitrogen, three parts in ten thousand of carbon dioxide, and small amounts of argon, helium, and other gases. Fire and explosions presented major hazards to coal miners. Methane, ethane hydrogen, carbon dioxide, carbon monoxide, and hydrogen sulfide are found in underground mines. Methane (CH_4), known as "swamp gas," "marsh gas," or "light carburetted hydrogen," is a nonpoisonous, tasteless, ordorless, and colorless gas. Methane is the result of the decomposition of organic matter in the mine and the amount of methane is dependent on its depth below the surface. Deep mines, as a rule, have more methane gas than shallow mines. Methane is removed from the mine naturally by brisk air currents that are of sufficient quantity and properly directed. Methane, when it mixes with air, forms "fire damp," which is a highly volatile gas, easily ignited by an open flame or an electric spark. Fire damp is formed when a concentration of 5 to 15 percent methane mixes with air. Since this gas is one-half lighter than air it tends to gather along the top of mine workings. The proportion of "fire damp" may be determined by the fire boss with an accuracy of 99 percent or more by using a gas safety lamp. The height of the flame determines the presence of this gas. The fire boss was assigned the responsibility of checking gas and air quality in the mine. Carbon dioxide (CO_2) is a product of combustion from the breathing of men and draft animals, fires, decay of organic matter, and the explosion of coal gas or dust. Like methane, it is orderless, colorless, and nonpoisonous and will not support combustion by itself. This gas is called "choke damp," "black damp," or "stythe" because it excludes oxygen and settles close to the mine's floor. It received its name from the choking or distress caused by inhaling it. Miners subjected to air containing 3 or 4 percent carbon dioxide suffer headaches and choking. A concentration of 10 percent is fatal in a few minutes. Carbon monoxide (CO) is an orderless, colorless, and extremely poisonous gas. "White damp" gas is a mixture of mine gases which contains carbon monoxide in dangerous quantities. The inhaling of this gas can kill a miner immediately, even in small concentrations. "White damp" will burn and explode when mixed with air but the gas alone will not support combustion. "After damp" is any mixture of gases formed from a mine fire or an explosion. "After damp" may be an explosive or flammable mixture, or it may be nonflammable, but it is extremely poisonous.

Miners brought small birds (usually canaries) and animals with them underground to detect and warn them of the presence of lethal mine gases.[42] If the animals died it was a sign that deadly gases were present in the air. Some miners would burn hemp rope underground to detect the presence of methane. Methane was present if the rope burned brightly. Safety gas lamps were introduced to measure the presence of methane gas in the mine, beginning with the manufacturer of Sir Humphrey Davy's lamp in Great Britain on December 5, 1815.[43] The Davy lamp was one of the earliest safety lamps to successfully detect methane gas as low as 2 to 2.5 percent in air. The flame in the lamp is enclosed by an open wire gauze which allows the free flow of air in and out of the lamp. The wire gauze, being a good conductor of heat, conducts it away from the flame so that the heat of the flame is insufficient to ignite the gas unless the concentration in the mine is high and the flame burns too high. The early safety lamps burned naphtha, which is an artificial, volatile, colorless liquid distilled from petroleum. If the air in the mine was pure the flame in the Davy lamp was a clear yellow, while a blue flame indicated the presence of gases in the air. The light output of the lamp was poor, seldom

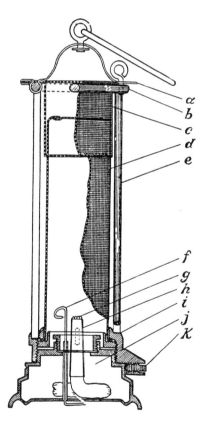

Components of Davy Safety Lamp: a) brass hood; b) brass ring; c) gauze cap; d) main wire gauze; e) three iron support posts; f) wick snuffer; g) cotton wick; h) wick holder; i) brass ring; j) brass oil container; k) key lock.
Mining Artifact Collector.

exceeding 13 percent of one candle power. The flame could also be easily blown out. A variety of improved safety gas lamps were introduced by several American and European manufacturers throughout the nineteenth century. The most popular safety lamps used by American miners were manufactured by Wolf and Koehler.

An old definition of a mine was "any excavation for the extraction of mineral and in which artificial light is needed." To this someone added the following remark, "and in which the air is always bad." There was no need for mechanical ventilation in the early, small drift-entry mine because a natural draft was created from the difference in the weight of air in the intake and the outtake. The outside atmosphere was dense and heavier than air in the mine in the winter, while during the summer these climatic conditions were reversed. Proper ventilation was important for the health and safety of the miners as coal miners went deeper underground to extract coal. Mine ventilation was required to remove the accumulations of gases and provide fresh air for the miners. Smoke from explosions of black powder, the reek of the miners' oil lamp, and coal dust from undercutting coal by hand or machine made air deep underground almost unbreathable. A well-ventilated mine was not subject to the build-up of poisonous or explosive gases and reduced the danger of underground explosion. Natural ventilation was not always feasible in slope and shaft mines located far below the surface. In early ventilating technique, a furnace was installed at the bottom of the shaft. An air intake shaft was sunk some distance away. The furnace shaft acted as a chimney by drawing the warm air up and out while fresh air entered the intake shaft to replace the air released. However, this method of ventilation posed a very serious fire hazard, and its widespread use was discontinued after such a furnace caused an explosion at the Steuben Shaft at Avondale, near Plymouth in Luzerne County. This new mine was constructed in 1868 by the Nanticoke Coal Company, a subsidiary of the Delaware, Lackawanna and Western Railroad, which was owned by three large coal, iron, and railroad corporations. A wooden breaker was constructed directly above the single three-hundred-foot shaft to the mine below. Sparks from burning wood, used in lighting the furnace, set fire to the wooden base of the shaft, which in turn ignited the wooden breaker at the top of the shaft entry and trapped the miners underground. There was only this one exit or egress from the mine, and therefore most of the trapped miners died of asphyxiation or smoke inhalation.[44] The Avondale disaster was the worst coal mining disaster to occur in the United States up to that time, claiming the lives of 179 miners on September 6, 1869. The tragedy prompted the immediate passage of mine legislation requiring that every mine in the Commonwealth have at least two exits and that no breaker be built directly over the shaft opening.

Early mine fans were powered by water, wind, and hand. The Guibal fan, named after the Belgian inventor who designed it, was one of the earliest mechanical fans. It was introduced in 1844 and was used extensively until the 1870s. The fan had a large metal blade revolving about fifty times per minute.[45] Larger mines adopted a system of multiple entry airways to provide adequate ventilation. A second drift entry or tunnel was driven parallel to the first and forty to one hundred feet distant. At intervals, as the two parallel entries are driven forward, cross tunnels are driven to connect the entries. As each new breakthrough is cut, the one behind is sealed. A fan is installed at the mouth of the ventilating entry, designed to force air through or to withdraw air from the mine. The movement of fresh air into underground workings extending for several miles required the development of a sophisticated system of barriers and trap doors to direct and maintain a constant supply of fresh air, because air current passing in a mine meets with resistance, called the "ventilating pressure," owing to the friction of moving air against the sides, top, and bottom of the airways.[46] A series of partitions called brattices, made of wood and cloth, closed off abandoned areas. These barricades separated different parts of the underground passages and controlled the flow of fresh air in the mine. Trapper boys, often less than ten years of age, were responsible for manually opening and closing these barricades as needed.

Array of hand-tools used by miners throughout the nineteenth century. Historical Society of Western Pennsylvania.

The collapse of the mine's roof was a leading cause in the death of workers during this period. A principal unpaid task of the miner was to maintain the roof. He checked the mine's roof to see if it was secure or needed to be timbered. A miner would "sound the roof" to see that it was secure by holding his hand up against the roof with the fingers spread out and tapping the roof with the end of the pick handle. A hollow sound indicated the roof was pulling loose, but if the sound was solid the roof was stable. If coal began to peel off the wall, the miners stopped working and corrected the condition. The mine roof was supported by wooden props or logs which were wedged between the roof and floor.

Miners had to perform preparatory and maintenance work prior to beginning the actual task of mining. All tasks in the mine, in addition to the work of cutting, shooting, and loading coal, were time-consuming activities. Miners called this work "dead work" or "unproductive work" because they were unpaid tasks and ancillary to production. They were paid only for the coal they actually mined. Some miners were often careless in protecting themselves against roof collapses, in detecting gas, and in insuring ventilation, and their inattention to this unpaid work proved fatal to many.

Coal mining was an arduous, tedious, and dangerous occupation. Miners used pick and shovel to dig coal, and mule-drawn wooden cars to haul it from the room to the surface, to the weigh station and tipple. The process of extracting coal entailed a series of distinct steps, and throughout this period there was little mechanization; instead, workers relied exclusively on human and animal power. Four basic tasks were involved in mining coal in nonmechanized mining: undercutting the coal face, drilling the face, blasting the coal, and loading the coal for transport out of the mine. Miners extracted coal from the seam by using a variety of primitive hand-tools, including a wedgepin, a sledgehammer, a variety of picks, an auger, shovels, tamping bars, and needles. The workplace at the seam ranged from two to twenty feet high, depending on the seam's thickness. The work of "undercutting" (undermining) coal was the most skilled, dangerous, and time-consuming activity in the premechanized mine. The undercutting of coal permitted the coal to break into chunks when it was knocked or shot from the seam. The skilled miner usually had to lie on his side and, using his sharpened pick, make a series of three- or four-foot horizontal cuts into the base of the face. When the cut was partway through the seam, the miner placed short wooden blocks or props known as "sprags" under the face of the coal after it was undercut to prop it up so that the overhanging mass would not fall on him, as he lay underneath to finish the job. This process is called "spragging." (The term is also used for a primitive system of braking mine cars.) The miner made two vertical cuts from the top of the seam down to the undercut, hammered iron wedges into the top of the outlined block of coal, and then attempted to remove the coal without shattering it. The advantage of a deep undercut was that more coal could be removed in a day's work. This task took up to several hours, depending on the miner's skill, the sharpness of his pick, and the hardness of the coal seam.

Coal that was too hard or difficult to remove by the undercutting and wedging method was later dislodged by explosives. Explosives were a last recourse because if improperly used, they broke the coal into fine coal, or "slack," which was waste since no market for it had been developed during this period. Miners became experienced in the use of a variety of explosives—first black powder and dynamite and later nitroglycerin. Explosives were first used to remove anthracite coal from the seam because the coal was too hard to extract manually. Blasting coal from the seam gradually replaced the wedging method, although there was some opposition to its widespread use. The extensive use of explosives was seen by some skilled miners as constituting an intrusion on their skill. A skilled miner, who opposed the use of explosives, wrote in 1875:

> A practical miner works all such coal without the use of [blasting] powder, unless there is some trouble in his place where he cannot take it down by the use of a wedge. On the other hand some Tom, Dick, or Harry that perhaps knows very little about coal mining brings a drilling machine and a keg of powder and by mere force of blasting puts out perhaps as much coal as another man, not caring who is suffocated by his powder smoke. The skilled miner has no advantage over the greenhorn.[47]

Black blasting powder, a slow-acting granular explosive, was originally used to remove coal from the seam. This powder was also used in stone quarrying, road and railroad grading, general excavating, and clay mining. Black blasting powder is a combination of three principal ingredients: saltpeter (potassium nitrate), 76 percent; sulfur, 14 percent; and charcoal, 10 percent. The powder, which does not freeze, was manufactured in two grades, "A" and "B." "A" powder contained potassium nitrate and was more water resistant, stronger, and quicker than "B" powder, whose active ingredient was nitrate of soda. Blasting powder was not completely water resistant and was unusable in wet work areas. Glazed (polished) and unglazed powder were manufactured. Glazed powder was the more popular type. Glazing was done by using a small amount of graphite to polish the outside of the grains. Black blasting powder, unlike dynamite, was manufactured in only one strength. The powder was packaged and sold in twenty-five-pound metal kegs, called powder kegs after 1874.

Undercutting coal by hand at the coal face. Carnegie Library of Pittsburgh.

The miner used the following procedure when blasting coal from the seam with squibs and black blasting powder: (1) drill a hole with a hand auger, a churn drill, or sometimes a breast auger strapped against the miner's chest from an upward angle after the coal is undercut (undermined); (2) place a black-powder cartridge in the bottom of the drill hole; (3) place a needle into the black-powder cartridge (the needle is a thin copper rod, usually five to six feet long, with a loop handle on one end and pointed on the other); (4) tamp stemming into the hole around the needle; (5) pull out the needle; (6) place a squib in the hole left by the needle, light the match end of the squib. The squib, also known as a match, reed, or rush, was used by miners to detonate the black powder. It was a small paper tube filled with quick-burning powder, with a slow match at one end acting as a fuse to permit miners time to evacuate the area. As soon as the squib began to fizzle the miner and his laborer ran for safety in a corner until the coal was blasted off the wall and the smoke cleared.[48] A single explosion could dislodge a ton or two of coal from the face.

The dislodged coal was shoveled with broad-billed, flat-backed shovels into small baskets or wooden cars by hired laborers. A brass check with the loader's identification number stamped on it was placed on the coal-filled car and the loader pushed it from the room into the main passageway, where the mule driver collected the car. The check identified which worker had loaded the car. Miners were paid piece rate based on the number of tons or bushels loaded daily. Coal was originally removed from the mine by miners using a crude harness that fit across their shoulders. Coal and waste were later removed from the mine on sleds and small wagons.

Oxen, horses, ponies, goats, mules, and dogs were all used to haul wooden carts or wagons underground and to the surface on narrow-gauge wooden tracks (later iron rails were used). Dogs were the animals first used by pioneer miners to haul their coal to the surface. A dog or a number of dogs were harnessed to a wooden cart while the miner pushed from behind. Michael Dravo, a mine owner near McKeesport, Allegheny County, is credited with first using mules instead of men and dogs for this laborious task during the 1830s. By the end of the period mules or horses had succeeded men and dogs in moving coal through the main mine tunnel. Coal carts were pulled on narrow-gauge iron railroad tracks, branching into the various rooms in which the miners worked alone or, more often, in pairs.

Surface buildings and structures at a typical bituminous coal mine of this era consisted of a wooden tipple, a loading platform, a small storage shed for mining supplies, and a stable if draft animals were used for coal haulage. The coal cars in the few shaft and slope mines operating during this period were delivered to the shaft bottom, where they were elevated to the surface by a steam-powered engine in the hoist house. These mines also had a fan house to enclose the fan that provided

Coke plant in Connellsville region, company housing in the background. Penn State University, Fayette Campus.

mine ventilation. Most tipples of this era were located on the river and extended over the river's bank. The tipple was a large tower-like structure at the entrance to the mine. Coal from the mine was delivered into the tipple, from which it was loaded into river barges or railroad cars below. This wooden structure was called a tipple because after the coal was weighed by the weighmaster, the contents of each car were "tippled" or dumped into the top of the structure. Coal slid from the car onto a series of screens in the tipple or directly into railroad cars or a barge. Coal was sorted into varying sizes by passing through a series of different-size screens—lump (largest), nut (one-inch) piece, and slack (one-half inch). Coal shipped to market after it was weighed but not sorted or processed in the tipple was called run-of-the-mine coal. During this period a majority of coal was shipped to market without much processing. Anthracite, unlike bituminous coal, was subject to extensive processing before its shipment. Gideon Bast constructed the first successful wooden "breaker" in the anthracite region at his Wolf Creek Colliery in 1844. In the anthracite region, the preparation plants were called "breakers," the name originating from the fact that hard anthracite coal is broken and sized in the plant. The breaker was an imposing, often windowless, wooden building where run-of-the-mine coal was broken up, washed, separated into uniform sizes, and cleaned of impurities, especially slate, by "breaker boys" before shipment to market.

Most contemporary mines had single entries and employed natural drainage and ventilation. Mine entries were usually the drift-entry type and located in the face of a river hillside. These mines were usually owned locally, employed fewer than fifty miners, and produced a daily output of about a hundred tons. Seven thousand workers were employed in the mines on the Monongahela Valley in 1868, producing 2.5 million tons of coal annually, of which two-thirds were barged out of the region to markets down the Ohio River as far south as New Orleans. A half-million tons of coal had been moved out of the Monongahela Valley to New Orleans by 1885. The Mingo Coal Works, Washington County, accurately described in George Thurston's *Directory of the Monongahela and Youghiogheny Rivers* of 1859, was a typical large mine of this era:

> The ground is leased. There are in the works 80 rooms, and 1,200 yards of entry. The improvements cost $15,000, and the works have a river front of one half mile.There are 100 hands employed at the colliery, who mined 550,000 bushels in 1858-1859, all of which was taken to Cincinnati. This company has 25 barges, worth $20,000, and one steam boat, the Mingo, worth $18,000.[49]

Miners' Organizations

Pennsylvania did not have a monopoly on production of bituminous coal as it did on anthracite between 1840 and 1880. There were fifteen bituminous coal producing states and 36,500 miners in the United States on the eve of the Civil War. Beside Pennsylvania, the principal bituminous coal producing states were Maryland, Illinois, and Ohio. The skilled British miners who came to dominate the American coal industry were militant industrial workers who had pioneered the labor movement in Great Britain. They founded the Miners' Association of Great Britain and Ireland during the 1840s, the first national mining union in the world. These transplanted skilled miners were imbued with a strong trade union tradition and were responsible for organizing local miners and establishing the early miners' unions in the United States. Miners, like other industrial workers, recognized that the only hope of advancing their shared economic interests lay in their collective and unified action. The complaints of the early coal diggers with regard to owners, aside from low wages, concerned the poor quality of company houses, high prices charged in the company stores, inadequate underground air ventilation, and the method used by coal operators in weighing their coal.

Miners were militant workers, although most had not organized into unions in the beginning of this period. They were willing to walk out of the mine, as a group, if they thought they were being exploited by coal operators. In 1848 Monongahela Valley miners struck against the operators, wage reduction from 2 cents to 1 3/8 cents a bushel. They returned to work at 1 3/4 cents after a three-week strike. These miners struck again in 1859 and demanded the installation of weigh scales at the tipple. Miners worked on a piece-rate or contract basis, earning wages based on the number of cars filled daily. This method was often inaccurate and cheated them, since the size of the carts and wagons used to transport mined coal was not standardized. This local dispute quickly spread throughout the coal districts of western Pennsylvania before the strike ultimately collapsed.

Many Americans looked upon labor unions with displeasure and hostility, viewing unions as a foreign importation, incompatible with American values of self-reliance and hard work. Labor organizers and leaders were denounced by opponents of unions as demagogues who were too lazy to work themselves and unwilling to permit self-respecting men to work. English-speaking miners who emigrated to the United States during the 1840s and 1850s were aware of the great upheaveal of the Chartist movement in England and were strongly influenced by these ideas.[50] The political principles of the Chartists' movement, published by the London Working Men's Association, were embodied in the following six points: an annual Parliament; universal manhood suffrage; vote by secret ballot; abolition of the property qualification for membership in the House of Commons; payment of members of Parliament, enabling those other than the wealthy to hold office; and uniform electoral districts.

The first attempt at unionization by miners occurred in the anthracite coalfields of eastern Pennsylvania, in Schuylkill County in 1848. The John Bates' Union, a local miners' organization, was named after John Bates, its founder and first president. Bates was an Englishman imbued with the lofty principles of Chartism. The anthracite miners demanded higher wages, better and safer working conditions in the mines, reduced hours of work, and abolition of the use of coal scrip. In 1849 they called an unsuccessful work stoppage that lasted from May 11 to June 21; the miners were forced to return to the pits by coal operators who redressed none of their grievances. Several unions followed, but like the Bates' Union, they also floundered and ultimately collapsed.[51]

The first attempt at organizing miners into a national union was made in Illinois by Thomas Lloyd and Daniel Weaver in 1860. These two skilled English miners had recently arrived in the coalfields of southern Illinois. Both Lloyd and Weaver, like John Bates, had participated in the Chartist movement, and they aroused Illinois miners to form a national miners' union. Miners from Maryland, Missouri, Ohio, and Illinois sent delegates to St. Louis who listened to Lloyd's passionate pleas for a miners' organization:

> Men can do jointly what they cannot do singly; and the union of minds and hands, the concentration of their power, become almost omnipotent. How long then, will miners remain isolated. Our unity is essential to the attainment of our rights and the amelioration of our present condition; and our voices must be heard in the legislative halls of our land. Come, then, and rally around the standards of union—the union of states and the unity of miners.[52]

The American Miners' Association was formed on January 28, 1861. The delegates elected Thomas Lloyd president, Daniel Weaver vice-president, and Ralph Green treasurer. The A.M.A. was the first miners' union to extend beyond a single state and published *The Weekly Miner*, the first official mine workers' journal. The delegates wanted to establish a uniform wage scale for all miners and demanded that the general assemblies of coal-producing states pass legislation to regulate the weighing of coal. The American Miners' Association had dues-paying members from the Pittsburgh district and Blossburg Field, Tioga County, by 1863, but union organizers did not recruit members in the anthracite region of eastern Pennsylvania. State branches of the A.M.A. were divided into districts, with each mining district having its own constitution and electing its own union official. The A.M.A. survived only a few years before collapsing in 1867 as a result of internal dissension. A number of local unions survived its demise in the Hocking Valley of Ohio, the Pittsburgh district, and in southern Illinois.

The anthracite industry was expanding by the 1860s and was being inundated by new workers, the number doubling from twenty-five to fifty-three thousand during the 1860s. Until the Irish came, most anthracite miners had been English, Welsh, and German, but with the shortage of labor, the ranks were filled by unskilled Irish-Catholic laborers who constituted 32 percent of the region's miners between 1860 and 1870. By 1870, there were forty-four thousand Irish immigrants in the northeastern counties, mostly concentrated in Luzerne and Schuylkill Counties. "Old stock" American miners saw the Irish-Catholic immigrants as lazy, stupid, drunken, and criminal. They were regarded as "foreigners" and described as unassimilable. English and Welsh disproportionately held the skilled jobs underground while the new Irish-Catholic miners were relegated to manual labor.

The Miners' and Laborers' Benevolent Association, formed in the anthracite region about 1867, was the next major miners' union. It spread from the anthracite fields to the bituminous coalfields of western Pennsylvania and Ohio. This union was subsequently absorbed by the Miners' National Association of the United States (1873-1876), which was organized in Youngstown, Ohio, in October 1873. John Siney (1831-1880), an anthracite miner from St. Clair, was elected its first president. St. Clair, Schuylkill County, was a small coal-mining town in the heart of the anthracite district of northeastern Pennsylvania. Historian Anthony F. C. Wallace wrote an excellent community study of St. Clair in 1981 entitled, *St.Clair: A Nineteenth-Century Coal Town's Experience with a Disaster-Prone Industry.*

The union achieved a measure of short-term success in its attempt to increase wages, reduce hours, and improve mine safety and preferential work assignments. By 1875 the union had established 347 lodges in thirteen states with thirty-five thousand members, or roughly one-quarter of all coal diggers. The conditions of workers were deteriorating as owners reduced their wages throughout the depression decade of the 1870s. Miners from the Tuscarawas Valley of Ohio, Connellsville coke region of Pennsylvania, and the Shenango and Mahoning Valleys on the western border of Pennsylvania and Ohio engaged in a series of hard-fought strikes before 1880. Miners declared "their willingness to eat grass, stone, dried leaves, etc., rather than to submit to mine coal for such prices."[53] To combat the new organization mine owners compelled their workers to sign "yellow-dog" contracts as a condition of their employment and refused to rescind wage cuts, reduce costs at the company store, or improve ventilation in the mine. The Miners' National Association could not weather the harsh depression of 1870s and soon dissolved.[54]

The Knights of Labor was formed on December 9, 1869, as a secret fraternal organization by nine tailors in Philadelphia. Uriah Smith Stephens (1821-1882), a tailor, Mason, and Greek scholar, was elected the union's first president and headed the organization until 1879, when he was succeeded by Terence V. Powderly, who held the post until 1893. The Knights of Labor was created, according to Stephens, "to secure for American workers the full enjoyment of the wealth they create," and the goal of the secretive labor organization was "the consolidation of all branches of labor into a compact whole." Their slogan was "an injury to one is the concern of all." The Knights was organized as an industrial union headed by a General Assembly to which workers belonged as individuals. The structure of this industrial union was organized as follows: Grand Master Workman, head of the General Assembly (national body); District Master Workman, head of the District Assembly; and Master Workman, head of the Local Assembly (five or more to a district). All gainfully employed workers regardless of sex, race, or color were admitted into the union. Bankers, lawyers, capitalists, gamblers, and stockbrokers were all forbidden membership in the secret organization. These occupations were representative of the new large-scale industrial enterprises and were regarded by workers as simply "parasitical" occupations.

The Knights remained a secret organization to protect its members from harassment by factory owners. This policy of secrecy had been abandoned by the late 1870s, in response partly to accusations from the general public and the Catholic

Church of members' participation in the Molly Maguires, a militant and secret organization of Irish-Catholic immigrants operating in the anthracite region of eastern Pennsylvania between 1862 and 1877.[55] The Mollies were one of a number of protective societies that Irish-Catholics had joined in their native country to fight British imperialistic control of Ireland. Branch organizations were established in the anthracite coal regions of northeastern Pennsylvania within the Ancient Order of the Hibernians (A.O.H.), the largest Irish protective and beneficial society in America. The center of Molly Maguire activity was the Schuykill coalfield. They advocated the use of violence and intimidation as legitimate weapons to improve the oppressive living conditions and poor wages of anthracite miners. Protestant Welsh, Scotch, and English miners maintained control over the most lucrative and skilled jobs within the mines while the Irish-Catholic workers were poorly paid and assigned the most difficult and hazardous jobs in the mine. During the Molly Maguires' nearly ten-year "reign of terror" its members were accused of committing at least 42 murders, 162 felonious assaults, and a myriad of destructive acts against coal and railroad properties. They destroyed collieries and assaulted or murdered mine bosses and company officials. Company officials were required to carry weapons and night watchmen were employed at the breakers by the companies to protect against vandalism. The railroad and coal companies, led by President Franklin B. Gowen of the powerful Philadelphia and Reading Coal and Iron Company, employed the Pinkerton Detective Agency to infiltrate and break up the secret organization. James McParland, an Irish immigrant, using the alias of James McKenna, successfully infiltrated the Mollies; later he was elected secretary of the organization. McParland provided the Pinkerton Agency with a list of 375 Mollies and some of these men were arrested and tried for their alleged crimes. Some twenty members were brought to trial in 1876 and 1877 and convicted. Ten were hanged on June 21, 1877, six in Pottsville and four at Mauch Chunk. Ten more men were hanged during the next two years, bringing the total to twenty. The A.O.H. revoked all its chapters in Schuykill, Carbon, Northumberland, and Columbia Counties in May of 1877. The era of the Molly Maguires was the bloodiest period in the anthracite regions of northeastern Pennsylvania. The organization disappeared from the anthracite coal region following the trials and was never heard from thereafter.[56]

The Knights became a public and national labor organization during the 1879-1893 presidential tenure of Terence Vincent Powderly (1849-1924), an Irish-American born in Carbondale, Pennsylvania, in 1849. Powderly worked on the railroad and served three terms as mayor of Scranton, elected on the Greenback-Labor ticket from 1878 to 1884. He was initiated into the secret order of the Knights of Labor in 1874, was elected its Grand Master Workman in 1879, and served until 1893. The organization grew in spurts through the 1870s and ballooned to seven hundred thousand members in 1886 after the Knights had won a number of major strikes in 1884 and 1885. District Assemblies of the Knights of Labor organized coal miners from the Pittsburgh district, Maryland, West Virginia, Ohio, Indiana, and Illinois between 1874 and 1879. Organizational drives for miners were undertaken by the Knights in the coalfields of Colorado and New Mexico in 1882.

Notes

[1] Sam H. Schurr and Bruce C. Netschert, *Energy in the American Economy, 1850-1975: An Economic Study of Its History and Prospects* (Baltimore: Johns Hopkins Press, 1960), p. 36.

[2] Ibid., p. 63.

[3] Howard N. Eavenson, *The First Century and a Quarter of American Coal Industry* (Pittsburgh: Privately printed, 1942), p. 138; "The Early History of the Pittsburgh Coal Bed, "*Western Pennsylvania Historical Magazine*, vol. 22 (September 1939). Eavenson was also a professor of mining engineering in Pittsburgh and a former president of the American Institute of Mining and Metallurgical Engineering. He was also a mining engineer and author who operated his own coal company in Harlan County, Kentucky.

[4] *U.S. Eighth Census, 1860 Manufactures of the U.S. in 1860* (Washington, D.C.: U.S. Census Bureau), pp. clx111-clxiv.
Production in Millions of Tons of Anthracite and Bituminous Coal in Pennsylvania

Years	Anthracite	Bituminous
1840	1,129,206	699,994
1845	2,625,757	1,130,000
1850	4,326,969	2,147,500
1855	8,606,687	3,429,700
1860	10,983,972	4,710,400
1865	12,076,966	6,372,900
1875	23,120,730	12,433,860
1880	28,649,812	21,280,000

[5] William Sisson and Bruce Bomberger, *Made in Pennsylvania: An Overview of the Major Historical Industries of the Commonwealth* (Harrisburg: Pennsylvania Historical and Museum Commission, 1991), p. 18.

[6] George H. Ashley, *Bituminous Coal Fields of Pennsylvania* (Harrisburg: Department of Forests and Water, 1928), p. 179.

[7] J. Sutton Wall, *Report on the Coal Mines of the Monongahela* (Harrisburg: Board of Commissioners for the Second Geological Survey, 1884), p. xxviii. Wall's study is the most thorough examination of individual mines in the Monongahela River district in the late nineteenth century.

[8] George Thurston, *Directory of the Monongahela and Youghiogheny Valley* (Pittsburgh: A. A. Anderson, 1859), pp. 253-266.

[9] James M. Swank, *Introduction to the History of Ironmaking and Coal Mining in Pennsylvania* (Philadelphia: published by author, 1878), p. 113; *Western Pennsylvanian* (Pittsburgh: James O. Jones Company, 1923), p. 35.

[10] Eavenson, *The First Century,* p. 496.

[11] Frederick Binder, *Coal Age Empire* (Harrisburg: Pennsylvania Historical and Museum Commission, 1974), p. 111.

[12] "The Century of Coal Production," *The Coal Industry* (January 1918).

[13] Sylvester K. Stevens, *Pennsylvania: Titan of Industry,* vol. 3 (New York: Lewis Historical Publishing Company, 1948), pp. 864-865; Helene Smith, *Export: A Patch of Tapestry out of Coal Country America* (Greensburg: McDonald/Saward Company, 1980), p. 102. The firm was operating mines at the mining towns at or near Yukon, Rillton, Irwin, Manor, and Hutchinson in Westmoreland County by 1925.

[14] Lola M. Bennett, *The Company Towns of Rockhill Iron and Coal Company: Robertsdale and Woodvale, Pennsylvania* (Washington, D.C.: National Park Service, 1990), p. 7.

[15] Swank, *Introduction,* p. 125.

[16] *Region in Transition: Report of the Economic Study of the Pittsburgh Region* (Pittsburgh: University of Pittsburgh Press, 1963), p. 260.

[17] Binder, *Coal Age Empire,* p. 83; Guy Mankin, *Power Handbook on Fuels* (New York: McGraw-Hill Publication, 1934), p. 11.

[18] Swank, *Introduction,* p. 72.

[19] William Nichols, *The Story of American Coals* (Philadelphia: Lippincott Company, 1897), p. 331.

[20] Robert D. Billinger, *Pennsylvania's Coal Industry* (Gettysburg: Pennsylvania Historical Association, 1954), p. 25.

[21] Peter Temin, *Iron and Steel in Nineteenth Century America: An Economic Inquiry* (Cambridge: M.I.T. Press, 1954), pp. 268-269.

[22] John Enman, "The Relationship of Coal Mining and Coke Making to the Distribution of Population Agglomoration in the Connellsville (PA) Beehive Coke Region" (Ph.D. diss., University of Pittsburgh, 1962), p. 87. The Clinton furnace operated continously from 1859 until its closure in 1927.

[23] Joseph D. Weeks, *Report of the Manufacturer of Coke* (New York: David Williams, 1885), p. 26.

[24] Fred C. Keighley, "The Connellsville Coke Region," *Engineering Magazine,* vol. 20 (1901). Keighley served as mine superintendent of Mammoth Number 1 mine, and later as general superintendent of the Oliver and Snyder Steel Company coke works near Uniontown. He was president of the Coal Mining Institute of America in 1900.

[25] Ibid.

[26] *Pittsburgh and the Pittsburgh Spirit* (Pittsburgh: Chamber of Commerce of Pittsburgh, 1927-1928), pp. 213-214; E. Willard Miller, *Pennsylvania: Keystone to Progress* (New York: Windsor Publication, 1986), p. 59.

[27] *Tenth Census of the United States: General Analysis of the Bituminous Coal Statistics* (Washington, D.C.: Bureau of the Census, 1881), p. 1208.

[28] George Littleton Davis, "Greater Pittsburgh's Commercial and Industrial Development 1850-1900" (Ph.D. diss., University of Pittsburgh, 1951), p. 246.

[29] John N. Boucher, *History of Westmoreland County* (New York: Lewis Publishing Company, 1906), p. 276.

[30] Margaret Mulrooney, *A Legacy of Coal: The Company Towns of Southwestern Pennsylvania* (Washington, D.C.: National Park Service, 1991), pp. 35-50; Keighley, "The Connellsville Coke Region." Jim Cochran was the father of Philip Cochran who owned the Washington Coal and Coke Company, which constructed the coal town of Star Junction, just south of Perryopolis, Fayette County, in 1893. Star Junction was one of the largest turn-of-the-century coke facilities with 999 beehive ovens. Coal was extracted for the beehive ovens at two mines, Washington No. 1 and No. 2. Sarah Cochran, widow of Philip Cochran, constructed Linden Hall, an opulent thirty-five-room mansion north of Perryopolis on State Route 51 South, at a cost of two million dollars in 1911-1912. The United Steelworkers of America purchased and restored the mansion in 1976.

[31] George Thurston, *Pittsburgh's Progress: Industries and Resources* (Pittsburgh: A. A. Anderson and Company, 1888), p. 181.

[32] Douglas Fisher, *Epic of Steel* (New York: Harper and Row, 1963), p. 110; George H. Ashley, *Bituminous Coal Fields of Pennsylvania* (Harrisburg: Department of Forests and Water, 1928), p. 216.

[33] Adam T. Shurick, *The Coal Industry* (Boston: Little, Brown and Company, 1924), p. 173; Joseph D. Weeks, *Report on the Manufacture of Coke* (New York: David Williams, 1885), p. 32; Franklin Ellis, *History of Fayette County, Pennsylvania with Biographical Sketches of Many of Its Pioneers and Prominent Men* (Philadelphia: Louis H. Everts and Company, 1882), p. 246.

[34] The furnace was named for Henry Clay, the prominent ante-bellum Whig politician and three-time presidential nominee of his party. The furnace was abandoned in 1861 by the owners. The Pennsylvania Historical and Museum Commission erected a historical marker on January 20, 1949, near the site of this furnace. The furnace was located two miles from the marker, which is located on U.S. 62 west of Charleston.

[35] Swank, *Introduction*, p. 77, 122.

[36] Priscilla Long, *Where the Sun Never Shines* (New York: Paragon Press, 1989), p. 6.

[37] Ibid., p. 7.

[38] Ibid., p. 60.

[39] George Korson, *Coal Dust on the Fiddle* (Hatboro: Folklore Associates Inc., 1965), p. 7.

[40] Enman, "The Relationship of Coal Mining and Coke Making."

[41] Dennis Brestinsky, et al., *Patch /Work Voices: The Culture and Lore of a Mining People* (Pittsburgh: University of Pittsburgh Press, 1991), p. 8; Smith, *Export*, p. 124; Bill Spence, "Anton Oil Wick Lamps," *Mining Artifact Collector* (Spring 1990).

[42] Muriel Earley Sheppard, *Cloud by Day: The Story of Coal and Coke and People* (Pittsburgh: University of Pittsburgh Press, 1991), p. 158; Richard Quin, "Indiana County Inventory." Washington, D.C.: National Park Service, 1990, p. 26; J. W. Koster, "Gases Commonly Met With in Coal Mines," *The Coal Industry* (February 1919).

[43] Elwood S. Moore, *Coal: Its Properties, Analysis, Classification, Geology, Extractions, Uses and Distribution* (London: John Wiley and Company, 1940), pp. 301-305; Jim Steinberg, "The Davy Lamp," *Mining Artifact Collector* (Fall 1989); Tony Moon, "The Standard Wolf Safety Lamp," *Mining Artifact Collector* (Fall 1990); Keighley, "The Connellsville Coke Region."

[44] Anthony F. C. Wallace, *St. Clair: A Nineteenth Century Coal Town's Experience With a Disaster-Prone Industry* (Ithaca: Cornell University Press, 1987), pp. 296-302; Alexander Trachtenberg, *The History of Legislation for the Protection of Coal Miners in Pennsylvania* (New York: International Press, 1942), p. 36.

[45] Shurick, *The Coal Industry*, p. 77; L. D. Harnett, "Origin and Development of the Mine Fan," *The Coal Industry* (January 1920).

[46] George H. Ashley, *Bituminous Coal fields of Pennsylvania, Part 1* (Harrisburg: Department of Forests and Waters, 1928), p.184.

[47] Long, *Where the Sun Never Shines*, p. 37.

[48] Brad Ross, "Squibs," *Mining Artifact Collector* (Fall 1989).

[49] Richard T. Wiley, *Monongahela: The River and Its Region* (Butler: Ziegler Press, 1937), p. 148.

[50] Andrew Roy, *A History of the Coal Miners of the United States* (Westport, Ct.: Greenwood Press, 1905, reprinted 1970), pp. 62, 79.

[51] Wallace, *St. Clair*, pp. 283-284; Donald L Miller and Richard E. Sharpless, *The Kingdom of Coal: Work, Enterprise, and Ethnic Communities* (Philadelphia: University of Pennsylvania Press, 1985), p. 151.

[52] Donald J. McDonald and Edward A. Lynch, *Coal and Unionism: A History of the American Coal Miner's Union* (Silver Spring, Md.: Cornelius Printing Company, 1939), p. 16.

[53] Ibid., p. 169.

[54] Eavenson, *The First Century*, p. 378.

[55] J. Walter Coleman, *The Molly Maguire Riots: Industrial Conflict in the Pennsylvania Coal Region* (New York: Arno Press reprint edition, 1969).

[56] Harold W. Aurand, *From the Molly Maguires to the United Mine Workers: The Social Ecology of an Industrial Union, 1869-1897* (Philadelphia: Temple University Press, 1961), pp. 96-114; Long, *Where the Sun Never Shines*, pp. 110-113.

The Golden Age of King Coal, Queen Coke, and Princess Steel, 1880-1920

Introduction

The "Golden Age of King Coal, Queen Coke, and Princess Steel" was 1880 to 1920. Its impact on the social and economic development of the nation and the Commonwealth, as the principal coal and coke state, was long term and profound. "Ole King Coal" had toppled wood as the nation's principal energy source during the mid-1880s. William Stanley Jevons (1835-1882), an English political economist and logician, while visiting Pittsburgh in 1882 observed, "Coal in truth stands not beside but entirely above all other commodities. It is the material energy of the country—the universal aid—the factor in everything we do. With coal almost any feat is possible; without it we are thrown back into the laborious poverty of early times."[1] Jevons had realized that abundant high-quality coal and coke near the Pittsburgh district had transformed the "smoky city" into the principal metallurgical center of the United States.

The United States Census Bureau issued a report in 1902 which, like Jevons, recognized the pivotal role coal played in the rapid economic expansion of "Smokestack America." Coal had supplied 72.5 percent of the nation's energy needs by this time. This fossil fuel drove the industrial revolution of America, and the miners who extracted it had fundamentally transformed the nation into an industrial juggernaut. As the report observed:

> coal has been coincident with the rapid advancement of this country . . . to the front rank among industrial nations of the world. Indeed the country's progress has been due largely to the abundance of its mineral fuels, chief among them was coal. Most of this development had taken place during the last two decades and has far exceeded the growth in population, indicating a rapid change from an agricultural to a manufacturing nation.[2]

The Commonwealth was the principal bituminous coal producing state throughout this entire period. Bituminous coal had supplanted anthracite as the principal mineral resource of the state during the 1890s. The state's soft coal industry employed more than 180,000 miners, who extracted an incredible 166.9 million net tons in 1920. The compact Connellsville coke district was the nation's principal coke-producing region. Beehive coke production was consumed at iron and steel companies in the Pittsburgh district, and eastern Pennsylvania, and Connellsville coke was shipped west by rail for use in the silver industry. Literally hundreds of new company-owned coal towns were constructed throughout the counties of central and western Pennsylvania to house the large and increasingly foreign-born labor force. Unorganized miners formed the United Mine Workers of America in 1890, and from this humble beginning the miners' union had evolved into the largest and most powerful labor union in the United States by the end of the era. The UMWA had successfully organized nearly one-half of all mine workers nationally by the end of this period. This brief era was indeed the "Golden Age of Coal" in Pennsylvania and the United States.

American Coal and Coke Production

The rapid expansion of the iron and steel industry and of the railroad system during the last quarter of the nineteenth century created energy demands for an abundant and inexpensive alternative to wood. Coal supplied this energy as the industry experienced a period of phenomenal growth in terms of production and employment. Wood had provided about 90.7 percent of the nation's aggregate

Per Capita Consumption of Coal in the United States	
1850	0.278 tons
1860	0.514 tons
1870	0.857 tons
1880	1.42 tons
1890	2.50 tons
1900	3.53 tons
1905	5.00 tons[5]

Coal Production in the United States and the World		
Net Million Tons	World Total	U.S.A.
1890-1894 (average)	538	156
1905-1909 (average)	1,053	392
1915	1,193	482
1920	1,319	597[6]

American Coal Production in Millions of Tons				
Year	Bituminous	Anthracite	Total	Percent Increase
1880	50,757	28,650	79,407	1870-1880, 96 percent
1885	71,773	38,336	110,109	
1890	111,302	46,469	157,771	1880-1890, 98 percent
1895	135,118	57,999	193,117	
1900	212,316	57,368	269,684	1890-1900, 70.9 percent
1905	315,063	77,660	392,723	
1910	417,111	84,485	501,596	1900-1910, 86 percent
1915	442,624	88,995	521,619	
1920	568,667	89,598	658,256	1910-1920, 31.2 percent[7]

U.S. Coal Production by States		
1880	25 states produced	71,481,570 short tons
1890	28 states produced	157,770,963 short tons
1900	28 states produced	269,684,027 short tons
1910	27 states produced	492,647,863 short tons

energy consumption, excluding wind, water, and human power, as late as 1850. Wood supplied nearly three-fourths of the country's energy supply (73.2 percent) as late as 1870, but in 1883 it provided only 47.5 percent of the nation's aggregate energy consumption while coal provided 50.3 percent (bituminous coal 33.4 percent and anthracite 16.9 percent).[3] Soft-coal production reached a staggering 110 million tons in 1885 and replaced wood as the nation's principal fuel. Great Britain was the leading coal-producing nation as recently as 1875 mining 46.6 percent of the world's production, while the United States produced only 17.1 percent.[4] The U.S. Department of the Interior observed that in each decade from 1850 to 1905 the output of coal practically doubled. Per capita consumption of coal also rose during this period.

The United States supplanted Great Britain in 1899 as the leading coal-producing nation and became the premier nation in both the production and consumption of coal during this era.

American bituminous coal production increased from 43 million tons in 1880 to 212 million at the turn of the twentieth century and to 569 million tons in 1920—more than a ten-fold increase. The number of miners increased from 100,207 in 1880 to 304,375 in 1900, and to 639,547 in 1920. Coal productivity, as measured per man-day in the industry, rose from 2.5 to 4 tons during this period.

There were twenty-five coal-producing states in 1880. Pennsylvania, Illinois, Indiana, and Ohio were the principal ones, accounting for nearly three-fourths of all bituminous coal production in the United States.

Each of the four states mentioned above increased their coal production by more than 130 percent between 1870 and 1880. The bituminous coalfields of western Pennsylvania measured about nine thousand square miles in 1881. Both production and number of workers employed increased continuously throughout the period. There were 666 establishments mining soft coal in Pennsylvania with an annual production of 16.5 million tons in 1880.

Eleven of the sixteen American counties producing over a half million tons annually of bituminous coal were Allegheny, Bradford, Cambria, Clearfield, Elk, Fayette, Mercer, Somerset, Tioga, Washington, and Westmoreland Counties, Pennsylvania.[8] Six of these eleven coal counties produced more than three-quarters of the state's annual production. The principal coal-producing counties in 1880 were Allegheny, 6.7 million tons; Westmoreland, 4.3 million tons; Clearfield, 2.8 million tons; Fayette, 2.7 million tons; Washington, 1.3 million tons; and Tioga, 1.2 million tons. Production in Pennsylvania increased tenfold, from 16.5 million tons to 166.9 million tons, while employment increased almost sixfold, from 33,391 to 184,168 between 1880 and 1920.[9]

Pennsylvania's annual bituminous coal production surpassed the state's anthracite production in 1897 for the first time. Bituminous coal output was 54,62,272 net tons in 1897 while anthracite output was 52,581,036 net tons. Combined bituminous and anthracite coal production of Pennsylvania was more than one-half of all coal mined in the United States between 1880 and 1899. The Commonwealth produced 66 percent of the entire coal output of the United States in 1880, and although this percentage decreased after 1900, Pennsylvania had the distinction of producing more than 50 percent of the nation's entire coal output until 1902. The violent and protracted 1902 anthracite strike reduced production, and the state's production was only 46 percent during the strike period. At least one-quarter of all bituminous coal mined in the United States during this entire period was extracted by Pennsylvania miners.

The counties surrounding the Monongahela River, the semibituminous "smokeless" coalfields of the Broad Top Mountain, and the North Central coalfields were the principal coal-producing regions in the state until this period. Demand for coal and coke from the iron and steel industry and the railroads was satisfied by the opening of new bituminous coalfields in Pennsylvania. The Second Geological Report of Pennsylvania, established in 1874, surveyed and identified the location of new coal reserves in the counties of central and western Pennsylvania from 1874 to 1884. The survey, under the direction of J. P. Lesley, the state geologist, published nearly 120 atlases and volumes, including geologic maps of the state's counties and one map of the entire state in 1885. The comprehensive survey helped "to uncover with precision the economic mineral resources of Pennsylvania in an age of great industrial expansion providing a firm foundation for later detailed studies produced by industry, government and academia."[11]

Pennsylvania's Bituminous Coal Production and Employment

Year	Production Net Tons	Employees	Percent Net Tons U.S. Total
1880	16,564,000	33,391	25.9
1885	23,413,692	44,145	23.4
1890	40,625,054	67,383	26.9
1895	51,818,112	84,976	26.0
1900	79,318,362	108,735	29.7
1905	119,361,514	164,941	30.2
1910	148,770,858	193,488	30.0
1915	157,420,068	187,734	29.7
1920	166,929,002	184,168	25.9[10]

The expanding railroad system of Pennsylvania was responsible for the opening of commercial coal-mining operations in a number of isolated and underdeveloped coal-producing regions. Though these counties had all produced coal for local consumption before the Civil War, large-scale commercial mining awaited the arrival of railroads and increased demand for coal. Many spur lines were constructed off the main lines of the Pennsylvania Railroad and the Baltimore and Ohio Railroad, connecting to the newly opened mines. The expanding coal industry contributed to the economic prosperity of these new coal-producing counties, and encouraged population growth as European immigrants arrived to mine coal and make their homes in the dozens of new mining communities that were rapidly being constructed during this period in Armstrong, Jefferson, Cambria, Indiana, and Somerset Counties.

Only Illinois, with 10,872 miles of railroads, exceeded Pennsylvania's 10,181 miles in operation by 1899. Railroads provided coal operators with low hauling rates to distant markets, since hauling coal was becoming the principal source of revenue for many railroad corporations. Railroads were the principal consumers of coal and became the principal owners of many of these newly opened mines and coal towns. The railroad and coal companies became interdependent with the establishment of "captive" mines of the railroad companies. The Rochester and Pittsburgh Coal Company and the Pennsylvania Coal and Coke Company, for example, were mines developed by railroad companies to supply their own need for steam coal. The coal development of Armstrong, Jefferson, Cambria, Indiana, and Somerset Counties replaced the semibituminous Broad Top Field and North Central Fields as principal mining regions. Coal was still extracted in these fields but their production was dwarfed in contrast to the large output of these new coal-producing counties. After the 1880s Pennsylvania's principal coal-producing regions were located in the Pittsburgh district, including Allegheny, Fayette, Westmoreland, and Washington Counties and the newly opened counties of southwestern Pennsylvania. The commercial development of the coal industry in these counties was delayed because of the plentiful, thicker coal seams located nearer Pittsburgh. Greene County was part of the Pittsburgh district but production in this county was modest until after 1910 when the rich Pittsburgh seam that underlay the county was opened for coal exploration.

The following is a brief description of the development of commercial mining in these counties, with an emphasis on the leading coal operators and coal mining towns constructed by them:

Armstrong County

Coal was Armstrong County's most important mineral resource in the twentieth century. Cannel coal was the first mined in the county and was used as "illuminating oil." This coal, which is very rich in volatile matter, especially hydrogen, is composed almost entirely of spores, spore cases, seed coats, and resinous or waxy products of plants that lived in the coal-forming swamps. The absence of woody material gives cannel coal a regular texture and grain not found in other types of coal. Cannel coal breaks like glass, with a conchoidal fracture, and because of its high content of volatile matter ignites easily, burning with great heat and a long flame.[12] Although the Allegheny Valley Railroad was completed into Armstrong County in 1856, large-scale commercial coal mining was delayed until 1899 when the Cowanashannock Coal and Coke Company erected the first company towns of Yatesboro and NuMine. The coal industry employed 4,290 workers and had produced more than 3.5 million tons of coal by 1910. The Yatesboro mine and company town, with rail connections, was erected in less than a year. A railroad spur line extended to Echo, where it joined the main line of the Buffalo, Rochester and Pittsburgh Railroad. Coal was shipped on this railroad to Great Lakes markets. This was the largest and most productive mine in the county by this period. The principal coal companies in Armstrong County were the Helvetia Coal Mining Company, the Allegheny River Mining Company, and the Buffalo and Susquehanna Coal Company.

Somerset County

Mining began in Somerset County during the last quarter of the eighteenth century. Coal was used by local blacksmiths, who hauled it themselves to their shops, probably before 1800. Coal mining remained essentially a local industry until the completion of the first railroad into the county in 1872. The railroad era brought a new era of land speculation and economic growth to Somerset County. The Baltimore and Connellsville Railroad constructed a route connecting Baltimore and Pittsburgh, via Cumberland, Maryland. In Somerset County the railroad followed the Casselman River from Turkeyfoot to Meyersdale, then passed through a mountain tunnel at Sand Patch to Wills Creek, Maryland. Irish and German laborers were imported to the region in 1870 to construct the railroad. The railroad was completed in 1871, when the final spike was driven near Casselman, and it provided passenger and freight service through the county. The railroad allowed the development of the county's rich coal and lumber resources and fostered European emigration. The Keystone Coal and Manufacturing Company was founded by Henry A. Stiles of Philadelphia, Henry T. Weld of Mount Savage, Maryland, and George F. and William J. Baer of Somerset in 1870. The company opened Keystone, the first coal town in the county in 1872. Keystone, known locally as "Stilesville," was named after Henry A. Stiles, who served as the first president of the company. Baer was a prominent lawyer, judge, businessman, and president of the Somerset and Mineral Point Railroad. The mine and village were located about two and a half miles southeast of Meyersdale in the eastern part of Summit Township. Coal was hauled from the Keystone tipple to Keystone Junction over a narrow-gauge railroad which was abandoned after the Salisbury and Baltimore Railroad was built in 1878. The Cumberland and Elk Lick Company established the Shaws Mine complex in Summit Township in 1875. Shaw was the second coal-company town constructed in Somerset County.[13] Beehive coke ovens were constructed at the mine, with coke production beginning in 1886. The principal coal companies by 1920 were the Consolidation Coal Company, the Berwind-White Mining Company, and the Quemahoning Coal Company. Daniel B. Zimmerman of Somerset County was an astute businessman who was a cattle dealer, agriculturist, and coal operator. Zimmerman, like William J. Baer, was one of the local pioneer coal operators in the county. He was president of the Quemahoning Coal Company that was founded in 1898. He developed numerous mines and the company mining towns of Goodtown, Wilson Creek, Ralphton, and Zimmerman, Somerset County. His coal interests in the county totaled 140,000 acres of coal lands in 1907, with additional holdings at Listie, Jerome, and Rockwood. These resources made Zimmerman the largest local and independent coal operator in the county. The communities of Cairnbrook, Boswell, Hollsopple, Hooversville, Jenners, Jerome, and Windber had their origins as coal-company towns.

Cambria County

Cambria County is one of the few counties in the bituminous coal regions of Pennsylvania that mine four important commercial seams. The Lower Kittanning coal is the most extensive seam and is located in the southern part of the county. The Upper Kittanning lies about 125 feet below the Lower Kittanning seam. This seam was mined for steam coal. The Lower Freeport seam, located in the northern part of the county, is an excellent coking coal with low sulfur content. This seam is located 120 to 190 feet above the Lower Kittanning seam. The Upper Freeport seam, the fourth principal commercial seam in the county, is mined in the area around Barnesboro. This coal was used in railroad locomotives and with varying results in coke production. A number of small coal mines were opened in the South Fork area of the county before the Civil War. A number of commercial coal operations had begun around Barnesboro and Patton in northern Cambria County by the 1870s. The Blacklick coalfields were opened in the 1890s by the Berwind-White Mining Company of Philadelphia in the Windber area on the northern boundary of Somerset County. Bakerton, Colver, Revloc, Spangler, Barnesboro, Vintondale, Dunlo, Nanty Glo, and Cassandra are all communities that owe their existence to the coal boom during this period. The principal mining companies were Barnes and Tucker Company, Berwind-White Mining Company, Heisley Coal Company, and Coleman and Weaver Company.

Indiana County

Coal was discovered in Indiana County as early as the 1760s. The Upper Freeport is the largest reserve of valuable and accessible coal in the county. The Lower Kittanning, Lower Freeport, and Pittsburgh seams are also mined commercially. Local coal played a pivotal role in the development of the pioneer salt industry in the county. The Blairsville division of the Pennsylvania Railroad was opened in 1856, and small coal shipments were made from the county after this date. The Western Pennsylvania Railroad traveled from Blairsville to the Allegheny River once it was opened in 1864. Glen Campbell, founded by the Glenwood Coal & Coke Company in May 1889, with a population of three hundred, was the first company town constructed in Indiana County.[14] The Rochester and Pittsburgh Coal Company, founded in 1881, acquired coal properties in Indiana in 1899 and 1900 and within a decade was the largest coal producer in the county. R&P operated numerous mines and constructed the company towns of Ernest, north of Indiana, and Lucerne Mines, Center Township near Homer City. Aultman, Clymer, Commodore, Graceton, Iselin, McIntyre, Rossiter, and Sample Run were all new, booming mining towns founded in the county during this period. The principal commercial coal companies in Indiana County were the Rochester and Pittsburgh Coal and Iron Company, Clearfield Bituminous Coal Corporation, and the Pennsylvania Coal and Coke Company.

Jefferson County

The Rochester and Pittsburgh Coal and Iron Company was organized in 1881 to tap into the new coal resources of Jefferson and Clearfield Counties. The corporation was financed by Walston H. Brown, a New York financier and the Adrian Islin family of New York. The firm acquired more than six thousand acres of coal land near Punxsutawney, Jefferson County, and constructed the earliest coal-mining towns in the county during the 1880s: Beechtree, Washington Township north of

Coal Production Growth of Central Pennsylvania					
Year	Armstrong	Somerset	Cambria	Indiana	Jefferson
1881	245,000	298,000	780,000	52,000	420,000
1891	299,945	441,070	3,073,098	539,628	3,600,052
1901	1,686,075	3,8898,738	8,614,49	815,659	6,034,656
1911	3,760,460	8,330,274	16,371,550	8,555,610	5,356,338
1921	3,386,951	9,141,045	15,713,730	6,192,613	2,639,997[15]

DuBois, in 1882, and Walston, outside Punxsutawney, in 1883. The company constructed houses at Beechtree; they cost $250 each and were rented by the company to miners and their families for $60 a year, while those at Walston cost $200 each, and were rented for $48 annually. Horatio, Adrian, Anita, and Eleanor were mining communities in the county. The Rochester and Pittsburgh Coal and Iron Company, Jefferson Coal Company, and the Anita Coal Company were the principal mining companies.

Industrial Expansion and the Burgeoning Coal Industry

The economic expansion of "Smokestack America" during the last quarter of the nineteenth century was predicated on increasing energy supplies and an enlarged labor force. Coal solved America's burgeoning energy demand. Every man, woman, and child was the beneficiary of coal during this period. The annual consumption of coal per capita rose nationally from one ton in 1870 to two tons in 1896 and to 6.6 tons in 1918.[16] The American coal industry expanded as a result of increased energy demands from the growing iron and steel and railroad industries. The Bessemer and the Siemens-Martin open-hearth processes, introduced during the second half of the nineteenth century, represented new steelmaking technologies. American steel production was previously measured in pounds, but the widespread use of these processes made it possible to mass-produce steel in large quantities quickly and at a reduced cost. Steel production surpassed iron production in the United States for the first time in 1892. The United States had become the greatest steel-producing nation in the world by 1900, when annual steel production was twice as large as Great Britain, its nearest competitor.[17] Per capita consumption of steel increased from about five hundred pounds in 1900 to about twelve hundred pounds in 1920.

The initial breakthrough of a viable steelmaking process came in the 1850s with the invention of the Bessemer process. Charles Henry Bessemer of Sheffield, England (1813-1898), and his American counterpart, William Kelly (1811-1888), a Pittsburgh native and kettle maker, both claimed simultaneously to have invented the pneumatic or Bessemer process of converting pig iron into steel. These inventors discovered that when drafts of air were blown through molten iron, it created a superior metal relatively free of carbon. In 1854, Bessemer began his experiments in designing a converter near his Sheffield cutlery works capable of transforming iron to steel by removing carbon. The first official news of Henry Bessemer's new steelmaking process in the United States came in an article of the *London Times*, August 14, 1856. Bessemer created steel in thirty minutes—not the usual three months—that was only a little more costly to produce than iron. He received an American patent for the pneumatic process in 1856. Kelly developed a crude pneumatic converter that transformed iron into steel in a matter of minutes at the Cambria Iron Company, Johnstown, Pennsylvania, in 1857 and 1858.[18] William Kelly's original converter now resides at the Smithsonian Institution in Washington, D.C. Kelly contested Bessemer's American patent on the basis of work undertaken by him in the 1840s and 1850s. He proved to the satisfaction of the United States commissioner of patents that he had the idea of applying air in a pneumatic converter in 1847. He received a patent for the invention that was renewed in 1870. In 1861, Kelly made the first American steel by the Bessemer-Kelly pneumatic process, employing the bottom-blown tilting converter at Johnstown. Bessemer took out the earlier patents on his converter but Kelly, in conjunction with David Morrell, president of Cambria Iron Company and others, secured later patents on the process. The Johnstown Iron Plant began operation in 1852. The facility has operated under the following ownerships since its formation: Cambria Iron Works, 1852-1855; Wood, Morrell and Company,1855-1862; Cambria Iron Works, 1862-1898; Cambria Steel

Company, 1898-1916; Midvale Steel and Ordinance Company, 1916-1923; and Bethlehem Steel Company, 1923-present. In 1866 the Bessemer and Kelly groups of patent holders combined their interests, ended their litigation, and joined forces. An agreement was concluded by which the Bessemer group received 70 percent of American royalties, while the Kelly group received 30 percent.

The first commercial Bessemer converter was erected in Wyandotte, Michigan, by the Wyandotte Iron Works in 1864 and produced steel of commercial quantity. The firm produced 2.5 tons of steel at a time—in one "heat." Alexander Holley was impressed by the importance of Bessemer's invention and after acquiring rights to Bessemer's American patents, erected an experimental Bessemer converter at Troy, New York. Holley started Bessemer steel production under these patents at the Troy plant on February 1865 and continued it at the Pennsylvania Steel Company facility at Steelton, near Harrisburg, where the first Pennsylvania Bessemer furnace was installed. Steel rails were rolled by the Cambria Iron Company from ingots produced at the Pennsylvania Steel Company in 1867, and the company erected its own Bessemer works in 1869. The Bessemer process dominated steelmaking during the nineteenth century, but the process gradually gave way to the open-hearth process in the twentieth century.

Cheap Bessemer steel revolutionized the iron industry. The pneumatic process introduced the mass production of steel. Steel rarely sold for less than $250 per ton before this new steelmaking technology was introduced. The production of inexpensive steel prompted Andrew Carnegie to note, "The day of iron is past! Steel is king!" In 1873 the Carnegie Company began construction of the Edgar Thomson Steel Works, named for the third president of the Pennsylvania Railroad, at Braddock, twelve miles down the Monongahela River from Pittsburgh. The mill was constructed on Braddock's Field, the site of the historic battle in 1755 between the British and colonial army commanded by General Edward Braddock and French and Indian forces. The Edgar Thomson Works, completed by Andrew Carnegie in 1875, received its first order for two thousand steel rails from the Pennsylvania Railroad. Pig iron was originally obtained from Carnegie's Lucy Furnace, built in 1871-1872 at Fifty-first Street and the Allegheny River, Pittsburgh, until its own blast furnaces were constructed in 1880. There were nine steel companies in the United States in 1890 with capacities in excess of a hundred thousand tons of steel annually, along with nearly five hundred smaller firms with steel capacities as small as three thousand tons. Carnegie's Edgar Thomson Steel Works was the largest steel plant in the nation, with an annual capacity of 450,000 tons in 1890. Carnegie's steel companies had an annual production of 3.5 million tons of steel ingots and over 3 million tons of finished-steel products at the turn of the twentieth century.

The rise of American steel production was meteoric during the next forty years. Steel production rose from a mere 22,000 tons in 1867 to 26,205,913 tons in 1906. The production of Bessemer steel, which was centered in southwestern Pennsylvania, had come to dominate the steel industry by 1880. Each Bessemer converter was larger than a typical anthracite furnace and provided a cheap means of producing large quantities of steel. Most of this steel went into the production of rails. Steel rails were preferred to iron rails because of their durability, their ability to support heavier freight cars, and their strength that permitted trains to travel at faster speed. Steel production was concentrated in Pittsburgh and the surrounding Monongahela Valley. Johnstown (Cambria Iron Company and the Johnstown Steel Rail Company), the Mahoning and Shenango Valleys, Youngstown, and Steubenville were other principal steel centers in the region. The steel mills in these metallurgical centers all used coke from the Connellsville district, shipped by rail or river. This expansion of steel capacity insured the growth and popularity of the Bessemer steel process during the last quarter of the nineteenth century and insured the expansion of the nascent coke industry of western Pennsylvania.

Pierre Martin (1824-1915), son of a French ironmaster, developed the open-hearth or Martin furnace in 1865 that used recycled steel scrap, improved the efficiencies of blast furnaces, and enabled manufacturers to produce a more refined steel. Abram S. Hewitt introduced the open-hearth process at Trenton, New Jersey, in 1868. This process removed sulfur and phosphorous and permitted the use of poor-quality iron ore. The percentage of steel made by the open hearth method exceeded that of Bessemer steel production in 1908; after this date a majority of American steel was produced by the open-hearth process. This process accounted for 36.5 million tons of ingots as opposed to ten million tons of Bessemer steel ingots in 1920.[19] The development of the coke industry was the foundation of the iron and steel industry of the United States. The increased production of steel placed new demands for coke that stimulated the coal industry.

The American Coke Industry

Raw bituminous coal, anthracite, charcoal, and coke were all used by iron manufacturers in Pennsylvania as fuel in their blast furnaces during this entire period.[20]

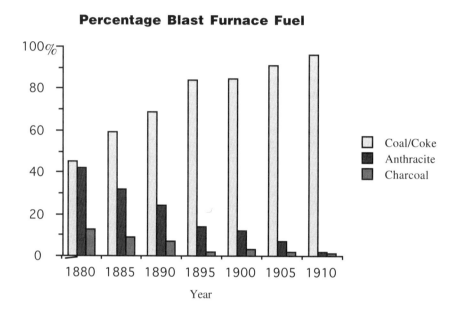

Iron blast furnaces, located principally in eastern Pennsylvania, continued iron production with anthracite although the quantities of anthracite iron production continued to decline with each passing decade. There were 158 anthracite or anthracite/coke furnaces in America in 1880, and 90 such furnaces in Pennsylvania, 15 furnaces in New York, and 12 furnaces in New Jersey in 1896. Their combined annual production was 3,156,487 tons. Eastern anthracite iron manufacturers had found it cheaper to import coke from western Pennsylvania to fuel their furnaces by the dawn of the twentieth century.[21] Anthracite was last used alone as a fuel in an iron blast furnace in 1900, and all use of anthracite in furnaces had been discontinued completely by 1923.[22] Ironmakers found it easier to get the oxygen in iron ore to combine with the carbon in coke, because the air holes permitted greater contact between the two elements.[23] The quantities of ingredients required to produce a ton of pig iron in a typical iron furnace during the 1880s were 1.78 tons of iron ore, 9.58 tons of coke, 4.44 tons of limestone, and 4.22 tons of air. This produced one ton of pig iron, .517 ton of slag, .063 ton of flue dust, and 5.82 tons of gas.[24] Coke provided more energy per pound at less cost than anthracite or charcoal. It produced a higher and more efficient operating temperature in blast furnaces because it was more porous than anthracite. Anthracite furnaces were smaller than contemporary western coke blast furnaces, which were on average two and a half times larger. Expansion of the coke-fueled iron and steel industries spelled the demise of the anthracite furnaces of eastern Pennsylvania.

The once dominant iron charcoal furnaces of Pennsylvania had become a relic of a former era by the turn of the century. The eight remaining charcoal furnaces in Pennsylvania were producing less than one percent of iron production by 1910. Charcoal furnaces hung on in other parts of the country and by 1890 charcoal production nationally was 703,522 tons of iron, of which nearly a quarter of a million tons were produced in Michigan. The charcoal iron industry thrived in Michigan and Wisconsin as long as timber was plentiful near the sites of the furnaces.[25]

A number of factors favored the concentration of iron and steel production in southwestern Pennsylvania and eastern Ohio after the Civil War. There was an ample supply of water for industrial uses, accessibility to markets, availability of Mesabi iron ore, and the proximity of excellent metallurgical coke. The best coking coal was located in Westmoreland and Fayette Counties, in the Connellsville coke region. Coal from this region was eminently suited for the production of coke because it was "thick, contiguous, clean, soft, friable, and advantageously positioned for cheap and easy mining."[26] Connellsville coal had the following traits required of a superior metallurgical fuel for the blast furnaces: (1) mechanical strength sufficient to withstand the crushing strain in the blast furnace; (2) sizes over two inches; (3) hardness or resistance to impact, a lack of brittleness permitting rough handling without undue fragmentation; (4) porosity, in order that the coke might expose a maximum surface for reaction with furnace gases; (5) low ash content, under 8 percent if possible; (6) low moisture, under 1 percent; (7) low phosphorous, under 0.02 percent; (8) low sulfur, under 1.25 percent; and (9) a high carbon content.[27]

There were 256 blast furnaces using either bituminous coal or coke in the United States in 1896. These furnaces were distributed in seventeen states and produced more than thirteen million gross tons of pig iron annually, which was four times the amount produced by all anthracite furnaces in the United States. There were 76 blast furnaces in Pennsylvania, 53 in Ohio, 39 in Alabama, 24 in Virginia, 17 in Illinois, 12 in Tennessee, 6 in Kentucky, 5 in Maryland, 4 each in West Virginia and Wisconsin, 3 in New York, 2 each in Missouri, Colorado, Georgia, and Indiana, and 1 in Minnesota.[28] A typical blast furnace of the period used about 3,330 pounds of iron ore, 1,200 pounds of coke, and 500 pounds of limestone to produce 2,000 pounds of iron.[29] Between 1871 and 1919, 88 percent of all iron in the United States was made with coke produced in the beehive coke oven.[30] In 1909 there were 579 coke plants with 103,982 beehive ovens in the United States.[31] By 1908 fully 98 percent of the total production of pig iron in the nation was made with coke, either alone or in combination with raw bituminous coal. Pennsylvania was the leading coke-producing state in the nation with its peak market share of 70.3 percent attained in 1907. The peak output of coke nationally was reached in 1916, when nearly forty thousand coke ovens were in operation converting 33,792,000 tons of bituminous coal into 22,486,000 tons of coke.[32]

During this period the annual production of coke, like coal, increased phenomenally from 3.3 million short tons to more than 51.3 million tons in 1920.

Large-scale beehive coke production spread to thirteen states and territories during the 1880s; in 1880 beehive coke ovens produced nearly 3.3 million tons of coke at 186 establishments nationally. Pennsylvania was the leading coke-producing state with 124 establishments producing 74.5 percent of the country's coke, valued at $16.3 million in 1880. The Connellsville coke district was the nation's premier coke-producing region. Fayette County produced 45.8 percent and Westmoreland County 27.4 percent of the state's coke production in 1880, while Allegheny County, a distant third in production, produced only 3.5 percent.[33]

As early as 1882 there were at least fifty individually operated coke plants in the region, utilizing several thousand beehive ovens.[34] From 1870 until World War I, the Connellsville coke district supplied a majority of all coke used by the iron and steel companies in western Pennsylvania, northern West Virginia, and eastern Ohio. The economic expansion of the district is reflected in increases in the number of

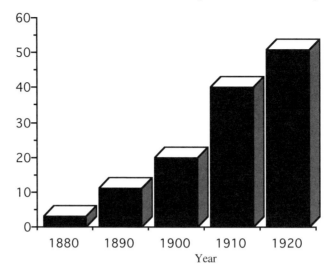

U.S. Coke Production (millions of tons)

beehive ovens, tonnage output of coke, value per ton of coke, and gross revenue for the district. From 1880 to 1900 the number of beehive ovens increased from 7,211 to 20,954, production rose from 2.2 million to 10 million tons, price per ton of coke rose from $1.79 to $2.70, and gross revenue rose from $3.9 to $27.4 million.[35]

The ever-increasing growth of the coke business in the region provided employment, spurred the growth of coal mining towns in the region, and necessitated a railroad building boom to transport coke north to the Pittsburgh district. A Pennsylvania official in 1887, observing the changes in the region as a result of this coke boom, noted that "this coke region [has] grown from 3,600 coke ovens in 1876, into a vast furnace of 13,000 ovens, while a poorly-paid, helpless band of workmen of ten years [has] grown into a vast army of 13,000 cokers."[36] Nearly 75 percent of all wage-earners in the region were employed in the coal-mining and coke industries about 1880. From 1900 to 1910, it is estimated that 80 percent of all capital investments in the region were in these industries.[37] For example, the population of Mount Pleasant Borough, Westmoreland County, grew from 1,197 in 1880 to 3,652 in 1890 to 5,810 in 1910. The coal company towns of Grace, Adelaide, Trotter, Leisenring, Paul, Leith, Wynn, Morrel, and Calumet were all constructed in Fayette and Westmoreland Counties between 1880 and 1889.[38]

The H. C. Frick Coke Company, formed in 1882 by a number of Pittsburgh iron and steel manufacturers, including H. C. Frick, was the largest coke company in the region and in the nation. The new firm issued forty thousand shares of stock at $50 a share, with capitalization valued at $2 million in assets and stock issue in 1882. These men and corporations held the following shares as recorded on May 5, 1882: Andrew Carnegie, 1,000; Thomas M. Carnegie, 500; Henry Phipps Jr., 500; H. C. Frick, 680; E. M. Ferguson, 660; Walton Ferguson, 660; Carnegie Bros. & Company Ltd., 2,500; and H. C. Frick & Company, 33,500.[39] The company owned 1,026 beehive ovens and 3,000 acres of coking coal in the year the company was formed. The company controlled thirty-five thousand acres of coal land and owned nine thousand beehive ovens in 1890. The principal competitors to the H. C. Frick Coke Company in the Connellsville region were the W. J. Rainey Company, the McClure Coke Company, Sample Cochran Sons and Company, Joseph R. Stauffer and Company, A. A. Hutchinson and Brothers, and James Cochran and Company during the 1880s. Frick purchased the McClure Coke Company, the second largest coke business in the district in 1895. It had operated fifteen coke plants constructed between 1871 and 1879 and when acquired it was operating eleven plants and nineteen hundred ovens.

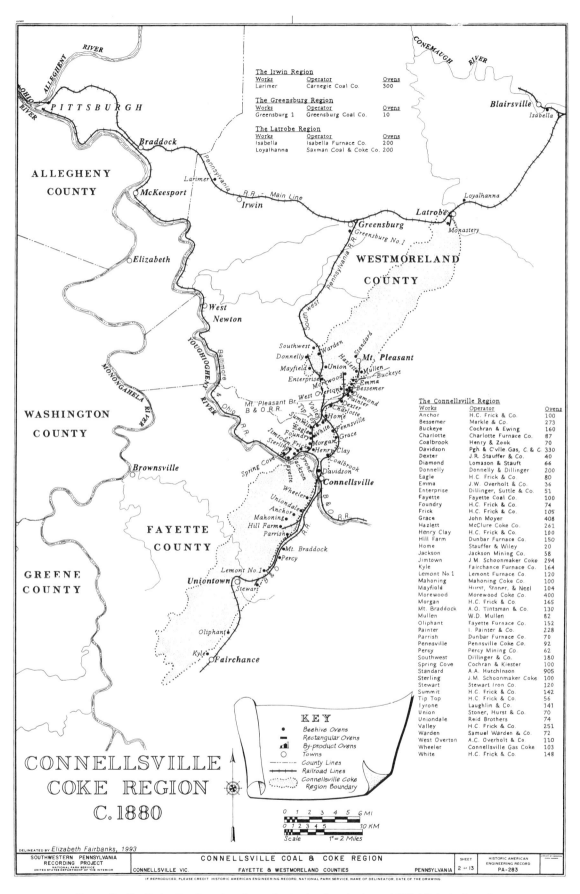

Connellsville Coke Region showing principal coke plants c. 1880. Historic American Engineering Record (HAER).

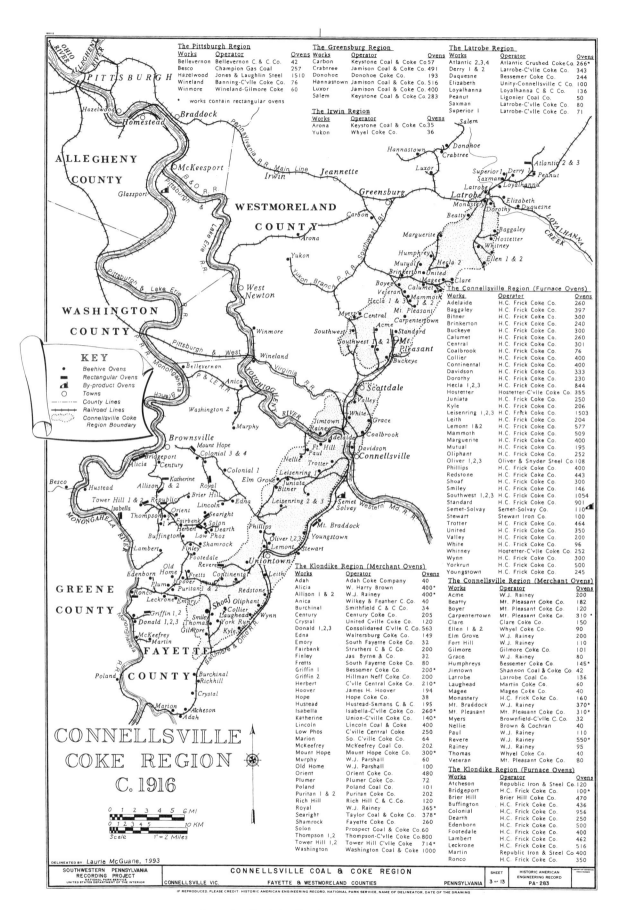

Connellsville Coke Region, c.1880. HAER.

Frick was elected president of the H. C. Frick Coke Company and the firm erected a new corporate office at Broadway and Walnut Avenues, Scottdale, Westmoreland County, in 1880. A second corporate office was constructed next door to the original office in 1904. Both offices are still extant and have excellent integrity. The vast coke empire of the H. C. Frick Coke Company was created by acquiring existing coke plants in the Connellsville district, the firm having constructed only twelve coke plants in the Connellsville coke region between 1871 and 1907—Adelaide, Standard Mine Number 2 (which replaced Standard Number One, destroyed by fire), Shoaf, Brinkerton, Henry Clay, York Run, Hopwood, Collier, Phillips, Bitner, Frick (Novelty), and Smiley. Management preferred instead to purchase established coke plants, and between 1870 and 1908 acquired nearly fifty plants. H. C. Frick Coke Company purchased four facilities between 1870 and 1879, seventeen facilities between 1880 and 1889, seventeen facilities between 1890 and 1899, and fourteen more coke plants between 1900 and 1908. The new American millionaire "Coke King" was operating ten thousand ovens with eleven thousand employees in the Connellsville district by the late 1890s.[40]

Andrew Carnegie became acquainted with the coke baron soon after Frick's marriage to Adelaide Howard Childs in December 1881, beginning a long-term and often tumultuous business relationship. Frick was entrusted by Carnegie in 1889 to reorganize and consolidate Carnegie Brothers Steel, and by the turn of the century had successfully consolidated the various companies into the Carnegie Steel Company. The new company produced between 25 and 30 percent of the nation's steel, 50 percent of the armor plate, and 30 percent of steel rails. Carnegie was associated with Frick because, according to him, Frick and Company "owned the best coal and coke property."[41] This business arrangement assured Carnegie a reliable and inexpensive supply of coke, which along with iron ore and limestone is the principal raw ingredient required in the production of steel. "What Is Steel?" a poem appearing in the *New York Herald* in 1893, identifies the principal ingredients of steel and the geographic advantage of the Pittsburgh district for the development of the nascent steel industry:

> The eighth wonder of the world is this;
> Two pounds of iron-stone purchased on the shares of Lake Superior and
>   transported to Pittsburgh;
> Two pounds of coal mined in Connellsville and manufactured into coke and
>   transported to Pittsburgh;
> One pound of limestone mined east of the Alleghenies and brought to Pitts
>   burgh;
> a little manganese ore mined in Virginia and shipped to Pittsburgh;
> And these four and one half pounds of material manufactured into one
>   pound of solid steel and sold for a cent;
> That's all that need be said about the steel business.

The major counties and coking seams of Pennsylvania are (1) Allegheny—Pittsburgh, Upper Freeport; (2) Armstrong—Lower Kittanning, Upper Freeport; (3) Butler—Brookville, Upper Freeport; (4) Cambria—Lower Kittanning (B), Upper Freeport (E), Upper Kittanning (C), Lower Freeport (D); (5) Clearfield—Lower Freeport; (6) Fayette—Lower Kittanning (B), Pittsburgh, Redstone, Sewickley; (7) Greene—Pittsburgh, Sewickley; (8) Indiana—Lower Freeport, Lower Kittanning, Upper Freeport; (9) Somerset—Brookville (A), Lower Freeport, Lower Kittanning (C), Pittsburgh; (10) Washington—Pittsburgh; (11) Westmoreland—Kittanning (B), Pittsburgh, Upper Freeport.[42] While a majority of Pennsylvania coke was produced in beehive ovens in the compact Connellsville coke district straddling parts of Fayette and Westmoreland Counties, there were sixteen counties in the state producing more than 3.5 million tons in 8,456 beehive coke ovens during the 1880s.[43]

Pennsylvania Coke Producing Counties in 1885

Counties	Tons of Coke Produced	Number of Ovens Operating	Number of Ovens Idle
Allegheny	19,416	95	107
Armstrong	10,311	66	
Beaver	438	2	6
Bedford	41,682	110	
Blair	91,459	196	20
Butler	5,015	32	68
Cambria	100,606	228	2
Clarion	7,057	58	12
Clearfield	49,552	200	16
Elk	3,438	36	
Fayette	2,074,734	4,701	1409
Huntingdon	62,838	241	
Jefferson	90,053	151	206
Somerset	5,382	32	
Tioga	16,100	75	100
Washington	900	4	
Westmoreland	1,001,768	2,189	3,443

Blair and Cambria Counties were the next most important coking region in Pennsylvania during the 1880s and 1890s. Coke production in these counties took place in the bituminous coal basins along the sides and near the summit of the Allegheny Mountains. The coking industry in Cambria County developed in the eastern and northern townships, with the largest beehive coke establishments located in the Cresson area, and in the vicinity of Barnesboro and Hastings. The Pennsylvania Coal & Coke Company was dominant in the county's beehive coke industry by the first decade of the twentieth century. The county's coke production experienced a rapid decline in the decades after World War I as the by-product ovens replaced beehive coke production.

Blair County had four coke-making establishments, 190 ovens, and 107 employees in 1880, while in Cambria County there were three coke plants, 119 ovens, and 45 employees. The combined coke production of both counties in 1880 was about 150,000 tons. The Bennington coke plant, owned by the Blair Iron and Coke Company, was one of the largest coke plants. It was acquired by the Cambria Iron Company in the 1870s and remained active until about 1884.[44]

Coke manufacturing also developed during the last quarter of the nineteenth century in the semibituminous coal region of the Broad Top Field, located in parts of Huntingdon, Bedford, and Fulton Counties. Two coke-manufacturing reports on this isolated semibituminous coal region were issued by the Second Geological Survey of Pennsylvania. John Fulton, author of these reports and the general mining engineer of the Cambria Iron Company of Johnstown, observed that "[Broad Top] coke is destined to become the leading fuel for blast furnaces, and to retain this position from its almost inexhausitable source of supply, its calorific efficiency, and its continued economy." The first large-scale coke operation in the region began in 1875 at the Rockhill Furnace Number 2, owned by the Rockhill Iron and Coal Company, Huntingdon County.[45] The Kemble Coal and Iron Company constructed beehive coke ovens at the company town of Riddlesburg, near Defiance, Bedford County, during this period. The ovens produced coke used by the company's blast furnace, which was constructed in the late 1860s. Riddlesburg was an active mining community between 1870 and 1952. Colonial Iron Company was the last operator of the Riddlesburg ovens, which were last fired in 1950.

The Introduction of Alternative Coke-Making Technology

Bituminous coal or coke had not surpassed charcoal usage in blast furnaces by 1869. Anthracite continued to be the leading blast furnace fuel nationally until 1875 when it was superseded by bituminous coal and coke. The beehive coke oven was firmly established as the universal method for the production of coke during this period. From 1871 to 1919 about 88 percent of all iron produced in the United States used coke made in these ovens. Beehive ovens were producing nearly one hundred percent of all coke used as blast furnace fuel in the nation until 1893, and as late as 1910 they were producing 83 percent of all metallurgical coke.[46] The American beehive coke industry attained its maximum peak in terms of the number of ovens in 1910, when 100,362 beehive ovens operated in nineteen states. The following states had at least one thousand beehive ovens by 1910: Pennsylvania 54,360, West Virginia 19,792, Alabama 9,852, Virginia 5,389, Colorado 3,611, Tennessee 2,792, and New Mexico 1,030.[47] The peak production of beehive coke in the United States was reached in 1916 when, under the influence of war, coke production nationally reached 35,464,224 tons. Of that total, 27,158,538 tons were produced in Pennsylvania, the Connellsville coke region alone producing 21,654,502 tons. This represented 80 percent of coke production in Pennsylvania and 61 percent of the entire national production.[48]

The mid-nineteenth-century design of the beehive oven had changed very little by the dawn of the twentieth century. The number of beehive ovens continued to expand in the Connellsville coke district from 7,211 ovens in 1880 to 24,481 in 1910.[49]

Beehive Ovens in Connellsville District

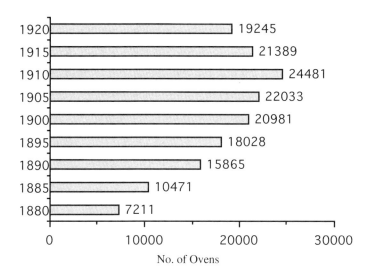

These ovens represented an increasingly archaic and wasteful technology although they still produced excellent coke. Hot coke was manually drawn by hand from them and this was a costly and time-consuming process. A well-constructed beehive oven converted about 70 percent of the burned coal into coke, but the remaining volatile matter was expelled from the oven's "eye" into the atmosphere as noxious smoke and gas. Their inefficient design permitted the loss of valuable chemical by-products—gas, ammonia, light oils, and coal tars. The Smithsonian Institution issued a report in 1924 that identified the wastefulness of the beehive coke oven design:

> In 1920, there were 24,000,000 tons of coal used to make 16,0000,000 tons of beehive coke. The principal by-products that were wasted from this source were: 216 million gallons of tar; 600 million pounds of ammonium sulphate, which could be used as fertilizer; and 120 billion cubic feet of gas; which could be used, the same as manufactured gas, for public utility.[50]

Two alternative coke-producing ovens that challenged the supremacy of the venerable beehive coke oven were developed during this period. They were the by-product or retort oven developed by European engineers, and the rectangular, Belgian, or Mitchell oven. The by-product or retort coke oven represented a revolutionary coking technology designed by engineers in England, Belgium, and Germany during the 1880s. The by-product oven would replace the beehive coke oven as the principal source of blast furnace coke in the United States at the end of this period. These ovens were used with marked success in Europe for about a decade, although their adoption was initially resisted by American coke manufacturers. They were satisfied with the excellent coke obtained from the beehive ovens and were reluctant to embrace this alternative.[51] The high cost of their installation and the lack of markets for by-products also slowed their initial development. There were no adequate markets for such coke by-products as benzol, ammonia, naphtha, tar, and pitch. The beehive oven was an inexpensive installation in contrast to the costly by-product oven. A comparison of these two types of ovens in 1912 indicated the cost of an individual beehive oven at seven to eight hundred dollars, while a single by-product oven cost twelve to eighteen thousand dollars, depending on the type of oven constructed.[52] By-product coke ovens were primarily experimental ovens in the United States during the 1890s because of the hesitation to embrace this new coke-making technology.

The by-product coke oven was described by a proponent as "one in which the gas evolved by distillation of the coal, in externally heated air-tight ovens, is withdrawn and saved."[53] These ovens, besides producing excellent metallurgical coke, were designed by European engineers to capture and recycle the chemical by-products of the coking process. The principal products are gas, ammonia, light oil, and tar. These recycled gases and chemicals would become the foundation of the chemical and plastics industries of the United States in the near future. The original by-product ovens of the 1890s saved some hundred thousand cubic feet of gas. Some seven thousand cubic feet were used for heating the ovens, and the remaining gas was clear profit for the coke operator. Each ton of burned coal in these early ovens produced 1,300 to 1,500 pounds of coke, eight to ten gallons of coal tar, three gallons of light oil, five to six pounds of ammonia and 9,500 to 11,500 cubic feet of coal gas. Better-designed ovens increased by-products saved during the next decade. The average amount of by-product recovered from every ton of coke rose in 1912 as follows: ammonium sulfate 31 pounds/net ton of coke; tar 10.7 gallons/net ton of coke; and surplus gas of 7,143 cubic feet/net ton of coke.[54]

The first by-product ovens were introduced in the United States in 1887, with actual production beginning in 1892. The first by-product coke in America was made in a small battery of twelve ovens built by the Semet-Solvay Company at the Solvay Soda Ash Works at Syracuse, New York, in December 1892.[55] The facility produced 12,850 net tons of coke the following year. The dozen ovens recovered ammonia for use in making soda ash and caustic soda. The ovens had been developed ten years earlier by the parent Solvay Company in Belgium. A single battery of fifty ovens was first constructed by the Semet-Solvay Company in western Pennsylvania at the Dunbar Furnace Company of Connellsville at Dunbar, Fayette County, in 1894. C. M. Atwater of Syracuse, who directed the construction of the by-product ovens at Syracuse, constructed the Dunbar coke plant. An additional sixty ovens were constructed in 1903-1904. Run-of-the-mine coal from the Pittsburgh and Upper Freeport seams was used in the ovens to make the coke burned at the adjacent blast furnace. The Dunbar ovens yielded 6 to 8 percent more coke per ton of coal than neighboring beehive ovens. Gas recovered from the ovens was shipped to chemical companies for further refining, pitch was sold locally, and ammonia was shipped to Syracuse. The ovens had horizontal movable flues for the recovery of ammonia. The Dunbar system included gas washers, an ammonia container, and two engines. The American Manganese Company of Philadelphia acquired the Dunbar Furnace Company on July 1, 1914. The firm manufactured manganese alloys, ferro-manganese, spiegeleisen, high manganese iron, and various grades of pig irons.[56] The American Manganese Company enlarged the original facility by erecting more ovens and constructing a new draining bin, a washery, an oil house, and a stir house. The firm employed 328 workers in 1916 and 523 workers in 1922. The facility ceased operation following the protracted 1922 national coal strike and in February 1924 was demolished by the owners.

The Dunbar location was unique because subsequent by-product ovens, unlike the beehive coke ovens, were erected near steel plants, usually located some distance from the coalfields. In 1895 the Cambria Steel Company of Johnstown was the first American steel corporation to construct by-product ovens to make coke in conjunction with a blast furnace. The firm constructed sixty Otto-Hoffmann by-product ovens at its new installation at Franklin Borough, northeast of Johnstown. The plant consisted of gas coolers, ammonia washers, ammonia stills, and ammonium vats. Tar and sulfate of ammonia were recovered from the ovens. The number of by-product ovens was increased from 160 to 260 ovens between 1901 and 1904 at the Franklin facility. Coke and by-product gases were produced from coal extracted from the firm's Rolling Mill mine. This "captive" mine, located at the foot of Westmont, Johnstown, had been operated by the Cambria Iron Company since its opening in 1856.

There were numerous manufacturers of by-product ovens and their designs changed over time; therefore, design and size specifications were not standardized. H. Koppers Company, Semet-Solvay Company, Wilputte, United Otto, and Otto-Hoffmann were the principal makers of these ovens (Illustration 1 & 2, Types of By-Product Ovens).[57] Each by-product oven was about thirty-six to forty feet long, seven to ten feet tall, and twenty feet wide. The ovens were usually constructed in batteries containing fifty or sixty units, with the largest plants having six hundred or more ovens. Each oven was a steel chamber surrounded by flues and lined with silica brick. About eight tons of coal were fed into the top of each oven by a charging machine, and each was then fired from seventeen to twenty-four hours, depending upon the type of coke required.[58] Coal was converted into coke in half the time required by the beehive oven. While coal was burned without any external fuel in the beehive oven, the coal in the by-product oven was burned by gas from the outside. On each side of the oven were three horizontal flues that carried the gas that heated the oven. The flues were made of tiles, two inches thick, through which the gas traveled from other ovens that had been burned. The ovens were fired from heating flues in the walls. Each oven had a door opening at each end, both doors being closed during the coking operation. When the coking process was completed, a ram operated from a car at the rear forced the coke up onto a cooling platform. Coke was quenched with water and was now ready to be loaded onto cars and shipped to the blast furnaces.[59] Gas from the oven was transferred to the by-product house where it was cleaned and the by-products obtained. The oven had a coking chamber from which the gases from the heated coal were drawn through uptake pipes into water-sealed collecting troughs. The gases were then drawn through a series of condensers and scrubbers from which the various by-products were deposited.

The real beginning of the American by-product coking industry was in 1906 when Dr. Heinrich Koppers of Essen, Germany, was brought by the United States Steel Corporation to design and construct 280 by-product ovens at its Illinois Steel Company at Joliet, Illinois. Dr. Koppers had designed by-product coke ovens in Germany that were capable of producing fifteen hundred pounds of coke from each ton of coal carbonized. These ovens were able to recover valuable by-products formerly wasted as noxious smoke. United States Steel executives were satisfied with the quality of coke produced from the new Koppers ovens, and later installed 490 Koppers ovens at their new Gary, Indiana, works.

The H. Koppers Company was incorporated in Chicago in 1912 and engaged in the construction of by-product ovens for a number of steel companies. A group of investors from Pittsburgh, led by H. B. Rust, purchased the company in 1914 and renamed it the Koppers Company. The company moved its corporate offices from Chicago to Pittsburgh to be nearer the coalfields of western Pennsylvania.[60] By 1915 the Koppers Company had become the largest designer and builder of these chemical-recovery coke ovens in the United States.[61] A typical by-product oven built by the company in 1930 "consists of a silica brick chamber which is maintained at a temperature—about 2000 °F. This chamber is tapered 17 to 19.5 inches in width, 10 feet in depth, and 37 feet in length and holds 12.5 tons of coal."[62]

Dr. C. W. Saleeby, a leading proponent of these ovens, described coal "as an inexhaustible treasury of infinite and manifold riches" because of the gaseous by-products created by the "cooking" of coal.[63] Coal has been called a "black diamond" because of its valuable by-products. Dyes, inks, antiseptics, aspirin, and saccharin are just a few of the many products derived from refining coal. The burning of coal in the by-product ovens created a number of principal materials that are used in the production of other goods. Coal tar is the basic material for hundreds of compounds, including naphthalene, heavy oil, pitch, resins, and explosives. Tar is a thick, brown to black, viscous liquid obtained by the distillation of coal, wood, peat, and other organic materials. The bulk of the ammonia goes into the production of ammonium sulfate, which is used in the manufacture of nitric acid, explosives, and fertilizers. Light oil contains benzene, used in tanning fluids, aniline dyes, motor fuel, plastics, and synthetic rubber. Toluol is the basic ingredient in explosives, particularly TNT (trinitrotoluol). It is also used to make antiseptics, fingernail polish, printing ink, saccharin, and detergent. Xylene is used to make motor fuel, gasoline solvent, herbicides, and solvent naphtha used as a rubber solvent, electrical insulation, linoleum, varnish, and a variety of products which are used in the manufacture of dyes, drugs, and chemical reagents (Illustration 3, Chemicals derived from the by-product ovens).[64]

A second advantage of these new, alternative coke ovens was that they yielded more coke per ton of coal. The beehive coke oven produced 66.8 percent coke per ton while the by-product oven averaged 73.6 percent.[65] The new by-product oven design also permitted the use of inferior-quality coal, although it had to be crushed, sorted, and washed to remove impurities, especially sulfur.

The growth of the by-product oven, as an alternative to the beehive oven, was a slow and evolutionary process. *Engineering Magazine* compared the two coke ovens in 1899: "Dr Johnson once defined a weed as a plant of which the use had not yet been discovered; perhaps we might equally define a by-product as something valuable which most people willingly permit to run to waste, and in these days of close competition the difference between waste and economy frequently means all the difference between commercial failure and success."[66] Seventeen steel firms installed by-product ovens at their facilities from 1895 to 1903. Additional by-product batteries were erected by Bethlehem Steel Company at their Sparrows Point, Maryland, facility in 1903 and the Pennsylvania Steel Company at Steelton near Harrisburg in 1907. There were twenty-nine coke by-product plants in the United States producing about 5.6 million tons of coke by 1906. There were three by-product facilities in western Pennsylvania operating 442 by-product ovens by 1910.[67]

The proximity of the beehive coking industry in the Connellsville district was largely responsible for the delay in the adoption of the new by-product coke ovens by the steel companies of the Pittsburgh district. *The Weekly Courier* noted in 1914 that "few operators in either the Connellsville or Lower Connellsville regions have sufficient coal acreage to warrant replacing the beehive installation they now have with the expensive by-product oven."[68]

The H. C. Frick Coke Company became a subsidiary of United States Steel Corporation when it merged with the American Coke Company, the Continental Coke Company, South West Connellsville Coke Company, and the United Coal and Coke Company on April 1, 1903.[69] United States Steel management, unlike rival steel companies in western Pennsylvania, declined to invest immediately in the new by-product technology. The firm had major investments in tens of thousands of beehive ovens located in the Connellsville and Klondike coke districts and management used beehive ovens as long as they were profitable.

United States Steel erected its first by-product coke plant in the Pittsburgh district at Clairton, about twenty miles downstream on the Monongahela River from Pittsburgh, in 1916-1917. The St. Clair Steel Company, a subsidiary of the Crucible Steel Company, built the Clairton Works in 1901. The plant was sold in 1902 and became the Clairton Steel Company. The plant had three four hundred-ton-per-day blast furnaces, twelve open-hearth furnaces (eight basic and four acid) sixty-ton heats each, a forty-inch blooming mill, and a twenty-eight-inch three-stand billet mill. The plant was sold again in 1904 to the United States Steel Corporation. The physical size of the Clairton steel facility continued to grow during the second decade of the twentieth century. A coke by-product plant at Glassport, Allegheny County, was the only such facility in the Pittsburgh district until the construction of by-product ovens at the Clairton plant. The Chicago, Youngstown, and Birmingham districts produced more by-product coke than the Pittsburgh district before 1917 and the construction of the Clairton Coke Works.

The Clairton by-product coke ovens were put into operation in 1919 and consisted of 640 ovens with a daily coal capacity of 10,500 tons. Ten more by-product coke oven batteries were added. Six of these batteries, containing sixty-one ovens each, were put in operation in 1924. The other four batteries, containing eighty-seven ovens each, were put into operation in 1928. Clairton was the largest by-product coke plant in the nation and its completion made the Pittsburgh district the largest by-product coking district in the United States. Raw bituminous coal was shipped daily by means of one-thousand-ton barges on the Monongahela River from the Connellsville coke district.[70]

Coal was burned in the by-product ovens at Clairton for about nineteen hours to drive out volatile gaseous matter, then the oven doors were opened and the coke was removed and pushed into the quenching car. Water was sprayed for forty-five seconds to quench the coke. The watered coke should not have contained more than three percent water, and once watered down it was allowed to drain for one minute. The coke was then dropped at a wharf from which it was fed by a rotary belt and pushed through chutes into railroad cars for delivery to the blast furnaces at the Edgar Thomson Works at Braddock, and the Homestead, Duquesne, and Clairton Works.[71]

By-product coke output was controlled by the demand for coke, which in turn was dictated by the demand for steel. American involvement in World War I was a powerful catalyst for the expansion of by-product coke production. The war had restricted the import of coal tar derivatives used to manufacture many essential dyes, drugs, and explosives products. America was still dependent upon Europe for part of its supply of creosote, benzol, and similar coal derivatives. A shortage of toluol, benzol, and sulfate of ammonia needed in the production of explosives developed with the closing of trade with Europe. These by-products were traditionally imported from Germany and Great Britain, which had developed sophisticated and extensive chemical industries since the 1890s, unlike the small coke-residues market in the United States. Four new by-product plants were built between 1915 and 1918 at Clairton (768 ovens), Cleveland, Ohio (180 ovens), Lorain, Ohio (208 ovens), and Gary, Indiana (140 ovens) to meet this demand of the munitions industry.[72]

A serious reappraisal of existing coking methods was undertaken during World War I. The Pennsylvania Department of Mines in 1918 urged coke operators to embrace the more productive by-product oven instead of the wasteful and inefficient beehive oven:

> The value of the by-products such as tar, ammonia, gas, benzol, and the varied dyestuffs was not appreciated until the war in Europe revealed our dependence upon foreign nations, particularly Germany, for these materials. . . . The trade in by-products may be considered established on a really large scale, and the many millions of dollars formerly wasted every year will be saved.[73]

By-product coke production initially developed very slowly. Although these alternative coke ovens supplied less than 0.1 percent of all metallurgical coke manufactured in 1893, coke production doubled from 11,219,943 tons in 1914 to 25,997,580 tons in 1918.[74] By 1912 one-fourth of the annual coke production of the United States was being obtained from by-product ovens; the dominance of the beehive oven had been lost permanently by 1919 when by-product coke ovens produced 56.9 percent of all coke, compared with 43.1 percent of coke in the beehive ovens. According to the eighteenth annual report of the United States Steel Corporation,

	National Coke Production in By-product and Beehive Ovens				
Year	By-Product Ovens Coke (Net Tons)	Beehive Ovens Coke	Total Coke Production	Percent Production By-Product Ovens	Beehive Ovens
1880	0	3,338,000	3,338,000	0 percent	100 percent
1885	0	5,107,000	5,107,000	0 percent	100 percent
1890	0	11,508,000	11,508,000	0 percent	100 percent
1895	19,000	13,315,000	13,334,000	0.14 percent	99 percent
1900	1,076,000	19,457,000	20,533,000	5 percent	90 percent
1905	3,462,000	28,769,000	32,231,000	11 percent	89 percent
1910	7,139,000	34,670,000	41,709,000	17 percent	83 percent
1915	14,072,895	27,508,255	41,581,150	34 percent	66 percent
1920	30,833,951	20,511,092	51,345,043	60 percent	40 percent[76]

for the fiscal year ending December 31, 1919, the firm's subsidiaries mined more than 28 million tons of coal in 1919. The subsidiaries, located chiefly in the Pittsburgh district, produced 15,463,649 tons of coke, of which 5,933,056 tons were produced in beehive ovens and 9,530,593 tons in by-product ovens.[75] United States Steel management, like that of other steel companies that operated "captive" mines, was embracing the by-product ovens and the alternative method of coke production by the end of this era. The increased popularity of the by-product coke ovens meant the decline of coke production in wasteful beehive ovens.

Seven steel corporations were operating by-product coke plants in the Pittsburgh district alone by 1920. The American chemical industry underwent a period of rapid growth during the 1920s as more uses for these chemical by-products were developed. The following figures identify the primary products and their yields per net ton of coal during the period between 1917 and 1924: coke 1,386 to 1,424 pounds, tar 7.0 to 8.6 gallons, ammonium sulfate 17.8 to 22.8 pounds, light oil 2.4 to 2.9 gallons, and gas 10.4 to 11.6 million cubic feet. Steel companies' profits from by-product chemicals from the by-product ovens came to equal those made from coke. The widespread use of these ovens by steel companies removed most of the economic advantages of the Connellsville coke district. The change from the beehive to the by-product method of coke production ranks as one of the most revolutionary changes in the coal industry. The demise of the beehive oven was imminent by 1920, and these ovens ceased to exist as a viable method in coke production during the 1930s. Several advantages of the by-product ovens accounted for their success in replacing the beehive ovens:

(1) The yield from the by-product oven was greater. A ton of Connellsville coal coked in a beehive produced on average .67 ton of coke while a ton of coal in the by-product oven yielded .72 ton of coke.

(2) A purer coke—lower in ash, phosphorous, and sulfur—was produced from coke made in by-product ovens. This made it possible for steel companies using by-product ovens to use poorer quality coal in the production of coke.

(3) Coals that would not produce good metallurgical coke in beehive ovens would produce coke of adequate quality in by-product ovens. This meant that harder and more-sulfurous coal from other coal tracts, especially those in the upper Monongahela River, could be used. Ordinary "steam" coals could be used in the by-product ovens by a blending and washing process that removed their high sulfur content. This process ended any shortage of metallurgical coke.

(4) The by-product ovens produced a variety of valuable by-products. Each ton of coal produced 130 pounds of breeze (pulverized coke, used in the domestic trade), 24 pounds of ammonium sulfate, 9.5 gallons of tar, 3 gallons of light motor oil, and 1,100 cubic feet of gas.

(5) Larger-capacity by-product ovens were designed and the coking process was shortened. This improved the economies of production in addition to the utilization of the by-products (including benzol and toluol).[77]

The Rectangular Coke Oven

The rectangular oven, developed in Belgium, was the second type of coke-producing oven introduced during this period. This European-designed coke oven was introduced in the Connellsville coke district by Thomas Jefferson Mitchell, superintendent of the W. J. Rainey Coke Company. This Cleveland-based coke company, founded by W. J. Rainey (1833-1900), was the chief competitor of the H. C. Frick Coke Company in the district. Rainey was born at Martins Ferry, Belmont County, Ohio, of Scotch-Irish immigrants. He became involved in the coal industry when he mined coal located under his father's Ohio farm. He opened his first mine in the Connellsville district in 1879 when he purchased the Fort Hill Works near Vanderbilt, Fayette County. The W. J. Rainey Coke Company, from this humble beginning, expanded to become the second-largest coke operator in the Connellsville coke district. This Cleveland-based company had at least a dozen mines and coking plants throughout Fayette and Westmoreland Counties. The company also owned the Cleveland Rolling Mill of Ohio.

This company, unlike its chief competitor the H. C. Frick Coke Company, aggressively installed rectangular ovens at its coke plants and soon became the principal coke operator in the Connellsville district employing these alternative ovens. T. J. Mitchell was the brother-in-law of W. J. Rainey and the general superintendent of the firm's numerous mines, some three thousand beehive ovens, and thousands of acres of coal land. He wanted to develop a coking oven that conduced to mechanized operation and still yielded high-quality coke. Beehive oven coke was still drawn by hand and labor costs were expensive. *The Mining Journal* determined the production cost of a ton of coke by surveying three beehive coke plants in the Connellsville coke district in 1888, and found that coking costs ranged from $1.00 to $1.25 per ton with labor costs contributing about 85 percent of the total expense. It took each coke drawer about three hours to draw and recharge each oven. Each man was responsible for six ovens, and each oven was drawn every other day.[78]

The rectangular oven made coke in a manner similar to the beehive oven, but each oven was larger, occupying about twice the area of a beehive oven. The standard beehive oven was about 12 feet in diameter and about 7 feet high, while the rectangular oven measured around 30 feet long, 4 1/2 feet to 5 feet wide, and 8 1/2 feet high.[79] The rectangular oven's larger interior space permitted each to produce 5 1/2 tons of coke with each charge as compared to 4 1/2 tons per charge in an average beehive oven. Beehive ovens were constructed in either single or double banks, called block ovens, while the new design of the rectangular oven abandoned the circular shape and the single door of the beehive oven. The rectangular oven, unlike the beehive oven, was designed specifically for the mechanical drawing of coke. Each oven had a door at both ends for the mechanical removal of coke and, therefore, such ovens were constructed in single banks. The mechanical pusher moved on tracks in front of the ovens, pushing the coke out the opposite door to be quenched. The hot coke was sprayed with water to stop its baking and then loaded for shipment to market.

The first bank of rectangular ovens was constructed under Mitchell's direction at Rainey's Mt. Braddock Works, Fayette County, between 1905 and 1908. The experimental rectangular oven was thirty feet long, four feet wide, and had a vaulted, horizontal crown. This prototype oven proved successful after a number of technical problems were resolved. The company had erected fifty rectangular ovens at the Mt. Braddock Works by 1907. The company also operated about twenty-six hundred rectangular ovens at its Acme, Allison Number 1 and Number 2, Elm Grove, Fort Hill, Grace, Paul, Rainey, Revere, Royal, and Union coke plants.[80]

The early successes of these rectangular ovens prompted other coke operators in the Connellsville and Klondike districts to construct these ovens. The last beehive coke plant constructed in the Connellsville region was built at the H. C. Frick Company's Phillips mine near Uniontown in 1907, and from that year until 1910 rectan-

Rectangular coke ovens. HAER.

gular ovens were the only coke ovens constructed in the district. Jones and Laughlin Steel Company built ten rectangular ovens at its Pittsburgh facility in 1907, while the Connellsville Central Coke Company constructed one hundred rectangular ovens near New Salem, first fired in September 1907. The River Coal Company constructed one hundred rectangular ovens at its Bridgeport plant near Brownsville; the Tower Hill Connellsville Coke Company erected forty-eight ovens at Tower Hill Number 2 Works, and E. A. Humphries erected forty ovens at Bradenville, Westmoreland County, east of Latrobe. The greatest number of rectangular ovens in the Connellsville district was reached in 1914. There were 1,132 ovens in the Upper Connellsville district and 2,666 ovens in the Lower Connellsville district (Klondike district). In contrast, there had been some 24,071 beehive coke ovens in the region in 1908.[81]

The new rectangular-oven technology was employed by a number of coke operators outside of the Connellsville district. The Pittsburgh-Westmoreland Company built three hundred ovens at Bentleyville, Washington County; the Keystone Coal Company and the Atlantic Crushed Coke Company built rectangular-oven coke plants at their Greensburg plants; and the Jamison Coal and Coke Company of Greensburg constructed these ovens at their coke plant near Fairmont, West Virginia. The Cascade Coal & Coke Company of Buffalo, New York, constructed two hundred rectangular ovens at their shaft-entry Sykesville Mine, Clearfield County, in 1917-1918. Coke was produced from coal extracted from the forty-eight- to sixty-inch Lower Freeport seam. There were two hundred beehive-type ovens already in operation at their mine. The rectangular ovens were constructed during the Great War by the firm in an attempt to reduce high labor costs. The new rectangular ovens were laid out in three parallel rows approximately sixteen hundred feet long. Each oven measured 5 1/2 feet wide by 32 feet deep and had a charging hole located in the center of the top. These ovens required twenty-six men to operate them while the same number of beehive ovens employed forty-eight men. The men working at the rectangular ovens were employed as follows: one man for each larry car, leveler, pusher, and loader; four men on the watering machine; four men on the daubing machines; five men cleaning up coke breeze after the loading machine; two men placing oven doors in position; two men shifting railroad cars; one man cleaning up the railroad track; two men attending the ashes; and two masons.[82]

The rectangular oven had a number of disadvantages that restricted its widespread use in the coke fields. First, the oven was more costly to construct than a beehive coke oven. Many coke operators, including the H. C. Frick Coke Company, the largest users of beehive ovens, had too much invested in beehive ovens to embrace the new and more expensive ovens. The Frick Coke Company owned 16,700 ovens in the Connellsville region with a daily capacity of about thirty-six thousand tons of coke in 1903.[83] There were already 24,071 beehive ovens when the first rectangular oven was constructed in 1908.[84] Second, there was the widespread belief that coal exhaustion in the district was imminent and therefore it was prudent for operators to maintain the existing beehive ovens. The Pittsburgh seam, which underlies the entire Connellsville coke district, produced between 10,000 and 13,500 tons of coal per acre. One hundred beehive ovens consumed about nine acres of Pittsburgh coal annually.[85] This continual use insured the inevitable depletion of the coal reserves in the original Connellsville district. The Connellsville coke district was extended southwest of Uniontown as new coalfields were opened in German, Menallen, Georges, Nicholson, and South Union Townships, Fayette County, in 1899. This new area was dubbed the "Klondyke" Field or Lower Connellsville region because the area was opened to extensive coal mining at the turn of the century, shortly after the gold rush in Alaska. Its rapid development by the coal companies and coal speculators resembled the gold rush in the Yukon. This region was predominantly agricultural, but was transformed in a relatively short time into a major coke-producing region. The development of the coal and coke industry in the region turned the local inhabitants "coal crazy" as "farms that had been considered only heirlooms of dead fathers and grandfathers suddenly blossomed into gold. Options were taken on every acre of coal land on the southern end of the county. These options were sold and resold again, till, finally, the coal seam alone with mining rights brought as much as two thousand dollars an acre!"[86]

Coal companies, coal speculators, and steel corporations from Pennsylvania and Ohio began acquiring coal lands here and in neighboring Greene County during the 1890s. T. J. Tuit, a pioneer investor of coal properties, purchased his first property in Greene County from S. Sealy Bayard, a farmer in Cumberland Township, 257 acres for $12,885 on May 28, 1897. A major portion of the rich coal reserves of the Pittsburgh seam of Greene County had been acquired by coal speculators, coal companies, and steel companies by 1907. Land that originally sold for less than $25 an acre during the 1890s was commanding from $100 to $600 an acre in 1907.[87] Some 2,033 beehive coke ovens were constructed in the region, producing 385,909 tons of coke valued at $792,880 by 1900.[88]

A major disadvantage of the beehive oven, which made the rectangular ovens attractive to some coke operators, was its inability to mechanically unload coke. Efforts had been made for many years to reduce the high cost of handling coke as discharged from the beehive ovens. Labor costs were high and the number of ovens that could be fired and unloaded daily was restricted. Machinery was invented at the turn of the century to draw coke mechanically from the beehive oven. This new technology doomed the further construction of the rectangular ovens.[89] These machines had a number of advantages over hand loading. The mechanical coke extractor, manufacturered by the Covington Machine Company, Founders and Machinists, of Covington, Virginia, was the most widely used electric coke drawer and mechanical loading machine and was the industry's standard. The company was formed in 1892 and operated a number of large machine shops between Newport News, Virginia, and Huntington, West Virginia. The firm manufactured power and

Covington coke extractor loading coke direct from ovens into waiting railroad cars. Pennsylvania State Archives.

hand punches, shears and rolls, and iron and brass casting of every description. The firm manufactured coke drawing and loading machines and coal leveling machines for beehive coke ovens. Each Covington machine increased daily coke production because each machine could draw about five beehive ovens per hour or about thirty-six ovens a day. Each electric coke drawer could level from thirty to forty ovens per hour and required only one man for its operation by 1913.[90] The use of the mechanical remover was also an advantage for a coke operator during strikes or at times of labor scarcity. The principal disadvantage of the machine was its high cost and the need to redesign the doors on existing beehive ovens. The original beehive coke oven doors were too narrow to permit coke removal by the mechanical extractor and, therefore, had to be redesigned and enlarged by the coke operators.

The Covington machine consisted of two parts—an extractor for drawing the coke out of the ovens, and a conveyor for screening and loading the coke. The Covington Machine Company issued a catalog to prospective coke operators describing in detail its electrically driven coke machine and its operation as follows:

> The machine, as it is made today, has an extension on the conveyor which runs along in front of the ovens, so that while the machine is drawing oven No. 2 the small amount of coke remaining in oven No. 1 may be pulled directly onto the conveyor. While the foregoing description mentions electric power only, we are prepared to furnish steam-driven machines, also steam being furnished by a small boiler carried on the machine. The steam-driven machine, however, requires an extra man to look after the boiler, and we recommend, when possible, that electricity be used.[91]

The Covington Machine Company installed the first electrically driven coke machine at Continental Works Number 1, Fayette County, H. C. Frick Coke Company, on October 1904. Frick's managers and engineers calculated the operating cost of using the Covington machine, including labor, electricity, and depreciation, at about fifty cents per oven, while the average cost of drawing coke manually was calculated at ninety-two cents per oven. The electric coke drawer had proved successful in its task at Continental Works, No. 1, where three machines were in operation and two machines were working at the Oliver plant by 1906. The H. C. Frick Coke Company installed additional electric coke-drawing machines at Baggaley, Fairchance, Hecla Number 1, Leisenring Number 1 and Number 3, Lemont Number 2, Oliphant, and Standard coke plants in Fayette and Westmoreland Counties. There were 150 electrically driven coke-extracting machines in use throughout the Connellsville district by 1908. It is unknown if these mechanical coke-extracting machines were employed in the other coke producing regions in Pennsylvania. The Covington Company sold coke-drawing machines in the coke regions of West Virginia at the Kanawha, New River, and Pocahontas fields. The Low Moor Iron Company at Kay Moor, West Virginia, introduced three Covington machines at their coke plant in 1901.[92]

The New "American" Miners

Rapid industrialization in the decades following the Civil War was accompanied by a remarkable increase in the nation's population, from 23 million in 1850 to 76 million in 1900. Some 23 million immigrants came to America from 1860 to 1910. Pennsylvania was second only to New York State in attracting immigrants between 1860 and 1920. The foreign-born population of the Commonwealth was 15 percent (430,000 people) in 1860 and 16 percent (1,400,000) in 1920. European immigration to Pennsylvania between 1901 and 1919 was 1901-1905, 226,007; 1906-1910, 259,979; 1911-1915, 273,473; 1917-1918, 16,470; 1918-1919, 11,257.[93] Immigrants were "pulled" to the United States for a variety of reasons, although most came to escape grinding poverty and to improve their wages and living conditions. Many industrial companies sent agents to Europe to fill their growing demand for cheap and unskilled workers. Advertisements by American capitalists promoting America as a land of milk and honey and opportunity for good jobs successfully lured millions of hopeful immigrants from southern and eastern Europe. Operators paid the cost of the trip from southern and eastern Europe to the mine or factory and then deducted the cost of the voyage from the worker's pay. A second-generation coal miner's wife from Osceola Mills, Pennsylvania, noted:

> I guess my family and my husband's family thought it would be milk and honey when they came to America. People told them how it would be the promised land over there. They weren't prepared for what they found.[94]

The expansion of the steel and coal industries created a series of long-term fundamental changes in the social, economic, and organizational structure of the state's bituminous coal industry. The increased demand for coal and coke created a severe labor demand within the Commonwealth that the existing labor force was unable to supply. Surplus labor from the neighboring farms and villages was simply too small to fill this acute labor shortage. The availability of jobs in the expanding coalfields of western Pennsylvania acted as a powerful magnet for a large number of workers from southern and eastern Europe. This immigrant and largely unskilled labor force found employment in the hundreds of isolated coal towns hurriedly constructed throughout western Pennsylvania. The "new" wave of immigrants represented numerous and diverse ethnic and religious groups, including citizens of the Austro-Hungarian Empire, Russia, Italy, Greece, and Poland. Most Americans, including the mine owners, regarded them as representatives of the "beaten races of the world" and collectively labeled them all as "hunkies."

These new immigrants created dramatic and fundamental changes in the ethnic and religious composition of the state. Between 1881 and 1890, 72 percent of European immigration to the United States arrived from the nations of northern and western Europe, while the remaining 18 percent came from the nations of southern and eastern Europe. The character of American immigration changed after 1890,

Miners wearing sunshine lamps pose for the camera. Carnegie Library of Pittsburgh.

and of the nearly 12 million immigrants who came between 1900 and 1914, a majority came from the nations of southern and eastern Europe. The pattern of European immigration was nearly reversed from 1901 to 1910, during which time northern and western European immigrants accounted for only 21.7 percent of all immigrants while 70.8 percent arrived from the nations of southern and eastern Europe.[95]

Before 1890, most miners in the bituminous coalfields of Pennsylvania were American born, English, Scottish, Welsh, Irish, or German. Magyars (Hungarians), Italians, Poles, Slovaks, and Greeks replaced them after 1890. The United States Immigration Commission of 1911 observed that miners and coke workers in the bituminous coal regions of Pennsylvania during the 1870s had been "Americans or representatives of English, Scottish, Welsh, German, and Irish races. The employment of immigrants from southern and eastern Europe began in about 1880. The arrival of unskilled, immigrant workers, the bulk from southern and eastern Europe, swelled the ranks of coal diggers. The Slovaks were the first arrivals that immigrated in considerable numbers. The great bulk of all immigrants from southern and eastern Europe, however, has occurred within the past eight or nine years."[96] Slovaks began working in the coal industry of Pennsylvania in the early 1880s and were soon joined by Croatians, Magyars, Italians, Poles, Russians, Ukrainians, Lithuanians, Bohemians, and other ethnic groups from southern and eastern Europe. Nearly one-half of all coal miners were foreign born, and one in seven was the American-born son of a foreigner in 1910.[97] Fewer than 8 percent of the foreign-born mine workers were English, Irish, Scottish, German, or Welsh. Nearly one-quarter were under the age of twenty-five; and one-half of all miners were under thirty-five years of age. The social composition of the coal-mining population nationally, according to the 1920 United States Census, was 62.9 percent native-born and 37.1 percent foreign-born: native-born white 56 percent, native-born "colored" 6.3 percent, alien 19.2 percent, first paper taken out 6.1 percent, foreign-born American citizen 11.8 percent.[98] Poles were the largest single foreign-born group of miners, numbering about fifty thousand by 1920. They worked in both the bituminous and the anthracite fields of Pennsylvania. Some two thousand Poles moved west and worked in the coalfields of Ohio, West Virginia, and Illinois. Italians were the second-largest group of foreign-born coal workers.[99] The marital status of coal miners was 35.5 percent single, 62.7 percent married, 3.7 percent divorced or widowed, and 5.1 percent with wives elsewhere.[100]

This immigrant and generally unskilled labor force was employed in both the bituminous and the anthracite fields of Pennsylvania. *The United Mine Workers Journal* observed in 1892 that "in the mines surrounding Scranton it was found that nine-tenths of the miners at present employed are Hungarian, Italian, and Slavs. Five years earlier the mines were nearly all American."[101] The Welsh, English, Scottish, and Irish who were the dominant ethnic groups among coal miners in Pennsylvania were soon replaced by these new unskilled workers in the mines. A survey of the nationalities of coal miners conducted by the Pennsylvania Department of Mines in 1912 identified more than thirty-five different nationalities working in the state's mines. The table below, derived from this study, identifies the major nationalities of both underground and surface mine workers employed in Fayette and Westmoreland Counties at this time.

Underground Workers		
Nationality	Fayette County	Westmoreland County
American	2,453	3,072
Slavonian	3,079	1,852
Italian	1,157	3,207
Polish	961	1,337
Hungarian	1,151	720
Total	8,701	10,188

Surface Workers		
Nationality	Fayette County	Westmoreland County
American	1,613	1,444
Slavonian	1,139	423
Italian	752	533
Polish	544	248
Hungarian	261	120
Total	4,859	2,758[102]

The influx of European immigrants provided coal operators with a surplus and cheap labor force. A Welsh anthracite miner complained in 1895 that "labor is so plentiful that operators can do just as they please. Pennsylvania is swarming with foreigners—Poles, Hungarians, Slavishs, Swedes and Italians . . . who are fast driving out English, Welsh, and Scottish miners out of competition."[103] Some skilled

British miners moved to the newly developed coalfields of the West and Southwest. Many English-speaking miners left the dangerous mines and became mine bosses, superintendents, or mine owners, while others were appointed coal inspectors or found employment in the state Bureau of Labor Statistics. And finally some British and Irish miners simply left the dirty and unsafe mines and found employment in the burgeoning industrial cities.

The new immigrant mine workers, unlike the antebellum English, Irish, and Welsh miners, had less mining experience or no skill or experience in coal mining. A majority of the new mine workers were farmers or agricultural workers, urban industrial workers, or skilled artisans with little or no mining experience. The Dillingham Commission of 1911 noted 7.7 percent of the southern Italians, 13.7 percent of the northern Italians, 9.8 percent of the Poles, 10.9 percent of the Magyars, and 10.7 percent of the Slovaks who emigrated after 1880 had been coal miners in their native lands; in contrast, 82.2 percent of the Scottish and 55 percent of the German mine workers had mining experience in their native lands.[104] Most of these "novice" mine workers were drawn to the coal towns for the same reasons as the immigrant-born Pittsburgh-area miner who wrote in his journal:

> I heard about America from my cousins. My cousins were working in the coal mines and I thought I could do that. I came across in 1900. I was 16 years old. Passage cost $105. I stayed a few days in Wheeling. Then I went up to the mining camp where they gave me a number (check). The next day I started to work.[105]

In contrast, few of these immigrants worked in the newly developed coalfields of southern Appalachia—Virginia, West Virginia, Kentucky, Alabama, and Tennessee. Rural mountain whites, along with a growing African American labor force, worked the southern coal mines, which began large-scale commercial production during the 1880s. Coal production in this region experienced a boom during World War I. African Americans, as slaves, had worked in the coal industry in colonial Virginia beginning in the 1750s. The development of the southern coal industry created severe labor shortages and presented new economic opportunities and an alternative to agricultural work as sharecroppers. In 1900, African Americans constituted 54.3 percent of Alabama's coal miners, 25.9 percent of Virginia's; 28.4 percent of Tennessee's, 23.7 percent of Kentucky's, and 22.2 percent of West Virginia's. Nine percent of all coal miners nationwide by 1910 were African Americans.[106] Approximately 70 to 80 percent of them were working in the nonunionized southern coalfields by 1920, concentrated chiefly in West Virginia, 25,000; the Birmingham district of Alabama, 17,000; and Kentucky, mostly in the western Kentucky coalfield, 10,000. An estimated 3,500 African American miners worked in the coalfields of Virginia and Tennessee. There were 12,000 African American miners in the northwestern and western field and more than 3,500 in Pennsylvania, chiefly employed in the Pittsburgh district and the Connellsville coke district.[107]

The Coal Company Towns

James B. Allen defined a coal-mining "company town" as "any community which has been built wholly to support the operations of a single company, in which all homes, buildings, and other real estate property are owned by that company, having been acquired or erected specifically for the benefit of its employees, and in which the company provides most public services."[108] Some miners called the single-industry village the "patch" or "patch town." The precise origins of these terms is unknown.

The mining town was financed, built, and operated by the coal company for the sole purpose of housing a labor force to extract coal or produce coke at a nearby mine. The practical purpose of the construction of company housing was to increase productivity and profits for the coal company by attracting labor, reducing labor turnover, and establishing control over the labor force. The General Assembly of Pennsylvania enacted a statute in 1854 permitting owners of extractive industries to develop their private property as they saw fit. The statute stated:

> At any time hereafter when any five or more persons, who may be joint owners, tenants in commons or joint tenants of mineral lands within this Commonwealth, may desire to form a company for the purpose of developing and improving such mineral lands, it shall be lawful for any company formed under this provision of this act to construct railroads in and upon their lands; also to erect dwelling houses and other necessary buildings; also all necessary machinery for raising, moving and preparing all minerals, found in their land, for markets.[109]

The statute granted complete legal, economic, and political autonomy to the individual mining corporation to construct and administer these communities as private and unincorporated entities. The first company mining towns were constructed in the anthracite region, but by the 1850s a number had been constructed in the bituminous coalfields of Pennsylvania. Some of the earliest mining communities were located in the Broad Top and North Central semibituminous "smokeless" coalfields. Robertsdale was constructed by the Rockhill Iron and Coal Company in the Broad Top field in 1873.[110] The Blossburg Coal Company, the Morris Run Coal Mining Company, and the Fall Brook Coal Company were the three largest coal companies in Tioga County before 1860. These firms all built a number of mining towns for their workers. The Tioga County communities of Morris Run (Tioga Improvement Company), Arnot (Fall Brook Coal Company), and Fall Brook (Fall Brook Coal Company), located in the rich Blossburg seam, were developed before 1880.

These coal towns were an almost instant creation. Their locations were established by geological considerations rather than transportation. They developed adjacent to the mining operation to minimize the walking distance of the miners. It was recommended that the work site be no more than fifteen minutes walking distance from the town, or thirty minutes by "dependable transportation." Mines were opening rapidly in unsettled, rural, and often remote townships with little housing available for the influx of workers and their families. The absence of towns and infrastructure required the coal companies to construct entire "camps" or "patches" to house and sustain their workers and their families. Housing, like mining buildings and structures, was simply part of the general investment in the mining enterprise. Even miners who could afford their own homes were simply unwilling to invest in them because they did not know the probable life-span of the mine. Most of these mining towns were constructed on the premise that coal mining at a particular site was a short-term business enterprise. Owners were aware that coal existed in a region, but the exact quantity and quality of underground coal at a particular location were often unknown. The sole purpose and lifeblood of the town was coal and when the seam was exhausted the town was quickly abandoned and often became a ghost town.[111]

The mining town was created as a single-industry community whose general layout was determined by a number of factors—topographic and physical setting, (whether the site was a narrow valley or an open flat plateau area), location with respect to other towns, size and probable life of the mine, class and nationalities of employees, and the conscientiousness of the company in community planning.[112] Depending on the size and location of the mine, coal operators constructed workers' housing and also furnished their labor force such necessary appurtenances as water, a retail store, a medical facility, a school, a church, and often a social center. They hired the teachers, the pastor, the storekeeper and the doctor. The coal company owned all surface property, including the mining buildings and houses, which were rented to the miners on a monthly basis with their rent deducted directly from their wages.

Panoramic view of Phillips, a H. C. Frick coal town near Uniontown. Pennsylvania State Archives.

The construction of these mining villages varied among the coalfields. Mine engineers, rather than architects or town planners, were employed by the coal companies to lay out the mining complex, including buildings and structures needed to extract coal or manufacture coke. Many mining villages were extremely primitive settlements consisting of the minimum accommodations needed to attract and house an increasingly immigrant labor force. They were constructed from simple designs with minimal attention to aesthetic qualities or town planning that could foster any sense of community. The usual mining village was composed of plain, serviceable buildings, often standing starkly in rows with little variation in their architecture. This construction gave most mining towns a drab and monotonous appearance. Housing was erected of inexpensive materials to minimize cost to the company. There were a wide variety of construction designs and materials, but generally the houses were built with cheap lumber and the exterior walls of clapboard. Some 5 percent of all housing was constructed of brick, tile, or stone. Mixtures of single- and double-family, and semidetached houses were often found in the same town. Single-family houses consisted of three, four, or five rooms. Duplex houses were eight, ten, and twelve rooms rented to two families. Multiple housing, including six or more units, was also constructed by some companies. Boarding houses and hotels were constructed to house single or transient workers. Many married immigrant miners traveled alone, intent on making sufficient money in the New World before sending for their families. These single workers crowded into boarding houses that rented rooms and provided meals and laundry service for single miners. The exteriors of the wooden houses were whitewashed or painted with cheap barnboard paint—lead gray, dull brown, or drab red. Flooring was a single layer of knotted or split board, permitting cold air through the holes, and was generally without carpeting. Few houses were constructed with cellars, and houses sat directly on the ground or were propped on stilts. They were generally constructed without electricity, water, and indoor plumbing. Electricity was used in mines for lighting and operating machinery after the 1880s but few houses built before 1910 were originally wired for its use. Light was generally provided by candles and kerosene lanterns, and the coal stove found in the kitchen was used for cooking and for home heat. Most houses had no piped water. Water for cooking and bathing came from outside pumps or hydrants in front of the houses, with one unit located every seventy-five to a hundred feet. Bathtubs were regarded as a luxury and were not generally found in company houses. As late as 1922, less than 3 percent of all miners' dwellings nationwide had bathtubs or showers. Mining was an extremely dirty occupation and miners returned home each evening often wet and always covered with dirt, grime, and minute particles of coal dust, called "bug dust." The inhaling of "bug dust" over many years was the primary cause of the debilitating black lung disease (pneumoconiosis).[113] Coal miners had to bathe themselves every evening in a large tub in the kitchen until mine companies were forced by state statute to provide a bathhouse. Miners asserted humorously that they were the cleanest men in the nation because they took baths daily. An important duty of the miner's wife was to have hot water ready for her husband's evening bath. Other duties of the women of the "patch" were purchasing food, selling surplus food, and taking in boarders to stretch the family's income. Wives took in boarders, usually of the same nationality, to make ends meet. Journalist W. Jett Lauch explained the cost of rent in the American bituminous coalfields:

Another Day by Frank L. Melega. Courtesy of M. V. Melega, Brownsville, Pa.

> The monthly rent for company houses in the different mining fields varies from $1.50 to $2 a room. The rent of a specified house being based on the number of rooms. It is apparent that by crowding, the rent payment for each person can be materially reduced.[114]

A variety of buildings were located at the rear of the houses. There were usually no sewers and every house had an outdoor privy—one for single houses and double ones for duplexes. Near each privy was an outdoor bin for coal storage. The miner, like any other coal consumer, had to pay for the coal that he carried home on his back in a sack from the tipple. Bread was a principal food consumed by miners and their families. Common baking ovens were located in the backyard for baking bread, pies, rolls, and the specialty dishes of the various immigrant groups.[115] Some mining companies permitted their workers to enclose the houses with fences so they might keep a cow, chickens, or pigs. Common land in coal towns was sectioned in lots, with garden plots ranging in size by type of crop grown.[116] Mining families were able to supplement their diet by raising their own vegetables and owning livestock, and avoiding these costly purchases at the company store. Many miners' wives cultivated large gardens, and kept chickens and occasionally a cow. Dorothy Schweider, in her study of miners' wives and their prominent economic role in the mining town, noted that wives sold surplus vegetables, eggs, and milk, which contributed from one-fourth to one-half of the family income.[117]

Company town. Pennsylvania State Archives.

Outhouses in company town. Indoor plumbing came very late.
Pennsylvania State Archives.

Coal mining was largely an immigrant industry with a majority of its operatives after 1890 being workers from the nations of southern and eastern Europe. All positions of authority were reserved for native-born men or immigrants from western Europe who had lived in the United States for many years and received American citizenship. The company-owned mining towns were multicultural communities that were usually socially stratified by race, ethnicity, and class. Ethnically segregated settlement patterns in these remote communities represented conscious attempts by some coal companies to fan and maintain Old World racial, ethnic, religious, and political antagonisms and keep, thereby, their workers divided. There appear to have been four types and areas of segregation, although not all are always clearly defined. First, an area in the village was occupied by members of management and by key personnel—the "bosses"—who usually were Americans or "Anglo-Saxon" stock. They lived in the best houses if they resided in the community at all; often they resided in a nearby town away from the dirt, grime, and smoke of the village. Irish, Scottish, Welsh, English, and native-born Americans were the foremen and superintendents at most mines. Housing for management—the mine superintendent, mine foremen, company-store manager, payroll clerk, and other office personnel—was often called "bosses row" or "silk stocking row" by foreign-born workers. The houses were generally larger and sturdier in construction than ordinary workers' houses. The bosses' houses, usually located a short distance from the miners' houses, were furnished with furnaces, indoor bathrooms, and hot and cold running water. The houses were often of singular design, in sharp contrast to the monotonous sameness that characterized most workers' houses. The houses had elaborate porches and were landscaped with lawns, trees, and shrubbery. The superintendent of the Crucible mine, owned by the Crucible Fuel Company, Greene County, lived in a fourteen-room house, accompanied by a tennis court and a swimming pool. A second segregated area housed white miners of American, Anglo-Saxon, or northern European stock. The English-speaking immigrants constituted a majority of miners and residents of the coal towns until the 1890s. The non-English miners, from southern and eastern Europe, called them "Johnny Bulls." The third segregated area in the town housed foreign-born white miners of eastern and southern European extractions. Most coal operators believed their new immigrant labor force to be inferior, derisively calling them "Wops," "Polacks," and "Hunkies." Each ethnic group resided apart in areas within the community called "Siberia," "Dago Hill," "Russian Hill," and "Hunkytown." African Americans occupied the lowest rung of the social and economic ladder in the company town.[118] Their housing was often mere shanties and was usually located some distance from the other miners' housing.

This policy of creating and enforcing the segregated company town was not employed by all coal companies. The Vesta Coal Company, a subsidiary of Jones and Laughlin Steel Corporation, was incorporated on December 22, 1891, under Pennsylvania law for "the purpose of mining, producing, transporting and selling its products."[119] The company operated seven "captive" mines on the Monongahela River between Allenport and Fredericktown on the boundary of Washington and Greene Counties between 1892 and the 1970s. Vesta No. 1 was located at Allenport; Vesta No. 2 was located near Roscoe; Vesta No. 3 was located at Coal Center; Vesta No. 4 was located at Richeyville, Smallwood, and Daisytown; Vesta No. 5 was located at Vestaburg; Vesta No. 6 was located at Denbo; while Vesta No. 7 was located at West Brownsville. The company had an informal housing policy that per-

Family vegetable garden.
Pennsylvania State Archives.

mitted any family to occupy a vacant house in the "patch." It was a common practice that relatives and fellow countrymen were encouraged by neighbors to move into the recently vacated house. Segregation in housing, based on ethnic and religious differences, was the practice at these villages, but it was self-imposed by workers and their families and was not part of a long-term systematic policy formulated by the company.[120]

President Warren G. Harding and Congress created the Coal Commission to investigate all facets of the coal industry after the especially violent and protracted 1922-1923 national strike. This eleven-month study, better known as the Harding Commission Report, was the most in-depth examination of community life and living conditions of miners in the bituminous coalfields that had been undertaken by the federal government. The Commission's findings were released in a voluminous multivolume report in 1923. The report was an excellent in-depth descriptive study of the physical composition of housing and daily life in coal-mining communities. Agents were sent to field-visit 713 company-controlled communities in the soft-coal fields of western Pennsylvania, West Virginia, and Ohio. The study estimated 70 percent of the nearly six hundred thousand miners lived and worked in the Appalachia Coal Region. Approximately one-half of all bituminous coal miners in the United States were living in company towns by 1922-1923. Eight out of ten miners in the bituminous coalfields of Pennsylvania lived in villages with less than twenty-five hundred population, and about one-half of them leased and lived in company-owned houses.[121] About two-thirds to four-fifths of all miners in the coal-producing states of southern Appalachia—West Virginia, Tennessee, Virginia, Maryland, and Alabama—lived in company housing. Eighty percent of West Virginia miners lived in company-owned housing, compared with fewer than nine percent in Indiana and Illinois.[122] As would be expected, the visited company-owned towns revealed a variety of living conditions. The report summarized the Commission's findings of company-owned housing as follows:

> In the worst of the company-controlled communities the state of disrepair at times runs beyond the power of verbal description or even photographic illustration, since neither words nor pictures can portray the atmosphere of abandoned dejection or reproduce the smells. Old, unpainted board and batten houses—batten going or gone, and boards fast following, roofs broken, porches staggering, steps sagging, a riot of rubbish, and a medley of odors—such are the features of the worst camps. They are not by any means in the majority; but wherever they exist they are a reproach to the industry and a serious matter for such mine workers' families dependent upon the companies for living conditions. Ninety-five percent of the company-owned houses in the 713 communities studied were built of wood. More than two-thirds were finished outside with weather board, usually nailed directly to the frame with no sheathing other than paper, and sometimes not even that. The weather board commonly used was plain overlapping siding, but in the northern coal fields a better sort of fitted weather board was frequently seen. Over two-thirds of the roofs were of composition paper. The houses usually rest on post foundations, with no cellars; but the double houses, especially in the vigorous climates, often have solid foundations, and occasionally excavated cellars. There are porches on nearly all except "shanties." Wood sheeting forms the inside of half the houses, plaster is 38 percent. Board and batten, the cheapest type of construction, was used in over a fourth of the dwellings, and in communities presenting a conspicuous range of general conditions.[123]

The Coal Commission devised an elaborate rating system to assess the quality of company housing and the quality of daily life in the village. Eight categories, on a scale from 0 to 100 percent, were used in comparing and ranking the coal communities. Housing, water supply and distribution, sewage and water disposal, community layout, food and merchandise supply, medical and health provisions, recreation, religion, and education were the eight categories used to rate the company villages. The communities were scored from the worst at 21.5 percent to the best village rated at 93 percent. The report concluded that sixty-six communities were of excellent quality receiving a score of 75 percent or above; eighty-two communities had substandard housing and scored less than 50 percent and were described as "dreary and depressing places in which to live." The majority of the company-owned towns were judged by the investigators as being average.[124] Ninety-five percent of the company-owned houses in the 713 communities were built of wood. Tile, brick, and stone were used in the construction of a minority of workers' housing. Most of these wooden houses had three to five rooms. There were also one- and two-room shanties and a minority of miners' houses having six rooms or more. The study noted fewer than 14 percent of the seventy-one thousand dwellings had indoor running water; bathtubs and showers were found in 2.4 percent of all dwellings, flush toilets in 3 percent, and running water in 14 percent. Two-third of the houses had electric or gas lights by 1920.[125]

Churches were found in a majority of the company-controlled communities. Buildings often served more than one denomination and services were often separate. Mine workers raised the money to build the churches in some villages, and the company contributed the land and some funds for their construction. Larger coal companies provided a doctor and dental services at the larger mining communities, while the smaller communities had no medical services or doctor. Professional facilities, including hospitals with medical services, were found in neighboring towns. The company trained first-aid teams to treat victims of mining accidents. Some large coal companies helped to subsidize wards for their workers at the local hospital. Minor medical services in the smaller "patch" towns were often provided by skilled midwives. Most miners of this period were practitioners of folk medicine. A variety of folk-medicine practices were practiced in the mining villages. These were employed as cures for a variety of ailments derived from mining coal and other ailments. Miners and their families took wild cherry bark to winterize their bodies, wore sacks made of asafetida (an offensive plant resin used in medicine) to prevent disease, sucked on a piece of coal to cure heartburn, rubbed coal dust on cuts to stop bleeding, wore red woolen underwear to cure rheumatism, carried an onion or garlic with them to fight a cold, drank kerosene to cure pneumonia, used chicken blood to cure shingles, wore a rattlesnake-skin belt to prevent lumbago, wrapped a red sock around their neck for a sore throat, set a pan of water under the bed for night sweats, and pierced their ears to fight weak eyes.[126]

The exact layout, size, and architecture of company towns varied from one town to another in the bituminous coalfields. Types of company housing with respect to the number and size of rooms, quality and type of building materials, and workmanship depended on the enlightenment and good will of the individual coal operator.[127] All the same, Margaret Mulrooney observed in *A Legacy of Coal* that most housing in larger coal towns of Pennsylvania shared five fundamental characteristics:

(1) Each town was financed, built, owned, and operated by a single company. Unlike other towns, the primary employer was also the primary landholder. In this dual capacity, the company town determined not only the economic character of the community, but the social, political, and cultural character as well.

(2) Houses in these towns tended to be two-story, wood-frame structures, whether detached or semidetached, with four or six rooms per dwelling.

(3) There was a clear hierarchy of architecture in each town that segregated management from labor and reinforced ideas of ethnic and occupational segregation.

(4) Houses within a given community were remarkably similar in style and materials since construction was carried out as cheaply as possible.

(5) Coal towns shared a similarity of spatial arrangement. In almost all cases, the location of the mine site and its associated buildings received primary consideration, while housing took a secondary role. Nevertheless, housing was always located near the work site to minimize travel time.[128]

Welfare Capitalism and the Construction of Model Industrial Coal Towns

As noted, many miners' houses were dilapidated or without weatherproofing, and many company towns lacked decent sanitary conditions and water supplies. These mining villages provided only basic necessities for miners and their families to sustain life. Few coal companies made provisions for social and leisure activities for their workers. Progressive reformers, who began visiting and examining industrial housing, deplored the squalor and overcrowded conditions that they witnessed. They demanded that industrial companies providing housing for their workers and their families make immediate housing reforms. This negative publicity by urban reformers about the living and working conditions of industrial workers persuaded some enlightened capitalists, including coal operators, to undertake improvement of their workers' environment. Paternalism is the best term to describe this response and the changing attitude of a small number of enlightened coal company executives. They insisted that their new interest in the welfare of their workers was motivated by sound business strategy, not by philanthropic considerations. They believed a well-treated labor force would create a contented labor force and hence more efficient and productive workers. In return, management expected their employees to be grateful and loyal for what the company offered. A coal industry spokesman clearly defined this changing attitude, stating that "we are at last beginning to rec-

Unidentified company town, western Pennsylvania. Pennsylvania State Archives.

ognize that people who live in pigsties are likely to act like pigs. If we want respectable and intelligent men and women to work for us in our plants, we must see that they have decent, healthy and comfortable houses."[129] A western Pennsylvania coal operator interviewed by a newspaper reporter in 1916 explained this new paternal attitude: "If you would make your business a success, you must get good service from your workmen, and if you would get good service from your workman; you must make it worth their while to serve you."[130]

Some larger independent coal companies and "captive" mine operators began the construction of a number of well-designed, planned, model industrial communities after the 1890s.[131] The company towns of Slickville (Cambria Steel Company of Johnstown), in Westmoreland County; Nemacolin (Buckeye Coal Company of Ohio, a subsidiary of Youngstown Sheet and Tube Company), in Greene County; Mather (Picklands, Mather & Company of Cleveland, Ohio), in Greene County; and Indianola (Inland Collieries Company, a subsidiary of the Inland Steel Company), in Allegheny County, were all examples of new towns reflecting corporate paternalism. These are all captive coal mines and communities constructed by steel companies.[132] Independent coal companies also constructed "model" mining towns including Windber (Berwind-White Mining Company), Somerset County; Jenners (Consolidation Coal Company), Somerset County; and Star Junction (Washington Coal and Coke Company), Fayette County.

The earlier "patch" towns developed in an ad hoc fashion during this economic boom period, with construction proceeding according to need. Their development over time accounts for the variety of building styles found in individual mining communities. The planned towns were designed by landscape architects and urban planners as completely integrated communities, consisting of a variety of social and institutional buildings besides the obligatory workers' housing and the company store. Daily life in many of the early mine camps was known for its drabness and monotony, which these "enlightened" operators attempted to address by building playgrounds, swimming pools, dance halls, recreational halls, and movie theaters. These new towns usually included a school, a water-filtration plant, and churches, and featured sewers and paved roads that were often tree-lined. *Coal Age* estimated monthly rental costs for these houses in 1923 as follows: superintendent's house, $25 per month; minor company official's house, $15; miners' housing $6 to $10.

These companies took meticulous care of the company houses. Each spring fences were repaired and whitewashed, and the leased houses were completely painted about every three or four years. They granted miners common ground on the outskirts of the village where animals could graze and vegetable gardens could be maintained. They permitted and encouraged miners and their families to plant their vegetable and flower gardens. Companies would award cash prizes for the best gardens in the village. Grindstone, Fayette County, owned by the H. C. Frick Coke Company, awarded prizes annually to the three best gardens. There was a five-dollar first prize, a three-dollar second prize, and a one-dollar third prize.[133] Each year some communities, for example Marianna, owned by the Pittsburgh-Buffalo Company, gave prizes to the family with the best-groomed yard in the town.

Superintendent's residence (location unknown). Pennsylvania State Archives.

The villages of Marianna, Washington County, and Mather, Greene County, are representative of these new, "model" company communities constructed for miners in Pennsylvania during this period. The Jones family of Monongahela City organized the Pittsburgh-Buffalo Company on January 4, 1904. The company was incorporated from a number of smaller coal companies and was capitalized at $6 million ($5 million common stock and $1 million preferred) with total assets valued at $7.9 million. The company owned thirty-thousand acres of coal lands in Allegheny, Greene, Washington, and Armstrong Counties. John H. Jones, president of the company, constructed a modern mine and "model" town at Marianna, Washington County, in 1907.[134] Marianna village was part of West Bethlehem Township until it was incorporated as a borough on January 24, 1910. The company erected 282 yellow brick houses, mostly single-family dwellings of four, five, and six rooms each, and a boarding house of fourteen rooms on the hillside behind the mine's shaft. It was one of the few mining communities in Washington County to have brick houses with indoor plumbing. James W. Ellsworth, a Chicago capitalist, laid out the model town of Ellsworth near Bentleyville in December 1899. He constructed brick cottages for his workers modeled after housing that he encountered during a visit to Wales. The houses were constructed with a yard and sufficient property for a small garden. Each miner paid $6.50 to $9.00 a month rent for his house, which could be applied toward the future purchase of the house. Ellsworth constructed the neighboring company town of Cokeburg about 1900. He sold out his coal properties to the Lackawanna Steel Company of Buffalo in 1907 and then left the region, never to return.

The Pittsburgh-Buffalo Company erected in Marianna workers' housing, a company store, a three-story public school, and an arcade (community hall). Inside the arcade was a drug store, an ice cream parlor, a bowling alley, pool and billiard tables, a dance floor, a skating rink, a movie theater, and a gymnasium. The company erected Saints Mary and Ann Roman Catholic Church. The large company

Playground constructed for miners' children at Phillips by the H. C. Frick Coal Company.
Pennsylvania State Archives.

store, located at the lowest part of town, was divided into a meat market, a section for groceries and dry goods, and an area for hardware and furniture. A railroad siding was located near the basement of the store to facilitate the transportation of food to the store. The company built a two-million-gallon, cement-lined reservoir to provide water for residents and to quench coke made at the beehive ovens. A playground for miners' children was constructed below Second Street above the beehive coke ovens. The town was constructed in a few short months and advertisements for mine workers appeared locally and in European newspapers. Marianna attracted White Russians and Italians as well as Scottish, English, and Slavic workers. The population of Marianna was over two thousand in 1910, while its population in 1970 was 872. The mine and community were acquired by Bethlehem Steel Corporation and the mine was operational until the 1980s when an underground fire forced its closure.

The company developed the Rachel and Agnes mines at Marianna as the safest and most up-to-date facilities in the nation. The new tipple, washer, and beehive coke ovens were the most modern installations at the time of their construction. The equipment and preliminary development at Marianna cost the Jones family more than a million dollars. The facility had a daily capacity of five to ten thousand tons. The Marianna village and mine were regarded by the Jones family as the most livable mining town and the best-built and safest mine in the nation. This assertion was ironic because the Rachel mine was the site of a disastrous explosion on November 28, 1908, killing 154 miners; only one survivor was pulled out of the rubble alive. Although the mine was noted as a gassy mine the disaster was caused by a coal dust explosion either from the flame of a miner's lamp or from an electrical spark. Thereafter, state regulations were passed to limit the accumulation of coal dust in the mine and require the use of sealed electric miners' lamps.

The village of Mather, Greene County, located on the south branch of Ten Mile Creek near the village of Jefferson, was constructed by Picklands, Mather & Company of Ohio in 1917. Mather, like Nemacolin, Bobtown, and Crucible, was a "model" company town constructed in Greene County during this period by an out-of-state steel company. These captive mines provided their companies with a steady supply of coal and coke for their blast furnaces. This Cleveland-based steel company acquired more than four hundred acres of coal from the Pittsburgh seam. Ground was broken for the first shaft on the old Moredach farm, near the village of

Jefferson, on August 7, 1917. The first shaft of the Mather Collieries was 348 feet deep, and 947 feet above sea level, sunk at the village of Mather, located on the Pennsylvania Railroad. The Mather mine supplied coal for its by-product coke ovens at the company's Toledo and Canton, Ohio, facilities, the Cleveland furnaces of the Allied Steel Companies, and the furnaces of the Steel Company of Canada (located at Hamilton, Ontario) as conditions from time to time warranted following its first shipment in 1918.

Baton & Elliott, consulting and contracting mining engineers of Pittsburgh, designed the mining complex and the village of Mather. They constructed nearly three hundred small but attractive and brightly painted, multicolored bungalows on shade-lined, paved streets. Each bungalow was equipped with a cellar, electric lights, and running water. The firm constructed some recreation facilities. A two-story stucco clubhouse included a living room, bathrooms, dining hall, barber shop, bowling alley, and billiard tables on the first floor. The second floor was occupied by a dance hall. Baton & Elliott built a filtration plant, several churches, a school, and a well-equipped playground for the miners' children. A tennis court and bandstand were later erected in the center of the village.[135]

A number of manufacturing companies provided coal operators with an alternative method of constructing workers' housing. They manufactured a complete line of prefabricated industrial housing ready to be assembled at a distant mining site. They also manufactured prefabricated hotels, stores, banks, churches, and bunkhouses that could be purchased and assembled along with the housing at a distant site. For a few thousand dollars per house a coal company received floor plans, pre-cut lumber, nails, paint, doors, and light fixtures all shipped by rail. There were a variety of companies involved in the catalog house business after 1890. The "Big Six" were Sears, Roebuck & Company, Montgomery Ward & Company, Gorden-Van Tine of Davenport, Ohio, and three companies located in Bay City, Michigan—Aladdin Company, Lewis/Liberty Manufacturing Company, and Sterling System Homes.[136]

The Ford Collieries Company of Detroit, Michigan, operated the Berry Mine at Bairdford and the Francis Mine at Curtisville, Allegheny County, mining coal from the Thick Freeport seam. The company town of Bairdford was constructed of prefabricated housing manufactured by the Aladdin Company. The unassembled houses were shipped by rail to Bairdford around 1920. At Indianola, Allegheny County, the Indianola Mine of the Inland Collieries Company, a subsidiary of the Inland Steel Company, acquired the property in 1916 formerly owned by the Indianola Coal Company. This was a 200-foot shaft entry mine that extracted coal from the Thick Freeport seam, located about 500 feet below the Pittsburgh seam and 120 feet above the Kittanning seam. This seam was mined extensively along the Allegheny and Kiskiminetas Rivers, with an average thickness of seven feet. The coal was used exclusively for the production of by-product coke. About one-sixth of the coal was being washed by 1925. The mine's tipple was equipped with picking tables, screens, and crushers for preparing coal before shipment. Thomas G. Fear was hired by Inland Steel in 1916 as the mine's first superintendent to design the mine and to construct a model town in Indiana Township. Fear was a graduate engineer from Birmingham, Alabama, who had constructed mining complexes in Tennessee and Alabama. Superintendent Fear built mining buildings, workers' housing, an office, a hospital, and a sewage disposal plant with water and sewer lines at Indianola.[137] The prefabricated miners' houses were shipped by rail from the Lewis Manufacturing Company, Bay City, Michigan, in October 1918 to Indianola. A local newspaper described the new community in 1919 as "the last word in coal producing and miners' home development." The homes featured electric lights, indoor bathrooms, and spacious backyards. The village houses, all with front porches, were constructed with their backs facing the roads around the cul-de-sacs. This was done to prevent coal miners, filthy after working in the mine, from entering and soiling the living rooms and furnishings. Instead miners entered their kitchens. The office and hospital were located next to each other near the tipple. The hospital had four beds, a treatment room, a dispensary for drugs, and a fully equipped operating room. The company donated land for the construction of the two churches in the village.

Prefabricated housing was constructed by other mining companies in other coalfields besides Pennsylvania. The Kaymoor Mines of New River, West Virginia, owned by the Lowmoor Iron Company, provided coal and coke for the company's iron furnaces. The company purchased nineteen "Redi-Cut Homes" from the Aladdin Company of Wilmington, North Carolina, in anticipation of a coal boom after 1918, which never occurred. The prefabricated houses were wooden bungalows, each holding two families, constructed from precut weatherboard with plastered-lath interior. The bungalows were shipped to the new mining town of Kaymoor One by rail.[138]

In summary, the quality of houses inhabited by coal miners and their families in Pennsylvania and surrounding mining states was uneven, from wooden shacks to roomy six-room houses equipped with the latest amenities, including indoor plumbing, electricity, and running water. They were diverse in size, method of construction, and material employed by the coal company in their construction. Four recognizable types of miners' housing emerged nationally:

(1) The Shotgun, typically a one-story, two-bay, wood-frame structure with a gable roof, post foundation, end chimney, and two or three rooms.

(2) The Pyramidal-Roof House, a one-story, three- or four-bay frame dwelling, often semidetached, with post foundations, central chimney, and four rooms.

(3) The Pennsylvania Miners' Dwelling, always a two-story structure, either detached or semidetached, with two bays per dwelling unit, a wooden frame, four or five rooms, front- or side-gable roof, end chimneys, and often a rear ell containing one or two additional rooms per unit.

(4) The Gable-Roof House, a one-story residence with two, three, or even four rooms, end chimneys, wood-frame construction, and occasionally a projecting one-story ell producing a T- or L-shaped plan.[139]

Company Store

Virtually all coal towns had company stores, which like housing were constructed by the coal companies as a convenience for their workers and their families. As in housing, there was a great variety in the size and architectural style of the company store. Except for prefabricated stores, each was custom built for each mining community of brick or wood. Many mining communities were isolated with poor roads connecting them to neighboring towns. This inaccessibility made the company store (known as the commissary in the southern coalfields) the commercial heart of the mining community. The store served as a gathering point, post office, bill-collection center, and shopping center. It provided all the daily material necessities of life in the isolated village. The miner and his family were dependent upon it for their food supplies in both good and bad times. They paid their bill for the company doctor, usually about one dollar a month, and received their mail at the store. The company doctor was hired and his tenure determined by the company, while his salary came out of the miners' wages. All stores stocked and sold foodstuffs and a variety of mining equipment. Coal miners, unlike most other industrial workers, were responsible for providing and maintaining their own tools and mining supplies. The store stocked black powder or dynamite, caps, miners' lamps and

Company store at Leisenring.
Private collection.

fuel for carbide and oil wick lamps, squibs (fuses), electric exploders, picks, shovels, flints, and machine oil.

A retired miner described the company store as "a variety store like Gee Bee. But it wasn't that big. It was a grocery, clothing, hardware, and gift store all in one. It was also the post office and gas station for the patch. It was two stories. The food was sold on the first floor and miscellaneous items on the second floor."[140] Small stores, usually constructed of wood, sold basic food items, such as groceries and meats, and also mining tools and supplies. The larger stores were similar to a conventional department store found in a larger town or city. The more spacious stores were often three stories, with different functions on each floor: the first level was used for the storage of goods, the second level for retailing and offices, and the third floor exclusively for offices.[141] The larger stores sold a greater variety of goods, including furniture, clothing, plumbing, hardware, and building supplies. If the item desired by the miner was not in stock, the store manager could special order it. Many stores employed an "order" boy who delivered store goods to houses in the "patch" if miners were unable to shop. The larger company stores operated ice cream parlors and soda fountains. Some larger company stores provided additional services including a laundry, millinery, and gristmill.

The Rise of Large Coal Corporations

The bituminous coal industry was composed of numerous small companies as late as 1880. These coal companies were usually family owned and locally controlled. Mines were of small annual capacity. A mine producing a thousand tons a day was considered a large operation. The United States Census of 1880 estimated that there were more than five thousand small "country bank" mines, and that together they contributed less than a million tons of coal toward the national total of some forty million tons. The average production of bituminous coal mines in the country was 14,269 tons a year in 1880. The largest output by any mine in Pennsylvania in 1880 was 332,056.[142] There were a few large bituminous companies at the beginning of the era but they represented a minority of all coal companies. Consolidation Coal Company, Westmoreland Coal Company, and Penn Gas Coal Company had absentee owners with corporate offices in eastern cities that controlled the daily lives of tens of thousand of miners in distant and isolated coalfields. For example, the Penn Gas Coal Company, a Philadelphia-based company founded during the 1850s, employed more than a thousand men and boys at their mining operations, located principally in Westmoreland County.[143]

Extractive facility with head frame, powerhouse, and associated buildings. Pennsylvania State Archives.

The rapid expansion of coal production and an enlarged and increasingly immigrant labor force created a series of organizational and technological changes within the bituminous coal industry. A number of large coal companies were formed during this period, but none of these corporations acquired monopolistic control of the industry. Both the ownership and the operation of the bituminous coal industry were widely diffused between 1880 and 1920. Edward T. Devine, a leading observer and author of an excellent narrative history of the American bituminous coal industry, noted that "there was no typical bituminous coal operator, but only a heterogeneous, unorganized, infinitely diverse, and hotly competing aggregate, with many representatives of almost any group that could be described, from a great corporation to an impecunious, struggling individual."[144] Coal mining companies ranged in size from the local "country bank" mine, employing as few as five workers, with coal production limited to a few hundred tons annually, to huge firms with corporate offices in Philadelphia, Baltimore, or New York. Their mines employed as many as a thousand workers with annual output exceeding a million tons. Twelve of the hundred largest industrial corporations in the United States in 1900 were mining companies. Many of the largest bituminous coal mining operations were increasingly owned and controlled by utility companies, railroads, and iron and steel corporations; they were called "captive" or "consumer" mines. "Captive" refers to coal production by these industrial firms to insure themselves an uninterrupted energy supply. Railroad corporations controlled more than one-fourth of the nation's coal production by 1909. "Captive mines" accounted for 21.4 percent of all mined coal in Pennsylvania in 1913 and 24 percent in 1919.[145]

There were more than five thousand coal mines nationally in 1905, and the largest 665 coal companies produced about 15 percent of the nation's 315 million tons of coal during that single year.[146] Major mining operations in the United States were increasingly owned by companies with corporate offices in New York, Baltimore, or Philadelphia. Twenty-seven Pennsylvania coal companies produced more than one million tons in 1910. These large companies employed from 1,432 (Shawmut Mining Company) to 19,406 workers (Pittsburgh Coal Company).[147] Westmoreland Coal Company, Consolidation Coal Company, Berwind-White Mining Company, Pittsburgh Coal Company, W. J. Rainey Company, Pennsylvania Coal and Coke Corporation, Keystone Coal and Coke Company, Rochester and Pittsburgh Coal and Iron Company, and the H. C. Frick Coke Company were the largest bituminous coal and coke corporations in Pennsylvania in 1920.

No company or combination of coal companies was able to establish a monopoly within the industry. From a national perspective no single company controlled more than 3 percent of bituminous coal production in 1905. Large bituminous coal corporations were simply unable to establish a monopoly or oligopoly in the industry as had developed in other industries during this period. There were 276 big-business combinations between 1897 and 1903 with a total capitalization of slightly more than $6 billion. By then some three hundred firms with assets totaling $20 billion controlled 40 percent of the industrial wealth of the nation. John D. Rockefeller's Standard Oil Company of Ohio formed the first monopoly of a natural resource in the nation during the 1870s. Standard Oil controlled nearly 90 percent of the nation's oil industry, prompting Rockefeller to quip, "Oil is my business." Henry Demarest Lloyd observed in *Wealth Against Commonwealth* (1894) that "Standard Oil did everything to the Pennsylvania legislature except refine it."

The creation of the United States Steel Corporation, under the direction of J. Pierpont Morgan, the New York banker, was the most widely recognized example of business consolidation to gain monopoly control of a particular industry. This New Jersey holding corporation was a merger of more than two hundred separate companies into the nation's first billion-dollar corporation, and was incorporated on February 25, 1901, capitalized at $1.4 billion, three times the annual budget of the federal government. The company owned seventy-eight blast furnaces, five hundred thousand acres of coking lands, more than a thousand miles of railroads, and large

Testing for accumulation of methane gas with a safety lamp at Vesta coal mines. Historical Society of Western Pennsylvania.

reserves of iron ore. The original member companies were Carnegie Company, Federal Steel Company, National Tube, American Bridge, American Tin Plate Company, American Steel Hoop Company, and American Sheet Company. Henry Clay Frick helped to organize and became a director of J. P. Morgan's U.S. Steel. The corporation controlled 40 percent of the steel industry in 1920. Monopolies or oligopolies in other industries were created by Standard Oil, which controlled 64 percent of the oil industry in 1911, and by International Harvester, which controlled 64 percent of the agricultural machinery industry in 1918.

A variety of factors made the bituminous industry highly competitive, despite the appearance of large corporate giants. High-quality coal was both abundant and distributed in some twenty-five coal-producing states covering thousands of square miles. High-quality coal deposits in Pennsylvania were located in more than two dozen counties and in an area covering more than ten thousand square miles. In addition, unlike other extractive industries, the amount of capital required to open a mine was small and, therefore, the larger corporations could never successfully eliminate small or medium-size companies. This inability to create a monopoly within the industry had catastrophic effects on the expanding bituminous coal industry.

Coal operators were aware of the competitive nature of the industry and the tendency towards over-development or undue expansion as early as the 1890s. Coal miners worked an average of only 214 days a year from 1890 to 1921. An overmanned coal industry meant reduced daily wages for miners and created a glut of cheap coal on the market. Government agencies considered 304 days a full working year in the industry. The maximum yearly employment was 248 days in 1918, the peak of the World War I period.[148] Coal operators were simply employing more workers than the industry could employ steadily throughout the year. Underemployment of miners was a conscious choice made by some coal companies, as documented by a survey undertaken in 1909 by the United States Census Bureau. The survey found that the true profit of many coal companies did not come directly from coal mining; instead, "it should be noted that many mine operators make a considerable profit by renting houses and selling merchandise to their employees at the company store."[149]

A surplus, inexpensive labor force and expansive coal reserves requiring little technology or capital to develop encouraged excessive production by coal operators, both individual and corporate. Seasonal demand and the inability to store coal also led to periods of excessive output. Railroads were competing for business and offered some coal companies reduced rates for long hauls to market. Economic enticements from railroads in the form of reduced freight rates constituted another reason for the expansion of the new southern Appalachia coalfields of West Virginia, Kentucky, and Alabama.

Some coal operators formed trade associations or merged with competitors to set prices and limit coal output and wages in an attempt to check the ruthless and competitive condition endemic in the industry. For example, more than 140 small to medium-size coal companies were operating up and down the Monongahela Valley by the 1890s. Fierce competition among these owners and against the new challenge from the southern coal producers reduced the price of mined coal below the actual cost of production. A futile attempt was made by these operators to limit production voluntarily to fix and maintain a minimum price for coal. These informal agreements or pools were generally ineffective. Cutthroat competition in the region was resolved temporarily by a series of consolidations that created two principal coal corporations in 1899. The Monongahela River Consolidation Coal and Coke Company was organized on June 9, 1899, under the leadership of J. B. Findley, a businessman and a native of Monongahela, who was the guiding spirit in the consolidation movement. The new firm was known locally as the "River Combine." The firm was originally capitalized at $30 million, one-third preferred and two-thirds common stock, and $10 million in bonds. The firm controlled dozens of coal mines located on the river. The Black Diamond, Catsburg, Coal Bluff, Cincinnati, Eclipse, Crescent, Knob, Vigilant, Beaumont, and Champion mines were owned by the new corporation on the Monongahela River. The corporation also owned towboats, barges, coal boats, river tipples and coal loadings, and docks on the banks of the Monongahela River.[150]

Pittsburgh Coal Company was the second firm formed from the merger of small Monongahela River mines. The company was founded in Pittsburgh and organized under New Jersey statutes as a $64 million corporation. The company controlled 140 coal-mining and distributing properties, including its own docks, loading and unloading facilities, railroad lines, and coal mines within a seventy-five-mile radius of Pittsburgh. Nine-tenths of its coal mines were located in Allegheny, Westmoreland, and Fayette Counties in western Pennsylvania.[151] The United States Industrial Commission identified it as the largest coal company in the United States in 1900. The Pittsburgh Coal Company secured the majority of the stock of the Monongahela River Consolidation Coal and Coke Company in October 1903. The two companies formally merged in 1916.[152] A. W. Mellon, Henry R. Rea, Henry W. Oliver, and H. C. Frick, all prominent Pittsburgh capitalists, were among the early directors of the Pittsburgh Coal Company. The company owned seventy mines located in Pennsylvania, Ohio, Illinois, and Kentucky, which produced 30 million tons in 1920. Sixty collieries were located in the heart of the Pittsburgh district; the company owned 152,745 acres of virgin coal land within the district.[153]

The Coal Association of Illinois and Indiana submitted a report entitled, *A Statement of Facts Concerning the Conditions in the Bituminous Coal Industry in the States of Illinois and Indiana* to President Wilson in 1914. The association was seeking relief from conflicting regulations of state and federal governments, which had rendered coal companies powerless to prevent cutthroat competition within the industry. An association spokesman, on the eve of the World War I, clearly identified the plight of the bituminous industry through uncontrolled overdevelopment:

> They [industry] have opened three mines where only two were needed; they have employed three men where only two were necessary. These mines and men can find productive work only during 175 instead of a possible 300 days in a year. This idle time of the miners is not confined to one season or period during which they can find employment. To the contrary, the men are always subject to call, for which they urge a greater daily wage that their annual income may be sufficient for their needs.[154]

Government was generally unresponsive to intervention and regulation of the coal industry by maintaining wages and hours or by establishing a minimum price of coal. The federal and coal-producing state governments were opposed to mergers of coal companies as unfair restraint of trade. The New River Consolidated Coal and Coke Company was fined under the Sherman Anti-Trust Act for establishing minimum prices of coal mined in the New River and Kanawha fields of West Virginia. Congress had passed the Sherman Anti-Trust Act in 1890. The act consisted of eight major provisions that declared that "every contract, combination in the form of trust or otherwise, or conspiracy, in restraint of trade or commerce, among the several states or with foreign nations, is hereby declared to be illegal." In 1911 the Supreme Court, under this law, found that Rockefeller's Standard Oil Company and the American Tobacco Company were monopolies or trusts and they were ordered broken up. However, the Court ruled by a vote of four to three in 1920 that the United States Steel Corporation was not a monopoly according to the provisions of the Sherman Anti-Trust Act.

There was no consensus among coal companies as to how to deal with overdevelopment, maintain a high price per ton of coal, and reduce wages. Some coal operators wanted limited government regulation, while other companies opposed any form of government regulation or intervention in the bituminous industry. It was suggested by some coal companies that the health of the industry could be improved by establishing a maximum and a minimum price at which coal could be sold, and then basing the miners' wages on the selling price. Others advocated that a law be passed taxing all the coal mines in the United States to provide for the families of miners killed in mining accidents.[155] The Berwind-White Coal Mining Company, like some other coal companies, opposed any attempt at corporate merger and opposed any form of government regulation of the industry. E. J. Berwind, president of the New York based company, expressed his company's position by insisting "he would never join any operators' association, holding that they are all right for ordinary operators but were beneath his dignity."[156]

Attempts at monopoly control of the bituminous industry by large coal corporations were dismal failures, the industry being still widely diffused in 1920. Coal output from mines ranged from a few thousand tons, or even less, to more than five million tons annually. The United States Geological Survey identified 12,122 corporations, partnerships, and individuals involved in the production of bituminous coal in 1920. There were 14,776 mines, including 1,440 "country banks" with a strictly local market and annual production of a thousand tons each, and 4,405 "wagon mines" shipping by rail nationwide. The remaining 8,921 "commercial mines" were divided among 6,277 producers. Only eighty of these "commercial mines" had an annual output of over a million tons.[157] "Snow-bird" or "fly-by-night" were terms used in the industry to describe tiny mines that opened with the first snow and disappeared with the advent of warmer weather. These mines were always small scale, inefficient, and crudely operated. They operated during winter when demand for coal was greatest and the price of coal was highest. These mines, often no more than a hole in the ground, were a convenient scapegoat upon which many of the ills of the industry were blamed, in spite of the fact that their total output was an insignificant part of annual production.

This national situation was reflected in Pennsylvania's own coal industry. There were 3,695 coal operators, both bituminous and anthracite, in 1909. These coal operators were organized as follows: individual 1,058 (28.6 percent), firm 664 (18 percent), corporation 1,942 (52.6 percent), and other 31 (.8 percent).[158] The Bureau of Mines of Pennsylvania estimated that there were 1,938 coal companies operating 2,584 mines employing 154,992 workers in 1919. There were 20 mines with production of over a million tons; 22 mines with production exceeding a half-million but less than a million tons; 112 mines with production of 200,000 to 500,000 tons; 144 mines with production of 100,000 to 200,00 tons; 174 mines with production of 50,000 to 100,000; and 118 mines with production of 25,000 to 50,000 tons; while 1,398 mines had production of less than 25,000 tons. Forty-four mines or 2.2 percent of the nearly two thousand mines produced a half-million tons or more annually, or nearly one-third of all coal output, and employed about one-third of all miners in the Commonwealth.

Size of Bituminous Mines in Pennsylvania in 1919

Tons of Coal	Number of Enterprises	Number of Mines	Workers
less than 25,000	1,348	1,452	17,860
25,000-50,000	118	138	5,354
50,000-100,000	174	229	14,417
100,000-200,000	144	220	21,629
200,000-500,000	122	216	33,151
500,000-1,000,000	22	74	14,770
one million plus	20	255	47,811[159]

Changes in Mining Technology

The trend to fewer and larger production firms affected the coal industry. The large coal corporations fundamentally changed the work process in their mines during this period. Most mines, even larger mines before the 1880s, represented a proliferation of small, organized underground workshops controlled by skilled miners. The craft process of mining coal took place within individual "rooms," about twenty feet wide. An unskilled laborer described work underground before the 1880s: "We were paid so much a ton about seventy-five cents, we'd be in a chamber (room) filling carts with coal. We didn't work in gangs together, we worked each one alone. I had a pick and I had a shovel and I had an auger to drill a hole—the auger could drill six feet in the face. There were eight hundred and some workers and each of them had their own room."[160]

The larger mining corporations introduced new management practices, including increased supervision, that were instituted to control the work process and make underground production more rational and efficient. Each miner was assigned a specific task, and this gave the company more control over its labor force underground and at the surface. This policy of increased supervision was disliked by most contemporary miners, and as one worker observed, "the large corporation are slowly but surely, riveting the chains of slavery around us."[161] The contract miner was becoming more like an operative in a iron or steel mill and less an independent artisan. Skilled miners, still known as "practical" miners, had historically controlled the work process underground. Coal operators had provided little direct supervision in the mining process; instead they had consciously abdicated control of the mine to the "practical miner." The skilled miner decided the amount of production, dictated the tempo of work, and was generally independent of direct supervision underground. This control and autonomy of the mining process by skilled miners was clearly described in the *Report of the Illinois Bureau of Labor Statistics* (1888):

> The skilled miner takes his tools into the pit and undertakes to deliver from the wall of mineral before him certain tons of coal for a certain sum per ton. He mines and drills and blasts and loads his own coal, timbers his own roof, takes care of his tools, and is responsible mainly for his personal safety and the amount of his output.[162]

The large coal corporations were organized in a hierarchical structure like other large, contemporary industrial companies. The president or executive supervised the entire coal operation. A number of specialized departments were created, each devoted exclusively to operating a particular facet of the mining business. These included the engineering, mechanical, electrical, accounting, sales, purchasing, medical, and safety departments. These firms employed a number of managerial supervisors to oversee coal production from its removal from the seam to its shipment to market. The general superintendent oversaw the operation of several mines, while the mine superintendent was in charge of the entire operation of a single mine, supervising all facets of the mining process. The superintendent acted as the connecting link between management and the miners. He was a former fire boss or mine foreman who was appointed to this position because of his practical expertise in the operation of the mine, or he was selected from the engineering department. He employed a number of subordinates, who were responsible directly to him, to supervise a part of the work process. Surface and underground foremen were hired by the superintendent to supervise the extraction, hauling, and tipple operations. The inside foreman would hire one or two fire bosses, who in turn hired a few assistants. The term "fire boss" is a misnomer because he was really a gas boss responsible for testing the mine air for the presence of toxic gas before the miners' entered the mine in the morning.[163] The fire boss was an experienced miner who entered the mine before the morning shift, as early as 3:30 A.M., and with his naphtha-burning safety lamp inspected every room in the mine and chalked his approval on the face and side of each examined room showing the date and time of examination and measurement of the presence of gas, especially methane.[164] The fire boss was also responsible for seeing that the brattices (barricades) were put up as required to regulate air ventilation in the mine. These barriers were operated by young boys called "trapper boys." The Bureau of Mines of Pennsylvania was requiring state certification by 1900, which the fire boss obtained by passing a written examination. His office was usually located underground.

Workers, besides the fire boss and shot-firer, who used electrical machines were all issued safety gas lamps to detect methane. They were issued by the lamp man, who was responsible for their cleaning and repair by the coal company. By 1910 the Pennsylvania Department of Mines had approved the use of safety gas lamps for gas detection, manufactured by the Clanney, Davis Deputy, Wolf, Schenk, Seippel, and Ackroyd & Best companies. These safety lamps were tested by the National Bureau of Mines at Pittsburgh.[165] Charles Wolf of Saxony developed the Wolf safety gas lamp, using naphtha instead of oil, in 1883. The Wolf lamp, manufactured by the Wolf Safety Lamp Company of America, New York, after 1913, was a self-extinguishing lamp in the presence of firedamp in explosive mixtures, or in the event it was overturned. The lamp was provided with a magnetic lock that could not be picked, had an igniter by which the lamp could be lit without opening it, and provided a good light.

The miner used to work alone or with his partner or "buddy" in a room, but after the 1880s large coal companies wanted their workers to be part of an interdependent, coordinated group of workers who were carefully supervised by management. This transformation of work from an individualized process to a complex division of labor was a major intervention in the workplace by management, and represented a major challenge to the skilled miners in their control of the work process. Under the new system, introduced first at the larger mines, the miner's tasks were delegated to a variety of specialized workers. This division of labor created a number of specialized jobs and resulted in the introduction of a two-tiered system of workers and wage payment. This division of labor had become the standard practice throughout the bituminous coal industry by 1900. Mine workers fell into two distinct and large groups, called "tonnage men" and "company men." Underground workers who were engaged in the actual extraction of coal from the seam were called "tonnage men." They were responsible for the drilling, shooting, and loading of coal in each room. The coal company paid their "tonnage men" piece-rate by the ton, the yard, or the cart of coal loaded daily. Tonnage men included the following categories:

Cutter ("machine man" or "machine runner")/scraper (helper): The operator of the mechanical cutting machine, his assistant, the scraper, and two loaders were assigned by the underground foreman to a group of rooms. These men moved the undercutting machine from room to room and undercut the coal seam at the bottom for a number of loaders. Each machine was operated by a single worker who cut a six-foot-deep channel under the working face, using a continuous chain with removable metal bits. The bits were sharpened daily at the blacksmith shop. The cutter was one of the highest-paying jobs in the mine, but it was dusty and dirty work and the constant inhaling of small particles of coal dust for years contributed to black lung disease. The scraper assisted the cutter by shoveling fine coal (bug dust) and debris aside or behind the machine. The scraper assisted the cutter by removing quantities of coal dust that lodged inside the cutting machine. This task was known as "pulling the slack." He was responsible for setting the pipe or jack so that the machine would start to cut the channel under the coal.

Coal loader: He loaded coal by hand or with a broad-bill shovel into mine cars after it was removed from the wall by the shot firer. He was responsible for separating impurities, especially shale, slate, and clay, from the coal. The carts were regularly inspected by the underground mine foreman and if the load was determined to contain "dirty" coal, the loader was denied credit for the wagon. He attached a small brass tag called a "check" with his number to each coal car after it was loaded. Each miner had a check with his number stamped on it and after the car was delivered to the tipple he was credited with the coal. An average loader filled two or three cars daily.

Shot firer/shooter/shot fireman: He was responsible for firing the blast of dynamite or black powder which knocked the coal loose from the seam after the holes were drilled by the driller. The mine wall from which the coal was blasted is called the working face, and the process of blasting coal with powder at the working face is

Shot firer at work preparing to blow coal from face. Hillman Library, University of Pittsburgh.

called "shooting coal." Coal was originally blasted from the working wall with black blasting powder. Dynamite and explosives were later used. This was the responsibility of the skilled miner in the premechanized mine, but the improper use of black blasting powder was blamed for many major mining explosions. While miners might still be required to drill the holes and take charge of permissible explosive cartridges, the shot firer performed the actual blasting. He moved from one room to another carrying an electric detonator and electric firing apparatus. The professional shot firer was required by state statute to shout three times, "Fire in the Hole," before he shot the explosives. He carried a safety gas lamp with him and measured for the presence of gas before blasting coal from the seam.

Pick man: He did retreat work using picks after the loaders had finished removing coal from the "room." In the room-and-pillar mining method, from 25 to 30 percent of the coal is usually removed in the first working. A higher extraction of coal requires eventually that most of the pillars of coal be removed. The coal pillars were often removed without adequate support of the roof. This work was called "robbing of the pillar" and occurred after the rooms of a heading or section had been advanced as far as possible. The coal pillar supports the mine roof, and "robbing" it is extremely dangerous work because of the possibility of roof cave-ins.

Pick miner: He removed coal from the seam in the premechanized mine using a variety of hand tools—picks, augers, and explosives. His future in the larger mine was diminished with the introduction by the 1880s of undercutting machines, such as the Harrison machine. By 1900 nearly one-quarter of all coal was undercut from the seam mechanically.

"Daymen" or "company workers" composed the second category of mine workers. This group was further divided into "inside" and "outside" workers. They were paid a fixed rate by the hour, day, or month. The exact number of workers and their various occupations, both inside and outside, at an individual mine was dependent on the size of the mine and its condition. The "daymen" were responsible for all activities except shooting, cutting, and loading coal. They were involved in maintenance, construction, repair work, haulage of coal, mine ventilation, and tipple operations. They were also responsible for propping up the roofs along the hauling ways with timber and masonry. "Daymen" laid iron tracks, repaired them, and kept them clear of coal and rocks which might have fallen off the loaded cars. If the mine was wet, a crew of pumpmen was on duty day and night to keep underground mine water under control. The largest group of company men was the tipple gang. They prepared raw coal for transportation to market by weighing, sorting, washing, and loading it. Company men included the following:

Weigh boss: He recorded the weight of each coal car as it came from the mine entry to the tipple. After hand-loading the cart each loader attached his brass check to it. These records were given to the mine clerk, who credited the miner's account. His office was at the tipple. The tipple was a large, tower-like building during this period constructed principally of wood. It resembled a large cylinder and averaged fifty to sixty feet in height and some twenty feet in diameter. It was located at the entrance to the mine; coal from the mine was hoisted and poured into the top of the structure directly from the mine. Coal was weighed and then sifted down the tipple through a series of different-size screens, sorting it into a variety of sizes ranging from one to four inches in diameter. Processed coal was then loaded from the tipple into railroad cars below, or onto river barges.

Checkweighman: His chief responsibility was to check the accuracy of the company's coal scales at the tipple. He observed the weighing of the coal, since the men were paid by the ton. Miners were often defrauded of part of their wage by deliberate and systematic "shortweighing" of their coal by unscrupulous nonunion coal companies. Nonunion miners had no legal redress against this economic exploitation. One of the first demands of miners when the United Mine Workers of America was formed was the employment of their own checkweighman to monitor the company weighman. The union weighman was paid by the miners themselves with a percentage of each man's tonnage.

Check boy: He removed the miner's brass check from the coal car and handed it to the weighman when it arrived at the headhouse scale. A loader who complained about conditions in the mine was often punished by the superintendent when his check, which he hung on each coal car to identify his load, would be "mysteriously" lost.

Grease boy: He sprayed lubricant on the iron axles of the coal-laden cars to prevent them from overheating. This was one of the many mining jobs done by young boys. There were forty-five thousand boys under the age of eighteen working in the coal industry nationally in 1920.[166] A majority of boys, as young as ten, entered the mine as their fathers' helpers and during this apprenticeship mastered the mining craft. They worked as check boys, trappers, grease boys, or spraggers, earning about fifty cents per day for their family. In Pennsylvania, a law enacted in 1903 forbade children under the age of fourteen to work underground, and under age twelve outside, but the law was routinely ignored by both miners and the coal companies. Large families and low miners' wages forced many families to lie about the ages of their sons to get them work. The company evaded the laws by not listing the boys on their payroll. Their wages were paid directly to their fathers by the company. The isolation of the mines made rigorous enforcement of the law impractical.[167]

Trip rider/mule driver: He operated the underground transportation system. Mines that used mules or horses to haul coal employed drivers to lead the draft animals to the surface pulling wagons loaded with coal. Each wagon or mine car held about fifty bushels of coal. A close relationship often developed between a mule driver and his mule. A number of regional minstrel songs were dedicated to these sure-footed animals, including "My Sweetheart's the Mule in the Mine":

> My sweetheart's the mule in the mines,
> I drive her without reins or lines,
> On the Bumper I sit, with my whip in my hand,
> I chew and I spit
> All over my sweetheart's behind.[168]

Brattice man: He constructed linings or partitions made of canvas, wood, or concrete, called brattices, which were used in mine passages to confine the air and direct fresh air into the working place of the mine.

Spragger: He traveled with the train of mine cars and inserted a "sprag" or rod between the spokes of the wheels to slow the downward momentum on the trip out of the mine. A sprag is a short, round piece of hardwood, pointed at both ends, measuring 1 1/2 feet in length and 2 1/2 inches in diameter. A sprag was also used by pick miners, set in a slanting position to keep the coal in place during the undercutting operations.

Hoisting engineer: He operated the steam-powered or electric hoist that moved men, supplies, and coal in the "man-trap" or cage in and out of the shaft and slope mines.

Fireman: He worked in the boiler house where coal was transformed into steam used to produce electricity in the power house.

Tipple crew: Men with a variety of tasks were employed at the tipple or plant washer, their jobs being to weigh, dump the mine cars, and prepare the coal for shipment by sorting it after it arrived from the mine.

Machinist: He repaired mechanical equipment, including coal cars, trolley locomotives, and other types of machinery used at the mine. The introduction of mechanical coal loaders kept machinists busy repairing these machines.

Dumper: He worked at the tipple and operated the latch on the pit-car which permitted the front of the car to open as it was tipped forward and empty the coal into railroad hopper cars.

Blacksmith: He sharpened the miners' tools daily, including picks, augers, bits, and the metal teeth on the undercutting machines. Miners paid him for these services at the company store with deductions from their pay. The blacksmith was called on to repair a variety of machines used in smaller mines. He shoed the horses and mules used for coal haulage.

Snapper: He worked with the motorman and took the cable which was on a reel on the front of the motor, went into each room, and hooked it onto the front end of the loaded car. He also coupled the coal cars.

Stable boss: In larger mines he cared for the mules or other draft animals used to haul coal from the mine. These animals were very valuable and many miners insisted that the companies cared more for their horses and mules than for their workers. If a draft animal died, somebody was usually fired; this did not happen when a miner was killed. Draft animals were housed each evening in the mule barn in drift- and slope-entry mines. In shaft mines and steep slope mines, mules or horses resided in underground stables and were brought to the surface only if ill or dead or during prolonged strikes. The General Assembly of Pennsylvania passed a statute in December 1965 making it illegal for coal companies to keep mules and other draft animals underground overnight.

Motorman: By the 1920s many of the bigger mines had replaced the mules and other draft animals. Electric or battery powered locomotives manufactured by the Jeffrey, Joy, Goodman, and Westinghouse companies were used to transport coal and waste from the mine to the surface. The haulage motorman drove these machines with loaded mine cars from the haulage ways to and from the tipple.

Brakeman: He was in charge of the coupling and uncoupling of coal cars, and applied the brakes. The brakeman, also known as a snapper, trip rider, or car coupler, also operated the track switches.

Trapper boy or door tender ("patcher" in the anthracite region): He was employed to open and close ventilation doors and to throw switches to allow the horse-drawn coal cars and their driver to pass. Opening and closing doors was called "trapping." The trapper also maintained and cleaned the switches. Trappers were usually young boys, who with experience worked their way up the mining hierarchy from trapper boy to mule driver, to miner's helper, and finally contract miner.

Tracklayer: He was an integral part of the mine operation, and his duties included the laying of new and the maintenance of existing track upon which mine cars traveled, both inside and outside the mine. Tracklayers usually worked in groups of four in sections of the mine to which they were assigned by the underground foreman.

Wire man: He drilled holes in the roof and attached wire-hangers, then strung and clamped the electrical wires for mining equipment, especially electric locomotives.

Pumper: He was responsible for keeping the mine dry and free of water. Some mines were as dry as dust but many mines, especially those located under subterranean streams, were prone to flooding. Water enters a mine from the roof and floor and sometimes through an adjacent abandoned mine. The amount of water in the mine varies widely with season; consequently a large pumping capacity must be provided by the company. The presence of water in a mine increased the cost of coal mining in proportion to the number of gallons and distance which water was pumped.[169]

Timber man: He was responsible for cutting, framing, and placing props (timber posts) used to support the roof of the underground mine. This was essential in underground operations because the roof collapses were the most frequent cause of mine injuries and fatalities. The term is also applied to any person who draws or recovers props or posts.[170]

By 1920 about 85 percent of bituminous coal miners were working underground and 15 percent on the surface. Sixty percent of mine workers employed in the bituminous mines were "tonnage" men, operating cutting machines, mining coal at the face with picks and shovels, and loading coal.[171] This division of labor in the extraction of coal, with its new emphasis on specialized jobs and intensified supervision and managerial control over the workplace, both at the surface and underground, is seen in the operation of the Leisenring and Standard Shaft Mines. These coal mines, although principally coking operations in the Connellsville coke district, are representative of the two-tier division of labor found in larger mines during the last quarter of the nineteenth century. John Leisenring of Philadelphia, owner of the Connellsville Coke & Iron Company, opened Leisenring Number 1 Coke Works and the company town of Leisenring, Fayette County, in 1880. Leisenring also operated Lesisenring Mine Number 2, opened in 1882 at West Leisenring, and Leisenring Mine Number 3, opened in 1886 at Monarch, Fayette County. Leisenring Mine No. 2 was the site of a major mining accident that claimed the lives of nineteen miners on February 20, 1884. The accident was caused by an explosion of gas and dust. The H. C. Frick Coke Company purchased the three Leisenring mines in 1890.[172]

Leisenring Mine Number 1 was a 385-foot-deep shaft, located one thousand feet above sea level. The mine produced 348,000 tons of coal used in 500 beehive ovens producing about 247,000 tons of coke in 1888. It employed 442 workers, divided among 252 underground workers and 170 surface workers.

The Standard Shaft Number 2 mine located near Mount Pleasant, Westmoreland County, was built by A. A. Hutchinson & Brothers of Pittsburgh who opened the mine in 1878. It was a shaft-entry mine situated in the Pittsburgh seam, eighty-four to ninety-two inches thick. A. A. Hutchinson & Brothers erected 150 company houses, a company store,

Underground	Surface
1 foreman	1 foreman
180 miners	6 blacksmiths and carpenters
23 company men	6 engineers and firemen
28 drivers	132 cokers and yardmen
5 miners' boys	19 company men
	6 office personnel, including superintendent, bookkeepers and clerks[173]

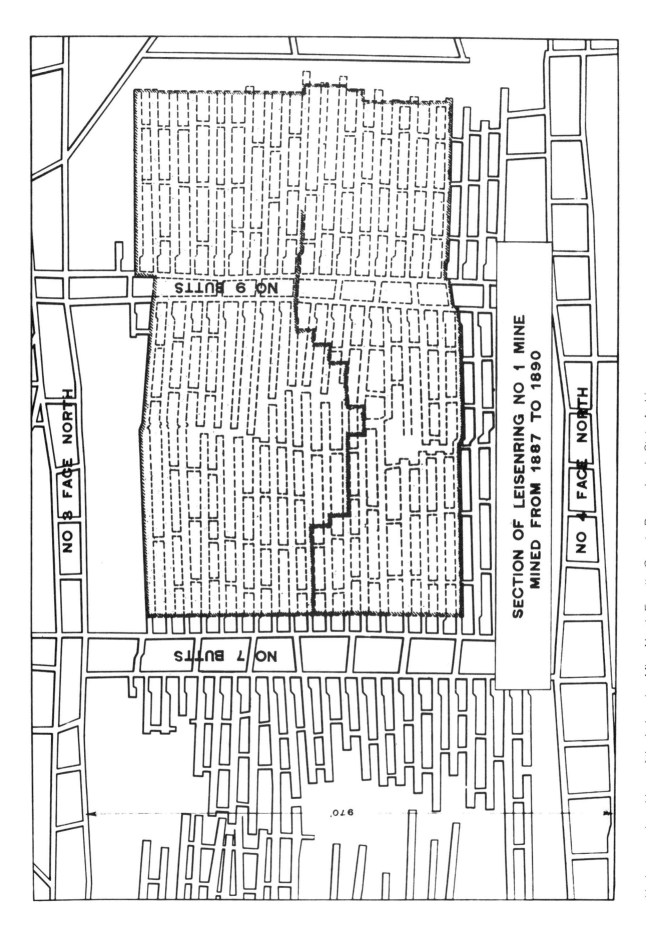

Underground workings of the Leisenring Mine No. 1, Fayette County. Pennsylvania State Archives.

and 509 beehive coke ovens. The mine's underground workings ran about four miles from the village of Standard to the village of Hecla. The H. C. Frick Coke Company bought the mine and coke plant in 1883 and immediately increased the number of coke ovens to 901, making it one of the largest beehive coke plants in the world by 1900. Standard Shaft mine produced 483,813 tons of coal, making it the second-largest mine in Westmoreland County in 1904. The Frick Company constructed a well-equipped underground mule stable at Standard, lighted by electricity. The firm, like other major coal companies, acquired its own railroad steel hopper cars. It purchased its own railroad cars to assure transportation without delay by avoiding the assignment of hopper cars from the railroad company.[174] By 1910 this was one of the largest coke plants in the world, with 908 coke ovens daily consuming over three thousand tons of coal and producing 125 carloads of coke.[175] The coke was transported by rail to the blast furnaces in the Pittsburgh region. Job classifications and the number of workers in the Standard Shaft Number 2 mine in 1880 are shown in table at right.

Underground	Surface
2 foreman	5 foreman
448 miners	14 blacksmiths and carpenters
35 company men	14 engineers and firemen
45 drivers or runners	340 cokers and yardmen
5 miners' boys	35 company men
16 doorboys or helpers	5 office personnel, including superintendent, bookkeepers and clerks[173]

As the scale of mining increased, individual mines became more complex extractive facilities. The number of surface structures and buildings located at a typical mine grew. There was no standard extractive mining complex in the bituminous coalfields. A large complex, by the turn of the century, had a variety of specialized buildings and structures. A mine could either include all or only a few of the following structures or buildings: ventilation-fan houses, mule stable (either surface or underground), tipple and scales, hoist house, head frame, supply house, repair sheds, blacksmith shop, machine or motor barn, carpenter shop, cap and powder houses, lamp house, mine office, motor barn, cleaning plant, power station, boiler room, substation, wash house, and perhaps coke ovens. A variety of factors determined the exact configuration of the surface structures and buildings at a particular mining operation. These factors included the type of mine entry, whether the mining facility was also a coke plant, the physical size of the operation, and the date when the facility began operation. A drift-entry mine, unlike a slope or shaft mine, did not require a hoist house since there was no need to haul coal, miners, or material to and from the mine. In 1900, there were 94 mines in the Connellsville coke region—38 drift mines, 32 slope mines, and 24 shaft mines. By 1909 there were 889 underground mines in Pennsylvania—they included 55 shaft-entry mines (6.1 percent), 76 slope-entry mines (8.5 percent), and 758 drift-entry mines (85.2 percent).[176] Most small drift-entry mines employed natural ventilation and usually did not require mechanical ventilation; therefore, there was no need for fan houses to contain large mechanical fans. Larger mines using electricity to operate undercutting machinery, haulage trolleys, and underground lighting needed a powerhouse, boiler house, and subgenerator shop to produce electricity and convert it from direct to alternating current and provide electricity for electric fans and power to the company town. Mines involved in the production of coke would require some type of coking ovens, usually constructed near the mine's

Head frame and powerhouse at unidentified mine in Connellsville Coke Region. Penn State, Fayette Campus.

entry to reduce transportation costs. Shaft-entry mines required a head frame located at the entry. The metal or wooden structure supported mechanisms used for hauling mine cars, coal, and miners to and from the mine.

The tipple (known as a "breaker" in the anthracite region) was a wooden structure where raw coal transported out of the mine was weighed, sorted, cleaned, and loaded for market. The structure was located near the mine's entry. After 1900 many of these structures were constructed of steel and concrete. Raw coal transported from the mine was being dumped into the tipple by a variety of methods by the 1920s—the bottom dump by which coal was released from the bottom of the mine car; the cradle dump by which coal was released by tilting the mine car to one side; and the end dump by which coal was released by tipping the mine car on end.[177] Waste materials, like slate and shale, were removed from the mine along with coal and placed in piles near the entrance of the mine. The gob pile, also known as the boney pile, ash pile, or the slate dump, was a mixture of slate, boney coal, and coal dust. The frequent deposits of mine waste over time made these heaps or hills quite large. The waste or slack pile of Pennsylvania's anthracite mines was called the culm bank.

Some coal seams contain solid coal from top to bottom, while other seams are split by partings of slate or other impurities of varying thickness. Certain coal seams contain high-ash coal called bone, boney, or bone coal, which has no commercial value and must be separated from coal. Slate is a dense, fine-textured metamorphic rock, extracted chiefly in Pennsylvania and Vermont. It has a variety of commercial uses, including blackboards, billiard tables, and grave vaults, but in coal mining it is waste material. Shale is a fine-grained sedimentary rock, fragile and uneven in composition. It is often incorrectly called slate. Slate has a clearer lamination than shale and often contains fossils of ancient ferns and plants.

A coal washer or preparation plant has the same function as a tipple, but it also washes the coal. A majority of bituminous coal during this period was shipped to market as run-of-the-mine coal without much surface preparation. Usually some impurities were removed underground by loaders, but little coal processing was done at the surface. A small minority of coal operators did prepare their coal prior to shipment to market. Their coal was sorted according to size, and impurities were removed and washed.

The Larimer Coke Plant of the Carnegie Coal Company was an early coke operation located midway between Larimer Station and Ardara, Westmoreland County, that cleaned and washed coal. Beginning in 1871, Andrew Carnegie experimented with making coke from fine coal or slack (bituminous coal one-half inch or smaller in size) at this coke plant, about two miles from Irwin. The Carnegie Coal Company's experiments proved conclusively that slack coal could be used for coke production if it was prepared before coking. Slack coal was washed to reduce its high concentration of ash and sulfur before placement in the beehive coke ovens. Carnegie at first used the "mound" process of coke production; then the company erected 122 beehive ovens at the Ardara site. Coke was made from slack coal shipped by rail from the neighboring mines operated by the Penn Gas Coal Company and the Westmoreland Coal Company. Carnegie Coal produced a good grade of coke continuously until 1900, when it abandoned coke production. The increasing demand for slack coal for steam made it no longer profitable to use it in the making of coke.[178] Slack coal, an excellent gas coal, was used in Philadelphia, New York, and other Atlantic Coast cities for gas manufacture for street lighting from the 1850s on.

Thirteen coal-washing plants with twenty-two washers were operating in the Pittsburgh district (Allegheny, Fayette, Washington, and Westmoreland Counties) by 1880, although few coal-wash plants were constructed after 1900. Coal companies concluded the added cost in cleaning coal was not justified economically, and therefore coal washers were seldom used in the bituminous coal industry until the 1920s.[179]

Mining Technology

Coal mining stubbornly resisted the introduction of machines and remained essentially a labor-intensive industry during its first century of development. Pick-and-shovel mining kept miners' productivity steady, as expressed in tons of coal mined per man per day. As late as the 1880s a majority of miners were still using the same crude hand tools and techniques employed before the Civil War. A number of factors had contributed to keeping machinery out of the mine and daily productivity low. Workers' resistance, lack of capital, technical difficulties in the development of machinery, and coal operators' passive indifference to mechanized mining. Contract miners and small individual mine owners shared the goal of high-priced coal. They believed the prosperity of the industry was made possible by limiting production to make coal scarcer in the market. This was a feasible strategy because a majority of the mines were still small, employing from a few miners to fewer than a hundred men. This small scale of operation also made investment in machinery an unsound business decision. Labor was cheap and abundant, so there was little incentive to invest in costly machinery.

Large coal companies believed that they could maintain profits by increasing daily productivity by introducing mining machinery. The undercutting machines were the first fundamental change in the work process since the beginnings of the coal industry in colonial days. Coal cutting or punching machines, powered by compressed air, were the first successful attempt at mechanized coal mining. Michael Menzies invented and patented a mechanical coal cutter in England about 1761. He proposed "to transmit power from an engine at the surface through a series of reach rods and chains passing over pulleys to a machine carrying a heavy pick, which was to undercut the coal."[180] To drive these reach rods he proposed to use a fire engine, a water mill, or a horse gin. During the 1860s miners were paid about four to five cents per ton to mine coal. An average miner could dig as much as one hundred bushels per day, but this required him to work long hours to earn a decent living.[181] The undercutting of coal was the most skilled and highly paid job in the premechanized mine. It was tedious, time consuming, and extremely dangerous work. There was a growing reliance by miners on a variety of explosives to remove coal from the seam, but the actual work of undercutting coal from the seam during the first century of coal mining had not fundamentally changed until the introduction of the mechanical undercutting machinery. The early "punching" machine was a horizontal jackhammer used to undercut coal. A mechanical undercutting machine was first introduced at a mine about two miles from Brazil, Indiana, in 1873. Brazil is located in the northern section of the Indiana coalfield which lies in the southwest part of the state.[182] The machine consisted of "an iron rim, four feet in diameter with movable steel teeth placed about one foot apart. The rim rested on small wheels which supported it and had cogs on its undersurface, which engaged on a shaft turned by the engine."[183] The machine ran on movable track and was fed by means of a screw working in cogs. The track ran parallel to the vein and about one hundred tons of coal could be cut in twenty-four hours. The machine was originally run by a five-horsepower engine and later by compressed air.

J. W. Harrison of Chicago patented a portable air-compressed undercutting machine in 1877 that was first successfully used by the St. Bernard Coal Company,

Henderson Coal Company, Hendersonville, Washington County. Coal Age, c1910.

Kentucky, in 1880. The original machine was a compressed-air drill puncher that struck the coal seam at an average of forty times a minute, which was about the same rate that a pick miner struck the coal seam with his sharpened pick. The Harrison machine mimicked the motion of the miner's pick, delivering a series of rapid blows intended to pulverize the coal. One man operated the machine by two handles, one held in each hand. The miner guided the machine and directed its blows on the seam, and with pressure from his foot forced the machine forward. A second miner shoveled away the fine coal (bug dust). Improvements on later models permitted the machine to strike the coal with a punching action of two hundred strokes per minute to the base of the seam at a depth of more than three feet. The pneumatic puncher offered great reliability and ease of control for untrained operators because it operated like a horizontal jackhammer. In 1876, the Lechner Machining Company introduced another type of coal-cutting machine at mines of the Straitville Central Mining Company in the Hocking Valley of western Ohio. The Lechner machine consisted of an engine operated by "a revolving bar into which sharp steel points were inserted. The bar was driven into the bottom of the seam, making a cut three feet wide and six feet deep."[184] The deep undercut allowed the coal to break cleanly away from the seam when the "shot-firer" fired the coal from the face. If the coal face was insufficiently undercut, the blast would pulverize the coal. Powdered coal, however, was of little commercial value and use during this period. These two machines were both operated by compressed air, produced by a compressor above ground and conducted into the mine in pipes. Air technology was safe and unlike the later use of electricity could not ignite a mine fire, although the air hoses made the puncher's use in the mine less flexible.[185] Compressed-air punchers weighing seven hundred pounds each were widely used underground in larger mines during the 1880s.

The Harrison machine and similar undercutting machinery were expensive during their first decade of use underground. Each machine cost about three hundred dollars. The machines were plagued by numerous mechanical breakdowns, and parts had to be ordered from the manufacturer or made by the mine's blacksmiths in the machine shop. The poor quality of these pioneer machines improved after the 1890s. The electric chain-breast cutter was an alternative to these earlier, air-compressed punching machines. The new machine was first manufactured in 1893. It was equipped with a thirty-five-horsepower motor and used a horizontal blade or cutter bar fitted with an endless chain capable of making a seven-foot-deep undercut. The front of the machine stuck out like a duckbill and was edged with a toothed chain. As the chain went around, its sharpened teeth gouged out a groove near the bottom of the coal face. The chain-driven machine was superior to the earlier pneumatic machines because it operated faster, used less energy, and produced less "slack" or pulverized coal in making cuts. The chain-breast cutter also spared the coal operator the expense of installing expensive compressed-air lines in the mines. A majority of coal operators had replaced compressed-air machines with electric undercutting machines by 1910. Undercutting machines were not used extensively in the anthracite mines of Pennsylvania because of the hardness of the coal, and because the highly disturbed condition of the strata in these mines made their use underground economically impractical.[186]

The use of mechanical undercutting machines continued to increase substantially with each decade after the 1890s. The quantity of coal undercut by machine depended on local variables, including the thickness of the seam, expertness of the machine operator, hardness of the coal, and the width of the working place. Undercutting machines provided coal operators a means to increase productivity. The machine elevated the individual miner's daily output from 2.57 tons in 1891 to 3.71 tons in 1914. Union operators in western Pennsylvania, Ohio, Indiana, and Illinois employed these coal-cutting machines in an attempt to boost productivity and lower their high operating costs. Northern operators were paying their union employees higher rates than nonunion southern mines, and the new competition from southern

Harrison undercutting machine. The pneumatic undercutting machine replaced manual methods.
Carnegie Library of Pittsburgh.

nonunion coal producers had motivated them to seek methods of reducing the cost to mine coal. The production of bituminous coal, nationally, stood at 4.28 tons per day per man in 1922, while productivity in machine-equipped mines was about five tons.[187]

The pneumatic undercutting machines were first introduced in the 1870s, but general production figures on mechanically undercut coal were not kept until about fifteen years later. Only 6,211,732 tons, equal to 6.6 percent of the total bituminous coal, was mechanically undercut nationally in 1891. Five years later the output of machine-mined coal reached 16,424,932 tons or 14.7 percent. In 1900, 52 million tons, about 25 percent of the production of coal nationwide, was machine cut. One-third of Pennsylvania's coal production and nearly one-half of Ohio's was machine cut by 1900.[188] In contrast, only 1.5 percent of British coal at this time was undercut by machinery.[189]

	Undercutting Coal By Machines and Picks in the U.S.		
	Percentage by Pick Mining	Percentage by Compressed Air Machines	Percentage by Electric Machines
1900	58.9	22.9	18.1
1905	57.7	22.4	19.8
1910	51.0	19.8	29.1
1915	44.6	9.7	45.5
1920	40.8	5.8	53.3[190]

Coal-cutting machinery was gradually replacing the miner's pick. These machines could undercut more coal in an hour than a pick miner could dig in a day, and their increased use displaced thousands of workers in the larger mines. One compressed-air cutter operated by a pair of workers could undercut four to six typical rooms in a ten-hour shift. This same work required twenty pick miners. As use of the mechanical cutter became more general, undercutting coal by hand ceased to be part of the miner's craft. A new group of mine workers called machine runners was created. The machine runner, along with a helper called a scrapper, took over the undercutting of coal. They were assigned a number of "rooms" by the underground foreman, and with a few loaders were responsible for the drilling, shooting, and loading of coal. Mechanical undercutters generally were not employed during this period by operators of the thousands of small "country banks" mines, where coal was still undercut using a pick.

After the coal was undercut from the seam, miners used an electric drill to bore a hole in the coal face into which a cartridge of dynamite was placed. Attached to the dynamite was a cap and two wires that extended for several feet. The "shot-firer" exploded the cartridge by electricity from a safe distance. After the blast the broken lumps of coal were ready to be picked up either by hand loaders or by a mechanical loading machine. The loaded coal was transported from the mine by animal-drawn carts or by electric locomotives called "electric mules."

The underground mine, besides being a workplace for the extraction of coal, was the site of an extensive and elaborate transportation system. Workers and supplies were conveyed to and from their work areas, and coal and waste materials required daily removal from the mine. A majority of coal was still being hand-loaded into wooden coal cars that held between one and three tons, and hauled by draft animals to the mine entry (head house). A variety of animals—dogs, ponies, goats, oxen, and horses—were used by coal companies to haul loaded coal cars to the surface, although after the Civil War, a majority of mine owners preferred the stubborn but sure-footed mule.

The growth in coal production after 1880 made many mines very large, and the increase in the size of underground mine workings made travel distances greater. This growth in size created a need for improved methods of underground transportion. Large commercial mines had underground workings ranging in size from a thousand to nearly a hundred thousand acres. For example, the Maple Creek Mine of the United States Steel Corporation on the Monongahela River between New Eagle and Monongahela City, Washington County, had underground coal reserves of approximately seventeen thousand acres of Pittsburgh coal seam situated in four townships of Washington County. The Renton Mine mined a hundred thousand acres of coal lands in Allegheny County, while the Shannopin Mine mined seventy-five hundred acres of its nearly thirteen thousand acres of coal reserves in Greene County. With existing mining techniques this reserve represented a twenty-year coal reserve. A variety of conveyances were used in transporting coal from the mines, ranging from baskets to wheelbarrows to coal cars. A variety of draft animals were used to haul coal and refuse, most commonly mules and horses, but also dogs, goats, or oxen, were used in small or thin-coal-seam mines. These draft animals were gradually replaced by electric locomotives, dubbed "electric mules," and other haulage machines. Gasoline, electric, and battery-powered haulage locomotives were all introduced in larger mines for underground haulage. Gasoline engines were installed in Kentucky mines in 1898. The gasoline engine gave off noxious gases, polluted the air, and posed a serious fire hazard, which restricted its widespread use. Electricity was introduced into many larger mines at the turn of the century and changed the work of miners. Electricity was used to operate cutting machines, air fans for ventilation, electric drills, shot firing, motor haulage, underground telephone and signal systems, and the opening

Compressed air locomotive.
Carnegie Library of Pittsburgh.

and closing of ventilation doors. The widespread use of electricity made the bituminous coal industry a more productive industry but it also made the workplace an unsafe environment.

The first electric-trolley haulage locomotives were introduced in 1887 and 1888 and were used in individual mines in Pennsylvania, Ohio, and Illinois. By 1900 larger companies were all using electric or battery-powered locomotives to transport coal, workers, and refuse from the mine to the surface. These locomotives were installed on the main haulage ways of larger mines while mules and mule drivers were still employed to gather loaded coal cars from the individual rooms. The electric-trolley locomotives were not used generally in mines where the air was gassy because of the danger of explosion caused by sparks between the trolley and its wires. When these electric trolleys were used for gathering cars from the working faces, they were equipped with an insulated cable that transmitted the electric current from the trolley line to the motor. The principal advantage of storage-battery locomotives was that unlike electric trolleys, they did not require direct current to operate and could travel underground wherever the track was laid, regardless of connection with an outside source of current. They were used extensively in gassy mines to avoid possible explosions caused by electrical sparks. The chief objection to these locomotives was that they needed to be charged regularly and that, therefore, they were out of service part of the time. The first storage-battery locomotives were employed in the Pocahontas Field, West Virginia. The storage-battery locomotives of 1914 ranged in size from two to ten tons. The introduction of these electric and battery-powered locomotives permitted the introduction of larger-capacity coal cars.[191] It was common practice in mines using both hauling machines in 1920 to employ the electric-engine trolleys with trolley wires for main haulage and the storage-battery locomotives for gathering.

Motorman operating a Jeffrey electric haulage locomotive.
Historical Society of Western Pennsylvania.

Life Underground: Where the Sun Doesn't Shine

Miners were becoming increasingly discontented with their daily lives underground and in the "patch towns" during the last quarter of the nineteenth century. They resented what they perceived as coercive and exploitive practices on the part of management. A majority of mine workers, who were still unorganized, toiled long hours underground for poor pay, often in deplorable and unsafe working conditions. Work in the industry was sporadic, as coal production often far exceeded demand. From 1890 to 1920 miners were idle an average of ninety-three working days each year. They were overworked to fill demands for home-heating coal in the fall and winter, while in the spring and summer there was much underemployment in the industry.[192] A typical miner during this period "worked in water half-way to [his] knees, in gas-filled rooms, in unventilated mines where the air was so foul that no man could work long without seriously impairing his health. There was no workmen's compensation law, accidents were frequent and there was no common ground upon which employer and employee could meet. They had no interest in common as they regarded each other with hostility and distrust."[193] Underground mining of

Miners with sunshine lamps.
Carnegie Library of Pittsburgh.

coal and other minerals by its very nature is performed under potentially dangerous conditions. The dangerous conditions in the mines were caused by poor or inadequate ventilation; use of open flames, especially miners' candle or oil lamps, in gassy mines; inadequately timbered mines; negligence in testing for mine gases; failure to control the accumulation of coal dust; and the improper use of explosives.

Coal mining has historically been a dangerous industry resulting in death and injuries to the men and boys who ventured underground to extract coal. That coal miners grew old before their time was proverbial. The workplace at the face could be from two to twenty feet high depending on the thickness of the seam, although it was usually shorter than the miner. John Brophy, UMWA District 2 president, explained in his memoir, *A Miner's Life,* that "one of the most exhausting things about mine work was the necessity of spending a ten- or eleven-hour day without a single chance to stand erect and stretch."[194] This continuous hunching for ten to twelve hours straight permanently altered a miner's posture. Inhaling stale air, gases and dust, hard and prolonged work in a stooping, strained position were all factors in aging coal diggers. The mine's atmosphere teemed with dirt and dust and miners suffered from a variety of occupational hazards, including black lung (pneumoconiosis) caused by their inhaling "bug dust." This medical ailment was referred to as "miner's asthma" or *phthisis pulmonalis nigra*. This ailment snatched the miner's breath away, leaving him to sit up at night gasping, unable to walk or lie down. Even after scientific evidence proved conclusively that breathing coal mine dust caused irreversible respiratory disease including black lung, coal companies denied the connection. Some operators maintained the inhaling of coal dust was good for miners. Tens of thousands of miners died or suffered from "miners" asthma while the industry and government ignored the horrible consequences of this debilitating occupational disease. Another seven decades would pass before pneumoconiosis was recognized by federal law as an occupational disease in the coal mining industry. The federal government provided black lung compensation for coal miners and their dependents following the passage of the federal Coal Mine Health and Safety Act of 1969.

The physical description of a miner in an 1886 issue of the *Union Pacific Employees Magazine* vividly describes the toll long-term mining had on the miner's body:

> Look at the man forty years of age, that has dug coal all his life is a deformed wreck, physically, if not mental. . . . Look at the number of miners with broken bones; the number with burns; with stooped shoulders; with weak and impoverished blood; with rheumatic pains from working in water; with affected lungs from working in bad air. A physically sound man fifty years of age, who has dug coal all his life, is almost impossible to find.[195]

Coal operators, like miners, had accepted the fact that mining was a dangerous and potentially unhealthful occupation throughout most of the nineteenth century. Miners were regarded as skilled independent contractors employed by the coal company and hence were responsible for their own safety and welfare while working at the mine. Most miners believed that coal operators evaded any responsibility in making coal mining a safer industry by simply dismissing all accidents and explosions as the result of employees' carelessness, neglect, or ignorance. Coal operators told state coal commissions that accidents in the industry were caused by miners who took foolhardy risks or imprudently attempted tasks that were surrounded by known elements of danger. Since they were paid by tonnage, many miners were negligent in assuring that the mine was safe. Instead of securing the roof and checking for dangerous levels of gases they immediately began mining coal. The Bureau of Industrial Statistics in Pennsylvania accepted the position of the coal operators. The agency asserted in 1908 that "more than half the accidents reported (in the industry) are chargeable to such causes." During this period management was lax in promoting any systematic safety programs in their mines in the face of a growing accident toll. Mine safety was given little consideration by most coal companies, and this callous indifference regarding the miners' health and welfare was clearly stated by a contemporary miner who angrily observed, "If I sell my labor to produce profits for my master at a bare existence for myself and my family, my master is responsible for any accidents that deprive me of the power to produce profits. I have given my all when I sell my strength."[196] A coal operator clearly expressed the attitude of a majority of coal operators when he told the Ohio Mining Commission that "the miner is free and can protect himself for he can engage in mining or not."[197] Operators regarded any government attempts to regulate mine safety as "unnecessary and unwarranted interference with the business of coal mining."[198] The UMWA, at its first annual convention held in 1891, passed a resolution demanding regular inspections of all mines by a state or federal government agency and the passage of legislation for "the adoption of a law, with heavy penalties attached, holding employers of labor liable for any and all accidents that may happen to their employees while in the line of service."[199]

Nearly fifty thousand miners died in American mining accidents from 1870 to 1914. This forty-year period was the most hazardous in the history of American mining. Fourteen states had experienced at least one major mine explosion by 1890. The expansion in production, the increased size of the work force, and the introduction of mechanical machinery and electricity were all factors contributing to the higher fatalities in the industry. Some miners accused the new, unskilled immigrants entering the industry in large numbers after 1880 of causing the rising fatalities. Their ignorance and inexperience as miners, according to their critics, contributed to the high accident rates in the mines. The new immigrant miners, unlike the earlier miners from Great Britain, had less mining experience abroad. The presence in the mines of skilled Scotch, English, Irish, and German miners was declining rapidly throughout this era as they were replaced by these less skilled workers. These original miners regarded the new immigrants as inferior and spoke of them in derogatory terms, dismissing them as "those ignorant foreigners." The UMWA district president of Pittsburgh in 1902, after the Rolling Mill Mine disaster in Johnstown, accused mine operators of neglect for hiring these unskilled immigrant miners.

> As long as they import foreigners by the hundreds, dump them into the mines without any instruction or training, and run along the theory that the mine boss, who gets his certificate from the State, is wholly responsible for the lives of hundreds underground, we will be greeted every few months by news of . . . appalling disasters.[200]

The United Mines Workers Journal felt they were justified in concluding that "the peril of the mines lies in the immigrant ship."[201]

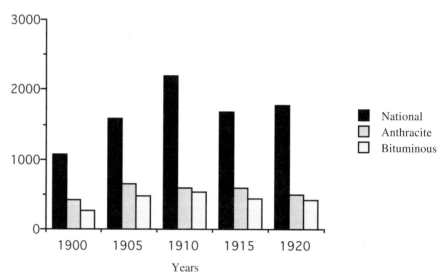

Most of these mine deaths were in ones or twos, although the spectacular underground mine explosions received the most media attention and reaffirmed to the general public the inherent dangers of the industry.[202] The major mine explosions of the era were caused by mine gases that came into contact underground with open flames. These explosions originated from the ignition of two materials commonly found in all coal mines—methane (CH_4) and the dense clouds of "bug dust" suspended in the atmosphere. Fine coal dust was extremely volatile and was produced by electric undercutting machines as they pulverized coal at the working face, and by the blasting of coal from the seam with dynamite or black powder.[203] Fine-powdered limestone dust, called rock dust, was later applied to the mine's roof ribs and floor as a safety measure to render explosive coal dust inert. Rock dusting allowed for better illumination in the mine. The widespread practice of rock dusting was instituted in most larger mines during the 1920s.

Open lighting was once universal in most mines and caused many explosions underground. The miners' open-flame sunshine lamp and the use of candles constituted a serious hazard in gassy mines. The acetylene lamp, called the "carbide lamp" by miners, made its appearance in the mines at the turn of the twentieth century. This lamp was a significant improvement as a source of underground illumination over candles and oil lamps. The carbide lamp burned with a light brighter than these earlier means. The carbide lamp, attached to the miner's canvas hat, was made of brass and composed of two parts. The lamp's upper chamber was filled with water, which was fed by capillary action into the bottom chamber containing carbide pellets (calcium carbide, CaC_2). The water acted on the carbide pellets producing acetylene gas, which was ignited by striking a serrated wheel acting upon a flint to produce a spark. Acetylene, C_2H_2, is a pungent gas formed by the action of water on calcium carbide. The carbide lamp burned for about four hours with a bright yellow-white flame that could be adjusted from one to three inches. The three major manufacturers of carbide miners' lamps were Arras of France, Seippel of Bochum, Germany, and the Wolf Company, with manufacturing facilities in England, Germany, and the United States.[204]

The carbide lamps were replaced by electric-battery lamps to reduce fire hazards from the lamp's open flame. Electric miners' lamps had been introduced by 1915 in some larger mines. The Edison battery lamp was one of the most popular models of the era. The lamp was attached to the front of the miner's cap, with a wire leading to a battery secured to a belt around the miner's waist. The electric lamps were charged overnight by miners or the lamp man at the lamp house. Although this electric lamp was safer because it was not an open flame, as late as 1940 only 53 percent of all miners nationwide were using them.[205]

Pennsylvania did not begin compiling accurate mine accident statistics until 1870. The Department of Mines of Pennsylvania has issued annual reports on the bituminous and anthracite coal industry since 1877. The agency routinely compiled such mining statistics as annual production, employees, fatalities (including the ethnic origin of the mine worker), production per fatality, fatalities per thousand employees, and fatalities per million tons produced. These grim calculations of human life were regarded as simply production variables by mine owners and the agency. A "major" disaster was defined by the agency as one in which five or more workers were killed. There were sixty-four major mining accidents in the bituminous coalfields of Pennsylvania from 1884 to 1945, using this arbitrary criterion.[206]

Five of these mining accidents claimed more than one hundred lives:

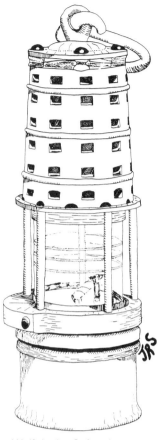

Wolf Junior Safety Lamp.
Mining Artifact Collector.

Interior of a lamp house with safety gas lamps.
Pennsylvania State Archives.

Mammoth Mine, Mt. Pleasant, Westmoreland County (1891), 109 killed
Rolling Mill Mine, Johnstown, Cambria County (1902), 112 killed
Darr Mine, Jacobs Creek, Westmoreland County (1907), 239 killed
Marianna Mine, Marianna, Washington County (1908), 154 killed
Cincinnati Mine, Finleyville, Washington County (1913), 113 killed

Fifty-four of these sixty-four mining accidents were caused by explosions of gas or dust; the remaining accidents were caused by roof falls, faulty mine cars, mine fires, and workers falling down mine shafts. From the founding of the UMWA in 1890 until 1907, over twenty-six thousand men died in coal mining accidents. Nationwide, 1907 was the worst year in the history of coal mining: 3,242 miners perished and tens of thousands were

Principal Bituminous Coal Mining Disasters in Pennsylvania between 1884 and 1945

Date	Mine	Operator
1884-Feb 20	West Leisenring	Connellsville C. & C. Co.
1885-Oct 27	Youngstown	Youngstown Coke Co.
1888-Nov. 3	Kettle Creek	Kettle Creek Coal Co.
1890-June 30	Hill Farm	Dunbar Furnace Co.
1891-Jan. 27	Mammoth	H.C. Frick Coke Co.
1897-Mar. 27	Berwind	Berwind-White Mining Co.
1898-Sept. 23	Umpire	Umpire Coal Co.
1899-July 24	Grindstone	Redstone Coal & Coke Co.
1899-Dec. 23	Sumner	Pittsburgh & Erie Coal Co.
1901-June 10	Port Royal No. 2	Pittsburgh Coal Co.
1902-March 6	Catsburg	Mon River Consolidated C. & C. Co.
1902-July 10	Rolling Mill	Cambria Steel Co.
1903-Nov. 21	Ferguson	Dunbar Furnace Co.
1904-Jan. 25	Harwick	Allegheny Coal Co.
1905-April 27	Eleanora Shaft	Rochester & Pittsburgh Coal Co.
1905-July 6	Fuller	Taylor Coal Co.
1905-Oct. 13	Clyde	Cylde Coal Co.
1905-Oct. 29	Hazel Kirk No. 2	Pbgh. & Westmoreland Coal Co.
1905-Nov. 15	Braznell	Braznell Coal Co.
1906-Oct. 24	Rolling Mill	Cambria Steel Co.
1907-Aug. 17	Sonman Shaft	Sonman Shaft Coal Co.
1907-Dec. 1	Naomi	United Coal Co.
1907-Dec. 19	Darr	Pittsburgh Coal Co.
1908-Nov. 28	Rachel/Agnes	Pittsburgh & Buffalo Co.
1909-Jan. 25	Orenda	Merchants Coal Co.
1909-April 9	Eureka No. 37 Upper	Berwind-White Mining Co.
1909-June 23	Lackawanna No. 4	Lackawanna C. & C. Co.
1909-Oct. 31	Franklin No. 2	Cambria Steel Co.
1910-Feb. 5	Ernest No. 2	Jefferson & Clearfield C. & C. Co.
1911-March 22	Hazel	Pittsburgh-Buffalo Coal Co.
1911-July 25	Sykesville Shaft	Cascade C. & C. Co.
1911-Nov. 9	Adrian No. 1	Rochester & Pittsburgh Coal Co.
1913-April 23	Cincinnati	Mon River Consolidated C. & C. Co.
1915-May 24	Smokeless No. 1	Smokeless Coal Co.
1915-July 30	Patterson No. 2	United Coal Co.
1915-Aug. 31	Orenda No. 2	Merchants Coal Co.
1916-Feb. 11	Ernest No. 2	Jefferson & Clearfield C. & C. Co.
1916-March 30	Robindale	Conemaugh Smokeless Coal Co.
1917-March 13	Henderson No. 1	Henderson Coal Co.
1918-Aug. 7	Harmar	Consumer Mining Co.
1920-June 2	Ontario	Ontario Gas Coal Co.
1920-July 19	Renton No. 3	Union Collieries Co.
1920-Aug. 9	Maryland Shaft	Maryland Coal Co. of PA
1922-Feb. 2	Gates No. 1	H.C. Frick Coke Co.
1922-March 20	Dilltown No. 1	Dilltown Smokeless Coal Co.
1922-Nov. 6	Reilly No. 1	Reilly Coal Co.
1924-Jan. 26	Lancashire No. 18	Lancashire Coal Co.
1924-July 25	Gates No. 1	H.C. Frick Coke Co.
1925-April 26	Hutchinson	Westmoreland Coal Co.
1926-Feb. 3	Pgh. Terminal No. 4	Pittsburgh Terminal Co.
1926-Aug. 26	Clymer No. 1	Clearfield Bituminous Coal Co.
1927-April 2	Mine No. 53	Ellsworth Collieries Co.
1928-Feb. 20	Kinloch	Valley Camp Coal Co.
1928-May 19	Mather Collieries	Picklands, Mather, & Co.
1928-Aug. 9	Hillside	Tunnell Smokeless Coal Co.
1928-Aug. 15	Irvona No. 3	Irvona C. & C. Co.
1929-March 21	Kinloch	Valley Camp Coal Co.
1933-Sept. 11	Oakmont	Hillman C. & C. Co.
1937-March 28	Kramer	NW Mining & Exchange Co.
1938-Jan. 12	Harwick	Harwick C. & C. Co.
1940-July 15	Sonman E	Sonman Shaft Coal Co.
1941-June 30	Kent No. 2	Rochester & Pittsburgh Coal Co.
1944-June 7	Emerald	Emerald C. & C. Co.
1945-March 12	Crucible	Crucible Steel Co.

injured. Of this national total 1,614 miners perished in Pennsylvania alone (806 in the bituminous coal industry and 708 in the anthracite industry). Five mine disasters in 1907 alone claimed a total more than eight hundred victims. The Darr Mine of the Pittsburgh Coal Company, located in Rostraver Township near Van Meter, Westmoreland County, on the Youghiogheny River, was the site of the deadliest coal mining disaster in Pennsylvania history and the fourth-worst mine disaster in the nation. Miners who worked at the Darr Mine lived in the surrounding communities of Van Meter, Jacobs Creek, Banning, and Wickhaven. On December 19, 1907, 239 miners were killed as the result of an explosion caused by an accumulation of gas and dust. The Monongah disaster of West Virginia was the worst mine disaster in American history. This occurred on December 6, 1907, "Black Friday" at Monongah Mines Six and Eight near Fairmont, West Virginia, when 360 men were trapped underground and a twelve-year-old trapper boy was killed at the surface.

Child labor was routinely used in bituminous coal mines. Boys, as young as seven, worked underground as runners, mule drivers, trappers, and couplers. Some coal-producing states enacted laws prohibiting children younger than twelve (later raised to fourteen) from working in the mines. These laws were rarely enforced. Pennsylvania coal operators were fined on average twenty-three cents for violating child labor laws in 1904. The low wages paid miners often forced parents to send their sons underground out of economic necessity. Federal legislation was passed in 1938 banning child labor in "hazardous occupations." The UMWA won a contractual agreement from coal companies in 1941 requiring a minimum age of eighteen to work in a mine.

Mining accidents caused by an accumulation of unsafe levels of gases (usually methane) or mine dust were shocking and sensational events and attracted public attention; however, explosions were not the leading cause of death and injury to coal miners, accounting for only a small percentage of the deaths in the industry. In Pennsylvania 1,954 mine workers were killed in sixty-four mining disasters between 1884 and 1945. Accidents caused by roof and rib falls, haulage accidents, and dangers from electricity posed greater daily health hazards to underground miners than death from the explosion of volatile mine gases or coal dust. The principal hazards to surface workers related to transportation, machinery, electricity, and falls. From 1913 to 1922 accidents killed 18,243 miners nationally at bituminous coal mines, a death toll of 4.3 per thousand full-time workers per year. The percentages of fatal accidents during this single decade were as follows: collapse of the roof and coal falls 50 percent, mine cars and locomotive related 18.1 percent, gas and coal dust explosions 12.2 percent, surface accidents 6.0 percent, electricity related 4.2 percent, explosives 3.7 percent, other causes 3.6 percent, and falling into the shafts 1.9 percent.[207]

The United States Geological Report of 1907 undertook a comparative study of mining accidents, casualties, and mine safety in the American and European coal industries. The study concluded that the death rate from mining accidents in the United States was three times greater than that in England and Belgium, twice the rate in Prussia, and almost four times the rate in France. American mines, according to the report, were inherently no more dangerous than contemporary European mines. The lower accident and fatality rates in Europe were due to safety legislation and rigorous enforcement. John Mitchell of the UMWA accepted the Geological Report's conclusion that "inadequate legislation and law enforcement more than any single cause was the essence of the safety issue."[208] Safety conditions and mining fatalities varied throughout the coal industry. He compiled fatality rates in the industry from 1907 and observed a strong statistical correlation between the number of fatalities and the extent of unionization. He observed 2.47 fatalities per thousand miners in unionized states, 5.07 fatalities per thousand miners in partially unionized states, and a phenomenal 9.49 fatalities per thousand miners in nonunionized coal-producing states.[209]

The human toll exacted by the coal mining industry during this era was staggering. The rise of underground explosions and increased fatalities and injuries

Mining Fatalities in European and Pennsylvania Mines Per 1,000 Employees		
Nation/State	1891	1897
Austria	2.54	.95
Belgium	1.40	1.03
Germany	2.80	2.27
England	1.50	1.32
France	1.67	1.07
PA (Anthracite)	3.08	2.84
PA (Bituminous)	3.18	1.72 [210]

prompted action which was brought about by union and public pressure. Some companies had unilaterally implemented their own safety measures by hiring their own mining inspectors. The prospect of losing their entire capital investment in a single explosion encouraged this new emphasis on mine safety. The H. C. Frick Coke Company, a subsidiary of the U.S. Steel Corporation, began a safety program at its numerous facilities with the "catchy" phrase, "Safety First, Quality Second, Cost Third." It trained rescue and first-aid teams who practiced continually, honing their skills. The company also posted signs in English and five other languages throughout the mines and at the surface, warning its workers of potential dangers.

Mining laws had been enacted by state legislatures of coal-producing states, beginning in the 1870s, and by 1920 an uneven collection of mine-safety laws existed across the coal-producing states. These laws were generally ineffective because they lacked meaningful enforcement provisions. Illinois and Pennsylvania, birthplaces of the nation's first miners' unions and the strongholds of the growing UMWA, were acknowledged as having the best safety laws in the industry, while West Virginia, largely nonunion at the time, was considered "the worst of the lot." State mining agencies with coal mine inspectors had been established by coal-producing states. The bituminous coal mining inspection force was created by a legislative act enacted in 1877 by the General Assembly of Pennsylvania. The act provided for the appointment of three mining inspectors by the governor. The bituminous coal region of Pennsylvania comprised thirty counties divided into three districts as nearly equal in size as possible. Five of these counties (Crawford, Erie, Forest, Potter, and Warren) eventually ceased to produce coal, while three other counties (Bedford, Greene, and Somerset) became active producers.[211] Pennsylvania's mine inspectors were directed to visit and inspect the state's mines to insure the safety of workers and mining. Each inspector was empowered by law to enter and inspect the mine and its machinery at all reasonable times, day and night. He was also charged with determining the causes of mine accidents and to do this was permitted to conduct investigations and issue subpoenas. Each inspector was independent, acting according to his own judgment, since there was no immediate superior to see if he was enforcing the laws. He was required by law to report the results of his inspections to the Secretary of Internal Affairs.[212] Moreover, Pennsylvania's state legislature granted state mine inspectors in the bituminous region a life tenure in 1915. The act was not applicable to anthracite-mine inspectors, who were still required to undertake a periodic examination administered by the state. As Pennsylvania's coal industry expanded, the number of mining inspectors was increased, from three inspectors between 1878 and 1880 to thirty inspectors after 1915. A long-standing complaint of some miners was that state mine inspectors were political appointees and therefore subject to operators' influence resulting in lax law enforcement.

Electric undercutting machine and trolley locomotives. Historical Society of Western Pennsylvania.

The United States Bureau of Mines (BOM) was established by an act of Congress on July 1, 1910, as a result of the rash of serious mining tragedies during the first decade of the twentieth century and especially in 1907, when the coal industry nationally recorded an appalling 3,242 deaths.[213] The agency's mandate from Congress was "the investigation of the methods of mining, especially in relation to the safety of miners."[214] The new bureau had no real power beyond gathering information. Joseph A. Holmes, the first director of the United States Bureau of Mines, undertook numerous experiments to reduce mine accidents and fatalities, especially those explosions caused by coal dust and the improper use of explosives in the mine. Mine dust was a serious hazard, especially in winter when the air and the mine were dry. A variety of explosives were used in mines without much regard to safety.

The Experimental Mine and Explosive Testing Center was constructed by the United States Bureau of Mines on a large tract of land near Bruceton, thirteen miles from Pittsburgh on October 11, 1911. A variety of safety measures instituted by the agency included the use of explosives and electrical equipment that had passed certain safety tests and been approved by the U.S. Mine Safety and Health Administration.[215]

The United States Bureau of Mines noted that the high death rate in American mines was not due entirely to the hazards inherent in the industry. The agency estimated conservatively that at least one-half of all deaths could be prevented if adequate precautions were undertaken by the coal operators. The fatalities per thousand full-time workers in American and European mines by 1920 indicated that American mines were still less safe than European mines. The average rate of 4.08 deaths per thousand full-time workers in the United States was more than three times the rate of 1.13 deaths per thousand in British mines. Coal mining was still one of the nation's most hazardous occupations. Nearly two thousand miners were killed annually, while injuries ran between fifty and one hundred fifty thousand per year during the 1920s.[216] An estimated seventy-two thousand more miners would die in accidents until the passage in 1952 by Congress of the next significant mine safety law. The federal Coal Mining Act for the first time established minimum safety standards that coal companies were required to meet. The BOM inspectors had been permitted to inspect mines, but until the passage of this act could only recommend safety practices to operators and could not enforce safe mining practices.

Daily Life in the Company-Owned Town

The coal companies played a dominant role in the everyday life of the mining community. Each company had substantial control over its employees and their families' personal lives in a manner virtually impossible in any other type of community in the nation. Miners were always vulnerable to pressures from the company and were increasingly unhappy with this control over all aspects of their daily lives, both in the "patch" town and at the mine. The company town, company store, and nonlegal tender (scrip) were all introduced as practical necessities, but these necessities over time were used by some coal companies to gain economic and political control of their isolated and captive labor force. Muriel Earley Shephard writes in her classic study of daily life in the Connellesville coke district, *Cloud by Day*, "They lived on feudal islands in the county but were not of it."[217] Neither the miner nor his family had the same rights or protection accorded most American citizens under the Constitution. Since many coal towns were rural and unincorporated villages, all political power was controlled by the coal companies which owned and operated the mining village as their private fiefdom. A former miner recalled this absolute political control held by the mine superintendent, noting that "he [the coal company superintendent] was mayor, council, big boss, sole trustee of the school, truant officer, president of the bank, in fact he was everything."[218] The miners who rented their houses in these villages had few tenant's rights. They rented their houses on the condition that they sign a lease agreement which could usually be terminated with only five days' notice. The lease permitted the company to evict them automatically when they ceased for any cause whatsoever to work for the coal company. The company constable needed no search warrant to enter the house at any time. Miners who rebelled or went on strike risked losing their homes. Striking miners and their families were routinely evicted from their rented houses by the coal companies. Strike activities caused so many evictions of workers and their families between 1898 and 1919 that the UMWA was forced to spend more than $16 million for their relief by constructing temporary housing, either barracks or tents, and providing food.[219] The United States Department of Labor concluded in 1917 that "a housed labor supply is a controlled labor supply."[220] Scholars have characterized this economic and political control exerted by the coal company as "a great anomaly in the midst of the free country."[221] Priscilla Long, author of *Where the Sun Never Shines,* observes that "in the land of the free the company town was isolated, remote, and anything but free."[222]

A Tioga County miner complains in 1882 that "the [mine] superintendent makes all the laws and those who are in his employ must abide by them."[223] The coal company employed a variety of private police to enforce its laws, including the dreaded Coal and Iron Police of Pennsylvania.

The monopoly held by the company in housing was duplicated by the ubiquitous company store with its notorious reputation for high prices, high interest rates, and inescapable debt caused by the extensive use of scrip. Most miners disliked the company store, derisively calling it "the pluck me" or "grab-all" store.[224] Miners resented the store for three principal reasons: prices were usually inflated in comparison to prices charged by independent retail stores; their supplies were deducted from their wages, called the "check-off," before they received their pay; and trading with the store was often compulsory. If a miner did not purchase sufficient merchandise from the store, he was punished with immediate dismissal, or assigned one of the dangerous or poorer-paying jobs underground. Store managers hired by the coal companies were expected to make a handsome profit. A miner from Bradford County told a state commission investigating alleged abuses in the mining industry in 1882 that "the employees in and about the mines where I work are compelled to deal in the company store, and have to pay a very high price for their goods." An Allegheny County coal miner told an investigator from the same commission that "if we do not deal in the company store we are not wanted at the mine, and are given a poor place to work. The company store system is a blot on the liberties of this country, and should be the concern of all whether in or out of the mine."[225]

Employees of company store operated by Pittsburgh Coal Company. Carmen DiCiccio.

Prices of Goods in Company Store and Independent Retail Store

Foodstuff	Company store	Independent store
Flour per barrel	$8.00	$7.75
Corn Meal	$1.50	$1.25
Butter per pound	$0.35	$0.30
Bacon per pound	$0.40	$0.10
Ham per pound	$0.13	$.11 -.12
Cheese per pound	$0.20	$.16 -.18
Tea per pound	$.60 - 1.00	$.25 -.75
Coffee per pound	$.28 -.37	$.25 -.30[226]

The Pennsylvania Bureau of Industrial Statistics conducted a survey in 1883 showing prices consistently higher in company stores than in neighboring stores. The practice of exploiting miners with excessive prices at the company store was not limited to the coal companies of Pennsylvania. The General Assembly of Ohio appointed a committee to investigate a miners' strike in the Hocking Valley in 1884 reported exploitive practices similar to those encountered in Pennsylvania.

The United States Census Bureau conducted an investigation of the coal industry in western Pennsylvania after a series of particularly violent strikes in 1910-1911. The study revealed that "many mine operators make a considerable profit by renting houses and selling merchandise to their employees."[227] The study investigated the earnings of the coal companies. It found that store earnings accounted on average for about one-quarter of the companies' total annual revenues.

Scrip was a certificate issued by an employer in lieu of cash wages, usually redeemable only for goods and supplies at the company store. In the coal industry the use of scrip was a controversial practice of the company. Its use was connected with the extension of credit, which often kept employees perpetually in debt to the coal company. Some companies paid their workers in coal scrip instead of legal tender. Scrip was contemptuously called "bogus" money, "flickers," "drag," "clacker," and "chicken feed" by discontented miners. The first scrip was issued in paper coupon form, hence the name. Scrip was later issued in brass, copper, and aluminum coins representing nickels, dimes, quarters, and half-dollars, and as paper money corresponding to legal-tender denominations. Rural isolation and poor roads made it difficult to transport money on a regular basis to pay miners, and so scrip was introduced as a substitute. It was first widely used during the 1880s and 1890s, in the isolated bituminous mining communities of West Virginia and Kentucky. Mine owners, however, did not give up its use when transportation improved and the money supply increased. Although West Virginia and Kentucky were the states in which scrip was used most extensively, mining companies in Virginia and western Pennsylvania also used it. The H. C. Frick Coke Company issued scrip, called "Frick Dollars," during the Panic of 1873 when cash to pay their employees was scarce in the Connellsville district. This practice was later extended in western Pennsylvania, especially by coke operators in Fayette County. There were at least sixty mines in Pennsylvania issuing scrip to their workers as a substitute for legal tender during this period.[229]

Coal Company Profits from Coal and Store			
Company	Coal Sales	Store Sales	Percent of Earnings
Keystone C&C Co.			
Arona	$331,126.88	66,566.90	20.1
Claridge	$206,528.67	56.694.34	28.4
Keystone Shaft	$250,903.38	50,938.00	20.3
Jamison C&C Co.			
Jamison No. 1	$304,576.22	91,445.21	30.0
Jamison No. 2	$448,611.59	95,071.51	21.3
Jamison No. 3	$244,391.69	83,178.19	34.0[228]

The use of coal scrip, like company housing and the company store, began as a necessity, but over time it became an abomination to miners and their families. The miner or a family member could draw scrip in advance of payday by visiting the payroll office, often located at the company store, to see if the company owed the miner money; if so, scrip was issued and accepted at the company store for food, clothing, mining supplies, and other essentials. The store manager would subtract these charges, called a "check-off," from the miner's take-home pay. Many families ran out of money before payday. Some company-store managers would encourage workers to make purchases on credit at the store. This debt was deducted from future wages, so by payday miners' families owed money to the company. Some debt-ridden miners on payday found three x's on their pay stub, meaning their wages had gone directly to pay off their store debt. Use of the check-off was a hardship because it forced families to shop exclusively at the company store, with its higher prices. It created a vicious cycle of continual debt and poverty. Often, miners were prevented by their debts from quitting their oppressive jobs in search of alternative employment in the city, or at a neighboring mine. Some retail stores near the mine accepted coal scrip, carrying large signs in their stores stating, "We cash scrip but at a discount." There were two sets of prices in these independent stores, one for cash and the other for scrip, with the latter discounted from 10 to 15 percent. Some unscrupulous shopkeepers discounted it by as much as 40 percent and took the discounted scrip to the company store and redeemed it for cash at face value. Proponents of the scrip system argued that it enabled miners to supply the necessities of life for their families without asking for the charity of friends and neighbors.

Coal companies' practices of inflating store prices, obligatory patronage of their store, and the use of scrip produced alienation and acrimony. Discontented miners did not passively accept these clearly exploitive practices. They staged wildcat strikes accompanied by periodic violence that prompted numerous state and federal investigations. In an attempt to minimize this continual exploitation of mine workers by coal companies, the General Assemblies of both Pennsylvania and Illinois passed statutes in 1891 prohibiting the ownership of the company store by the mining company. These statutes, however, were so loosely written that coal opera-

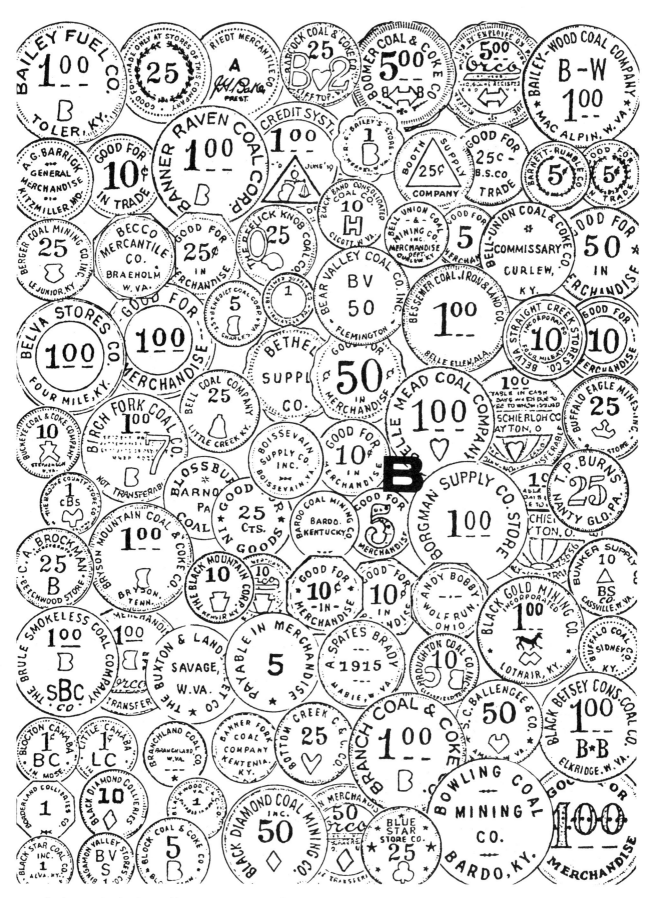

Scrip metal coins issued by coal companies. Scrip was also issued in paper denominations. Gordon Doodril.

tors were easily able to circumvent them. Pennsylvania law forbade mining companies to operate stores at their mines, but the law did not prevent them from organizing subsidiary companies whose stock was owned by the coal company. The Union Supply Company of the H. C. Frick Coke Company was established as a separate legal entity from the company's mining and coking operations. The firm operated sixty-three stores in Fayette and Westmoreland Counties in 1900.

Drive to Organize and the Founding of the United Mine Workers of America

Coal Companies and the Subsidiary Companies That Operated the Company Stores

Coal Company	Supply Company
Pittsburgh Coal Company	Federal Supply Company
Westmoreland Mining Company	Kiski Supply Company
W. J. Rainey Company	Rainey Supply Company
Pittsburgh Termininal RR and Coal Co.	Mutual Supply Company (Allegheny County)
	Sumner Supply Co. (Washington County)
Bethlehem Mines Corporation	Services Stores Corporation
H. C. Frick Coke Co.	Union Supply Co.
Berwind-White Co.	Eureka Stores Co.
Rochester & Pittsburgh Coal Co.	Mahoning Supply Co.
Jamison Coal & Coke Co.	Underwood Supply Co. (West Virginia)
	Hannastown Supply Co.[230]

The deteriorating conditions in the mines and life in the "patch" towns during the last quarter of the nineteenth century made bituminous coal miners increasingly angry and frustrated. Miners and their sons worked long hours underground at low wages, with little hope of any meaningful improvement in their status. Mary "Mother" (nee Harris) Jones (1830-1930), the charismatic and legendary labor organizer and friend of coal miners, repeatedly enjoined them "to pray for the dead and fight like hell for the living."[231] She was a brave and outspoken union organizer who was known to her antiunion enemies as "the most dangerous woman in America" and to coal miners as the "miner's angel" and the "Joan of Arc of American labor." Mary Harris Jones was born in Cork, Ireland, and immigrated as a child with her family to the United States via Canada during the 1840s. She worked as a school teacher in Michigan and was a proficient dressmaker in Chicago until her marriage to George Jones, an iron molder, in 1861. Her husband and four children died in a yellow fever epidemic in Memphis, Tennessee, in 1867. She became active as a lecturer for the Knights of Labor during the depression of the 1870s and as a Populist agitator during the 1890s. Mother Jones supported the miners in their strikes and protests against coal companies over low wages, long hours, and unsafe working conditions throughout the coalfields of Colorado, West Virginia, and western Pennsylvania from the 1890s until her death. She told coal miners that "I am not looking for office, I am looking for your interests and your children's interest." She celebrated her hundredth birthday on May 1, 1930, and died of old age on November 30. She was buried in a miners' cemetery at Mount Olive, Illinois. Several years before her death she asked to be buried at Mount Olive with the Virden martyrs. She wished to "sleep under the clay with these brave boys."[232] These coal miners were massacred in a violent union-management battle at Virden, Illinois, in 1898. The Chicago-Virden Coal Company imported 180 African American strikebreakers on October 10, 1898. The next day some forty miners were wounded and seven were killed during a bloody battle with hired strikebreakers and private police in Virden. The Progressive Miners erected an imposing granite monument to her memory that was formally dedicated on October 11, 1936. An estimated fifty thousand miners attended the dedication and stood in silent tribute to the "miners' angel."

Miners were extremely militant workers and were involved in some 709 strikes nationally between 1887 and 1894. These strikes involved 675,128 strikers, representing about 80 percent of the entire labor force.[233] Terence V. Powderly, after being ousted as president of the Knights of Labor in 1893, remarked why he believed coal miners were such militant workers able to endure lengthy strikes:

> In strikes, coal miners have always shown the most sublime fortitude and greatest endurance. From all I have witnessed, and from my studies of strikes, I am impelled to say that the miner can endure hunger and privation

until the front of stomach and his spinal column are about ready to lean on each other for support. The reason for this is that it requires the most heroic type of manhood to seek a living in the mine, and he who has the courage to make that step in the dark each day, which every miner does, must be made of the good stuff, must be endowed with great patience and capable of enduring privation and want.[234]

Miners had organized themselves in a number of unsuccessful labor organizations since the formation of the Bates Union, Schuylkill County, in the 1840s. Coal miners were represented by two national and competing labor organizations during the 1880s. The Knights of Labor had a broad-based membership policy that made it an eclectic labor organization. Farmers, small merchants, artisans, and all types of manual, semiskilled, and skilled workers were all welcomed into local assemblies. President Powderly placed little stress on immediate demands for wages and hours and generally opposed strikes; instead under his leadership the Knights advocated producers' cooperatives, trust regulation, currency reform, and the abolition of child labor. Powderly hoped to establish amicable labor-management relationships on a basis of cooperation and arbitration. Some miners departed from the Knights during the 1880s because they felt the Knights could not meet their particular social and economic demands. They wanted to establish a distinctive miners' union that would address what they believed were labor conflicts unique to their occupation. They were also unhappy with the Knights of Labor's preference for arbitration and were dismayed by Powderly's opposition to strikes. Disenchanted miners from Illinois, Indiana, Ohio, Pennsylvania, West Virginia, Iowa, and Kansas met at Indianapolis on September 12, 1885. They formed the National Federation of Miners and Mine Laborers, later renamed the Miners' National Progressive Union, in 1888. The union subsequently joined the new American Federation of Labor, founded in 1886 with Samuel Gompers its first president. The National Trade Assembly Number 135 was formed by the Knights as a rival miners' union the following spring. Both organizations aggressively competed to recruit unorganized miners. The two unions fought four years of disastrous internecine warfare for the allegiance of miners. The coal operators benefited from their bitter rivalry. The mine owners knew that it was only a matter of time until the disputes among the miners would destroy both organizations. The Miners' National Progressive Union represented ten thousand miners and had $139,000 in its treasury, while the District Association had fifteen thousand members, most of whom were workers in the coke fields of Pennsylvania.[235] Officials of the rival miners' unions had realized by 1889 that their continuous bickering was destructive to the general welfare of miners and worked only in the operators' interest. John McBride, president of the National Progressive Union, noted in December 1889: "[T]he discordant and demoralized state our forces were in, together with their weakness financially, seemed to court the destruction of conciliatory methods and invite a conflict with operators which could not but end in loss and disaster to us."[236] The worsening conditions in dealing with hostile coal operators prompted delegates from the competing unions to meet in Columbus, Ohio, in January 1890 to put an end to this disastrous infighting. Bitter and acrimonious disputes ensued among the 325 delegates during the next few days, but out of this conflict the delegates were able to set aside their long-standing differences and formed the "one and indissoluble" United Mine Workers of America on January 25, 1890.[237]

The new union had about 20,000 members out of some 255,000 miners and mine laborers nationwide. Anthracite miners and mine laborers were not organized by the UMWA until the strikes of 1900 and 1902, when John Mitchell convinced them to join the miners' union.

The UMWA functioned as an AFL affiliate although the UMWA charter had provision for at first allowing its two founding organizations to retain some of their essential features and administration. The UMWA was an industrial union embracing both skilled and unskilled workers, unlike Gompers's American Federation of Labor, which was organized exclusively as a craft union representing workers by trade or craft rather than industry. UMWA leaders, like John Mitchell, fought against the craft union principle as the exclusive basis of organization, and established the right of mine workers to organize as an industrial union. The UMWA, like the Knights, opened its doors to all workmen regardless of skill, nationality, or race, from the youngest trapper-boy to the skilled and experienced machine undercutter. The UMWA affiliated with the A.F.L. in 1890 and with the Knights of Labor until the miners union severed all ties with the fading Knights at their 1895 annual convention.

The delegates adopted the credo of "United We Stand, Divided We Fall." The miners' constitution has been frequently amended but the preamble stands as it was adopted in Columbus, Ohio, on January 25, 1890, asserting:

> There is no truth more evident than that without coal there could not have been such marvelous social and industrial progress as makes present-day civilization. Therefore, we have formed the United Mine Workers of America, for the purpose of the more readily securing the objects sought, by educating all the mine workers in America to realize the necessity of union of action and purpose, in demanding and securing by lawful means, the just fruits of our toil.[238]

The newly unified union elected John B. Rae of central Pennsylvania its first president; W. H. Turner, of Ohio, vice president; and Robert Watchorn, of Pennsylvania, secretary-treasurer. The salaries of the officers were as follows: for president $1,000 per year, vice-president $900, and secretary-treasurer $1,000. Rae had been born in Scotland and had worked in Scottish coal mines as a boy. He had been elected master workman of the Miners' Trades Assembly of the Noble and Holy Order of the Knights of Labor in 1886. The delegates also elected four men to constitute the national Executive Board of the UMWA, Patrick McBryde of Pennsylvania, William Scaife of Illinois, R. F. Warren of Ohio, and W. C. Webb of Kentucky.[239] The locals were grouped into districts and subdistricts, defined by state boundaries and the boundaries of individual coalfields. Jurisdiction over immediate daily affairs of miners in the various fields was delegated to district unions. The local districts were independent, having the right to elect officials, hold conventions, establish policies, and negotiate labor agreements within the district. The American coalfields were divided into twenty-one districts with each district required to supervise the activities of local unions within its geographic district. Pennsylvania was divided into five districts: District 1, Anthracite; District 2, Central; District 3, Low Grade (Greensburg Field); District 4, Coke Regions; and District 5, Pittsburgh District. Mining Districts 6 to 21 were Districts 6, 7, 8, 9, 10, Ohio; District 11, Indiana; District 12, Illinois; District 13, Iowa; District 14, Missouri and Kansas; District 15, Colorado, Washington, and the Territories; District 16, Maryland; District 17, West Virginia; District 18, Virginia; District 19, Tennessee and Kentucky; District 20, Georgia and Alabama; and District 21, Texas, Arkansas, and the Indian Territory.[240]

The UMWA constitution clearly articulated the diverse social and economic demands and greviances of mine workers throughout the coalfields during the last quarter of the nineteenth century:

(1) to secure earnings "fully compatible with the dangers of our calling and the labor performed,"
(2) to establish a system of payment in "lawful" money,
(3) to establish safe mining practices,
(4) to demand the eight-hour-work day,
(5) to provide education for miners' children,

(6) to seek favorable legislation for the protection of miners' health and welfare,
(7) to use all honorable means to maintain peace between ourselves and employers, adjusting all difference, as far as possible, by arbitration and conciliation, that strikes may become unnecessary.[241]

Two days after the formation of the UMWA, convention delegates voted to assess members' wages to build a fund to aid miners "who are locked out or on strike." Although the union's credo was "United We Stand, Divided We Fall," the members of the United Mine Workers of America were far from united and their influence extended over very little of the United States in 1890. The UMWA was founded in the depression decade of the 1890s, a difficult time to recruit new members to a fledgling union. The new miners' union remained a small and ineffectual labor organization during its first seven years. Coal demand nationally had slackened during the depression years and many miners found themselves either unemployed or underemployed, working three or four days each week. There were about twenty thousand union members in 1892; from 1893 to 1896, years of sharp economic depression, membership dropped to fewer than ten thousand.[242] Coal miners had withstood a number of severe wage reductions imposed by coal operators between 1890 to 1896. A sympathetic economist, commenting on how these wage cuts adversely affected the daily lives of miners, noted that "no one at all familiar with [these] conditions will deny that the miners' earning had reduced below the living point. Everywhere poverty and degradation were manifest."[243] By 1898 the union had observed signs that the depression was ending and a national strike was called for Independence Day 1897 under the direction of John Mitchell, the young charismatic vice-president of the union. Some 150,000 miners, one-half of the labor force in the bituminous coal industry, responded to his call for a nationwide shutdown.[244] The strike lasted three months and paralyzed coal production in the northern coalfields, but the strike failed in the newly developed southern Appalachia fields of West Virginia, Kentucky, and Alabama. The strike was a landmark event in mining history and in the growth of the UMWA. Northern coal operators from western Pennsylvania, Ohio, Indiana, and Illinois recognized the UMWA as the collective bargaining agent for their workers and negotiated an interstate collective contract with the miners' union for the first time in 1898. Miners were granted an eight-hour workday in union mines. The UMWA had at last secured a major victory for its membership and their victory excited many unorganized miners to join. Membership swelled from eleven thousand to thirty-three thousand within the next few years.

The newly formed UMWA represented about twenty thousand members, mostly native-born Protestant Americans or English, Irish, Welsh, and Scottish immigrants in 1890. The UMWA constitution, written at the convention and conscious of the changing social composition of miners, prohibited racial, religious, or ethnic discrimination. Miners of diverse ethnic and racial background were organized in the same locals, and in theory all were eligible for local and national union offices. The labor force of the industry since the 1880s was increasingly composed of unskilled immigrants from the polyglot nations of southern and eastern Europe. The original English, Welsh, Scottish, and Irish miners saw their numbers decline with the influx of Italian and Slavic (Polish, Lithuanian, Russian, and Slovak) workers. The UMWA leadership generally adhered to a nondiscrimination policy although strong ethnic, religious, and racial tensions continued among the rank and file. The United Mine Workers had an excellent record of organizing African American workers. There were five African American delegates to the UMWA founding convention and R. F. Warren, an African American miner from Ohio, was elected to serve on the union's first National Executive Board in 1890. By 1900, an estimated twenty thousand African American miners had become members, approximately one quarter of the union's membership. Much of this success belonged to a small but dedicated band of African American organizers, few of whom left written records. In 1907 the UMWA became one of the first American labor unions that allowed Asians to be members. Asian immigrants had faced severe discrimination in the mines and usually found work at the lower-paying surface jobs, such as sorting coal in the tipple.

John Mitchell, the powerful and dynamic president and organizer of the UMWA from 1899 to 1908, preached repeatedly to them that "the coal you dig is not Slavish coal, or Polish coal, or Irish coal. It is coal."[245] Miners who were proponents of racial equality within the union often proclaimed, "We're all the same color when we come out of the mine." Mitchell had experienced the suspicions and xenophobic prejudices that existed among various ethnic and racial groups while a miner himself at Braidwood, Illinois. Bigotry and racism were endemic in the mines and the coal towns, and repeated pleas by the UMWA for racial and ethnic harmony to unite miners were often ignored. Strong prejudice among mine workers was rampant, with epithets such as "greenhorn," "cheese-eater," "hunky," Polack," "dago," "harp," "wop," and "nigger" heard in the mines and most coal towns. The typical English-speaking miner looked down on the new immigrant miners whether consciously or not. Prejudices and dissension among union members were rampant, prompting UMWA President John McBride to address the membership in 1893: "The internecine strife occasioned by religion and nationality has . . . prevented your officers from effecting local organization [and] . . . in many instances has disrupted those already established. . . . [Y]ou cannot afford to harbor or countenance such a spirit of bigotry and intolerance."[246]

Coal operators from the principal northern soft coal states of western Pennsylvania, Ohio, Indiana, and Illinois formed the Central Competitive Field. This organization anticipated the possibility of benefiting from industry-wide unionization. The operators hoped that standardizing cost of production through a union contract signed by all operators would reduce existing competitive wage instability and bring order to the increasingly competitive industry. These operators saw the new miners' union as providing a mechanism for standardizing mining costs, reducing competition, and establishing a wage scale acceptable to miners. The union would also act as an enforcer of the contract, disciplining miners who staged wildcat strikes or reneged on agreements, and would also organize mines of recalcitrant operators who refused to bargain collectively. Coal operators hoped this new relationship with the UMWA would improve the productivity and profitability of their mines. F. L. Robbins, chairman of the Pittsburgh Coal Company, which was an original member of the Central Competitive Field and one of the nation's largest independent coal companies, wrote Mitchell in 1904 that "with honest, conservative men at the head of labor organization the liability of having trouble is decreased and it is a safer method of settling wage questions than dealing with the rank and file of employees."[247]

The early period of cooperative collective bargaining between the coal operators of the Central Competitive Field and the UMWA could endure only as long as a major portion of the industry operated on the closed-shop principle. The locus of production and economic power in the bituminous industry by the turn of the century had begun to shift slowly from the Central Competitive Field to the newly opened nonunion mines of central Appalachia. The coal industry was developing in the rural and sparsely populated mountainous regions of southern West Virginia, eastern Kentucky, and the coalfields farther south in Tennessee and Alabama. The ability of the UMWA to successfully organize miners in these coalfields would insure the continued viability of interstate union contracts and the economic strength of the UMWA in the immediate future.

John Mitchell (1870-1919) was born in Braidwood, Illinois, the son of John, a Scotch-Irish Orangeman, and Martha Mitchell. He was orphaned at an early age and at ten worked in local mines near Braidwood. He joined the Knights of Labor when he was fifteen and with the collapse of the Order took out membership in the United Mine Workers of America. He was elected vice-president of the UMWA in 1897 and was responsible for leading the successful 1897 strike. Under his aggressive leadership and the return of prosperity, the miners' union was involved in a number of strikes and work-stoppages to protest low wages, long hours, and unsafe working conditions. Mitchell reported to delegates at the annual 1898 miners' convention that the union had actively participated in 260 strikes of which 160 were won, 29 compromised, 36 lost, and 35 still pending.[248] He was elected the fifth president of the miners' union at the age of twenty-nine at the 1899 annual convention and served in this capacity until his resignation in 1908. The union led organizing drives west of the Mississippi and brought union protection to miners in parts of Colorado, Montana, Missouri, West Virginia, and Wyoming. The UMWA also signed union contracts for the first time in the coalfields of Alabama, Kentucky, Tennessee, West Virginia, and other states. In 1900, the UMWA organized only about eight thousand of the nearly one hundred fifty thousand miners employed in the anthracite coal fields. Anthracite miners of eastern Pennsylvania were organized after the 1900 strike and "the long strike" of 1902. As the winter of 1902 approached President Theodore Roosevelt intervened on October 16 and appointed a commission to mediate an end to the bitter strike. The strike was called off by President Mitchell and the UMWA on October 21. The Roosevelt Commission awarded the anthracite miners a 10 percent wage increase on March 22, 1903. Roosevelt felt compelled to intervene in the coal strike because mine owners refused an offer of arbitration, and the real possibility of anthracite fuel shortages in northern cities. Coal was so scarce during the anthracite coal strike of 1902 that the price of anthracite had risen to thirty dollars a ton by the time Roosevelt forced a settlement of the strike. Anthracite miners reached an agreement with the operators after a six-month strike, firmly establishing the UMWA in the anthracite coal region of eastern Pennsylvania.

The union, under Mitchell's aggressive leadership, attained a membership of 173,000 and a treasury of nearly one million dollars in 1903. Canadian miners in the provinces of Nova Scotia, British Columbia, and Alberta were organized by the UMWA in 1905, and consequently the official name of the miners' union became the International United Mine Workers of America. Mitchell was a strong advocate of trade unionism. The essence of trade unionism, according to him, was collective bargaining, which gave workers economic bargaining equality with management because it rid workers of fear, raised their efficiency, and established their citizenship in the new industrial order. When Mitchell resigned in 1908 as president because of ill health, the union reported an international membership of 330,000 dues-paying miners and was by far the largest and most powerful union in the nation.[249] He became a free-lance writer and popular lecturer on labor problems and in 1914 was appointed commissioner of labor for New York State, and later chairman of the State Industrial Commission. Mitchell died at the age of forty-nine in 1919. A memorial to him, designed by Peter B. Sheridan of Hazleton, stands on Courthouse Square in Scranton, in the heart of the anthracite region. A popular song, *Be Sure and Stick to Mitchell,* was dedicated to his memory and his role in organizing coal miners in eastern Pennsylvania:

> Be sure to stick to Mitchell boys
> Your faithful president
> For he's the one who won for you
> The gain of ten percent
> He made Truesdale, Baer, and Morgan
> Come down from the high stand
> And arbitrate with the working man
> Who holds the winning hand.[250]

The Bituminous Coal Industry, World War I, and the Coal Strike of 1919

The pinnacle of bituminous production in Pennsylvania and the United States was reached during the Great War.[251] Coal and coke in both domestic and European markets were needed in iron and steel mills, for railroad locomotives, for ships transporting troops to Europe, for electrical generators, for home and commercial heating, and to meet the needs of European allies. In World War I, bituminous coal furnished 69.5 percent of the nation's total energy, a level never attained since. Soft coal miners nationally produced 579,385,820 tons in 1918 alone. Throughout World War I, at the insistence of the Wilson administration, companies were continually asked to increase their coal and coke production to meet this new energy demand, at home and in Europe. Abandoned and idled mines were reopened and dozens of new shafts were dug in an attempt to satisfy this insatiable demand for coal and coke. The price of coal per ton on the commercial market had remained relatively stable between 1900 and 1916, averaging about $1.07 per ton. Compliance by the coal industry saw the average price of a ton of coal rise from $1.34 in 1916 to $2.26 in 1917 and $2.58 during the last year of the war. The price of coal peaked at $3.75 per ton in 1920, then plunged to $2.89 during the depression year of 1921.

The European War was a boon to the American coal industry. Coal Age.

Wartime inflation of coal prices and company profits meant more mines, mine workers, and production. There were 5,818 bituminous mines in the United States in 1910 and 8,921 a decade later. The number of mine workers rose about one hundred thousand nationally, while output rose by 162 million tons between 1900 to 1918. Bituminous coal output alone increased by 77 million tons nationally during America's involvement in World War I. The great demand for coal, combined with its rising price per ton, brought enormous profits and prosperity to hundreds of coal companies during the brief boom years between 1917 and 1920. At least 90 percent of 1,234 corporations reported an average net income increase of 16 to 26 percent of invested capital, before payment of federal taxes, and from 9 percent to 17 percent after payment of federal taxes, between 1917 and 1921.[252]

Pennsylvania, as the nation's leading producer of bituminous coal, shared in the coal and coke boom of the period. It produced 42 percent of the entire United States coal production in the five-year period from 1912 to 1916 and 36 percent in the five-year period from 1917 to 1921.[253] Pennsylvania's all-time-high production, for both bituminous and anthracite coal, was attained in 1918, when some 276.6 million tons were extracted from mines, employing some 329,904 miners. That year was also the historic high for coal production in a single state. The anthracite region extracted a record 100,445,299 net tons, employing about 148,226 miners in 1917, while the state's bituminous industry peaked the following year at 177 million tons of coal extracted by 181,678 miners. Bituminous production peaked nationally in 1918 when 579,385,820 net tons were extracted. Pennsylvania's own soft coal production of 177,217,294 net tons accounted for 30.8 percent of the nation's total pro-

Status of the Coal Industry of Pennsylvania in 1919				
	No. of Establishments	Capitalization	No. of Employees	Value of Product
Anthracite	254	$433,868,039	147,372	$364,084,142
Bituminous	1938	$648,626,800	154,992	$362,973,952[254]

duction. The wartime boom made the coal industry the largest employer in Pennsylvania, and second in terms of capitalization and value of products, by 1919.

The Democratic Wilson administration moved to stabilize coal production and improve working conditions in the coal industry to insure continued production and forestall labor disputes in the coalfields during the war. Congress passed the Lever Food and Fuel Control Act in 1917 to create the Federal Fuel Administration to insure industrial production and maintain labor stability during the duration of the war. The Lever Act also granted the president power to establish maximum prices for coal. Dr. H. A. Garfield, the elder son of former President James A. Garfield, was appointed the Fuel Administration's first director. The agency negotiated the Washington Wage Agreement with the UMWA and coal operators in October 1917. The agreement granted an increase of ten cents a day for pick-mining rates, a fourteen-cent raise to day laborers and monthly company men, and a 15 percent boost in yardage, dead rate, and room-turning rates. The Washington agreement bound the UMWA not to renegotiate the contract before the war officially ended or until April 1, 1920, whichever came first.[255] Strikes were forbidden under the agreement and all labor disputes in the industry were subject to arbitration by the War Labor Board or the Federal Fuel Administration. United States involvement in the war brought more than a year of labor peace to the bituminous coalfields.

Inflation nationwide ran at nearly 50 percent between 1917 and 1919 and rose an additional 20 percent in 1920 alone. From 1914 to 1919, the average annual earning of employed workers increased 87 percent, but their purchasing power increased only 5 percent. Miners' wages of 1917 had not kept up with wartime inflation. The U.S. Bureau of Labor Statistics reported in 1919 that an average American family needed an annual income of $2,243 for an acceptable standard of living, at a time when the income of an average miner was $1,194.[256] Angry and militant miners expressed their demands for wage increases at their international convention held in September 1919 by voting for an immediate 60 percent increase in tonnage and yardage rates, a six-hour day, and a five-day week. They had patriotically supported the war effort and proudly asserted they had "mined the tons (of coal) that beat the Hun." American labor unions, with the exception of some left-wing socialist groups and William Haywood's Industrial Workers of the World, had overwhelmingly supported the government's war effort against the Central Powers. Samuel Gompers, president of the AFL, fully supported the war effort and cooperated with the Wilson administration wherever possible. President Wilson placed AFL and UMWA representatives on national planning committees established to provide efficient allocation of scarce materials for war production.

The Industrial Workers of the World, who had never more than seventy thousand members, was a peculiar combination of socialists, syndicalists, and anarchists who challenged the existing labor-management relationship. They saw this conflict as a revolutionary class struggle. The "Wobblies" stressed the organization of workers that trade unions had usually ignored—agricultural workers, immigrant factory workers, longshoremen, and lumbermen. The I.W.W. was formed on June 27, 1905, at Brand's Hall in Chicago by a diverse group of two hundred political radicals and trade unionists, including Daniel DeLeon, Eugene Debs, Lucy Parsons (widow of a Haymarket Square "martyr"), "Mother" Mary Jones, and William "Bill" Haywood. Its membership was drawn primarily from four groups: the Western Federation of Miners (founded in 1893), the Socialist Party of Eugene V. Debs, the Socialist Labor Party of Daniel DeLeon, and a group of radical unionists. The enthusiastic gathering that launched the nationwide I.W.W., like the Knights of Labor, organized skilled and unskilled industrial workers and women, African Americans, and all ethnic groups into "One Big Union" as an alternative to the conservative, craft-oriented AFL.

William D. "Big Bill" Haywood (1869-1928), the dominant leader of the new radical union, had worked as a cowboy, miner, and prospector. Haywood was the former president of the Western Federation of Miners, which was the backbone of the new union. He declared that the aims and objectives of the new radical organization were "to bring the workers of this country into the possession of the full value of the product of their toil." The I.W.W. emphasized revolutionary programs including abolition of the wage system and the formation of industrial unionism. The Wobblies strongly protested American intervention in the European war and dismissed it and American involvement as simply a capitalist conspiracy. When I.W.W. members refused to support American participation, they were summarily branded as traitors by the Wilson administration. Federal and state officials suppressed I.W.W. publications and union halls were shut down during the war. The federal government in June 1917 indicted the union's top leadership under a number of wartime espionage laws. One hundred I.W.W. members were tried in Chicago before Judge Kenesaw Mountain Landis in 1918 for opposing American involvement in the war effort. All defendants were found guilty. A series of government raids and additional prosecutions under various criminal syndicalist statutes following the war, in which 150 I.W.W. members were jailed, led to the crippling of the union. The I.W.W. never recovered from these aggressive attacks by the federal government and soon faded from the national scene as an effective radical labor organization during the 1920s.[257]

Meanwhile, angry miners appealed in vain to Harry A. Garfield for an increase in their wages to offset rising costs in 1918. Garfield emphatically refused their wage demands, asserting that the 1917 agreement bound the UMWA not to renegotiate its terms until the war was officially ended or until April 1, 1920, whichever came first. John Llewellyn Lewis, who succeeded John P. White in 1919 as international president of the United Mine Workers of America, was pressured by the rank and file to call a national strike. A general miners' strike was called to coincide with the national steel strike. In September 1919, the employees of the U.S. Steel Corporation walked out in a drive for union recognition, to win an eight-hour day, and to obtain an "American living standard" from the company. The militant battle cry of the mostly immigrant steel workers demanded "eight hours and the union." The average work week in the steel industry was 68.7 hours in 1917. A typical unskilled steel worker earned on average about $1,400 per year, while government statistics estimated the minimum subsistence for a family of five to be $1,575. William Z. Foster (1881-1961), founder of the American Communist Party and a three-time presidential candidate, led the steel workers' strike for union recognition. The drive was brutally suppressed by the hostile steel companies after a lengthy and often bloody four-month strike.[258]

The 1919 national coal strike was the first of many strikes that President John L. Lewis called during his long and often tumultuous forty-year tenure as union president. President Wilson opposed the proposed strike asserting that the pending miners' strike was "not only unjustifiable but unlawful" and demanded Lewis withdraw his strike call.[259] Wilson sent William B. Wilson, his secretary of labor, to negotiate a new coal contract with the UMWA to avoid the impending strike. Congress had created the Department of Labor on March 13, 1913, "to foster, promote and develop the welfare of the wage earners of the United States, to improve their working conditions and to advance their opportunities for profitable employment." William B. Wilson (1862-1934), after serving in Congress, was appointed the nation's first secretary of labor by President Woodrow Wilson. The new secretary served from 1913 to 1923. Wilson was a former coal miner and a secretary-treasurer of the UMWA from Tioga County, Pennsylvania. William Bauchop Wilson (1862-1934) was born in Blantyre, Scotland, on April 2, 1862, the son of Adam and Helen Nelson Wilson. He worked at the mines at or near Arnot, Tioga County, beginning in 1871 at the age of nine. Wilson was a "half member" of the Miners' and Laborers' Benevolent Association at the age of eleven. A roadside historical marker erected by the Commonwealth of Pennsylvania on U.S. Route 15 on October 29, 1948, south of Blossburg identifies his farm home.[260]

Secretary Wilson met with John L. Lewis, William Green, secretary-treasurer of the UMWA (Green was later elected president of the AFL), and Thomas Brewster, president of the Coal Operators' Association. The C.O.A. had been formed in 1917 by the northern coal operators of the Central Competitive Field, although control of this owners' organization passed from union operators to nonunion operators during the 1920s. Negotiations at this meeting quickly collapsed and a national strike call was issued by Lewis for November 1, 1919. A government injunction was issued against the United Mine Workers leadership and Lewis was pressured by the government to cancel the national strike call. Lewis called off the work stoppage, telling the press, "I will not fight my government, the greatest government on earth."[261] Angry and independent-minded militant miners simply ignored Lewis's call to return to the pits and continued a wildcat strike curtailing nearly 75 percent of the national's coal production. Although Lewis was the leader of the militant miners' union, he was himself no radical labor leader. He was a long-time Republican who was fundamentally a conservative and cautious man, fearful of radical "doctrinaires" and "insurgents" in his union's ranks. He believed in competition and conflict in the market place as the basis of a free society. The strike, to Lewis, was not a weapon of class struggle but simply a viable union tool in the capitalist system. Unions were seen by Lewis as the reverse side of the corporate coin. The union and the corporation had similar functions—one organized sales and goods and the other organized labor. The government proposed a 14 percent increase but the miners refused this offer and continued their strike for higher wages. The coal miners' strike, unlike the Great Steel Strike of 1919, was successful. It was called to a halt in mid-December and a two-year settlement was signed in New York City in March 1920. The new contract gave union "tonnage men" a 34 percent wage increase and "day men" a 20 percent daily wage increase.

The bituminous coal industry appeared to both coal operators and miners as a healthy and robust industry as the decade of the 1920s commenced. The wartime experience clearly demonstrated to the world the essential role bituminous coal played in the functioning of a modern and complex industrial society. As Frank Keeny, West Virginia UMWA president, observed, "coal makes civilization possible."[262] "The Miner," a poem written by Berton Bailey for *Coal Age* in 1918, clearly states the invaluable and prominent role the miner and the coal industry had played in the recent war in defining the modern industrial society:

> Grimy, and caked with dust of coal he stands,
> Grasping his pick with his mighty hands;
> The arbiter of destiny and fate.
> Greater by far than the king or potentate.
>
> Shops may not run except at his behest,
> At forge and blast his strength is manifest.
> The rolls that rumble and the sheer and screams
> And all the million miracles of steam.
>
> Depend on him for fuel that will turn
> The wheels that urge them and the belts that churn.
> Guns that will shatter fortresses of steel,
> Ships that will plow the waves on steady keel. . . .[263]

A rosy vision of the future of the coal industry was held by most miners and the leadership of the UMWA. "King Coal" and "Queen Coke" had played an essential role in the economic development of the United States in the past and the present, and labor believed that coal would surely play a similar role in the immediate future. The UMWA celebrated its thirtieth anniversary in 1920 and John Llewellyn

Lewis had recently been elected its ninth president. The following nine men served as presidents of the UMWA between 1890 and 1960: John B. Rae (1890-1891), John McBride (1892-1894), Phil H. Penna (1895-1896), M. D. Ratchford (1897-1898), John Mitchell (1899-1907), T. L. Lewis (1908-1910), John P. White (1911-1917), Frank J. Hayes (1917-1918), and John L. Lewis (1919-1960). Lewis presided over the largest and most powerful labor union in the nation, representing 50.9 percent of all coal miners by 1920.[264] The union successfully negotiated an eight-hour workday, a forty-hour week, a standard daily wage for "day workers," and a standard tonnage rate for underground production workers, and it secured promises of improved safety conditions in the mines.[265]

The prognosis of continued economic growth of the industry engendered during World War I proved to be illusory as the bright vision of future prosperity began to fade during the 1920s. The wartime euphoria had masked a fundamental malaise in the industry. Overproduction and decreased demand for coal were becoming the curse of the industry as early as the 1890s. The bituminous coal industry had expanded production nationally ten fold from 50 million tons annually in 1880 to 568 million in 1920. Pennsylvania's bituminous coal production experienced similar growth, with output rising from 16.5 million to 166.9 million tons during this forty-year period. Production increases during World War I had simply exacerbated an already overdeveloped industry. The UMWA and bituminous coal companies would have to address the fundamental problems of too many miners mining too much coal in the 1920s.

Notes

[1] William Nichols, *The Story of American Coals* (Philadelphia: Lippincott Company, 1897), p. 1; Tommy Ehraber, "King Coal," *Pitt Magazine,* vol. 5, No. 1 (February 1990).

[2] Priscilla Long, *Where the Sun Never Shines* (York: Paragon Press, 1989), p. 117.

[3] Sam H. Schurr and Bruce C. Netschert, *Energy in the American Economy, 1850-1975* (Baltimore: Johns Hopkins Press, 1960), p. 36.

[4] James M. Swank, *Introduction to the History of Ironmaking and Coal Mining in Pennsylvania* (Philadelphia: published by author, 1878), p. 124.

[5] J. V. Thompson, *Coalfields of Southwestern Pennsylvania* (n.p.: Copyright John W. Boileau, 1907), p. 65.

[6] Anna Rochester, *Labor and Coal* (New York: International Publishers, 1931), p. 243.

[7] Schurr and Netschert, *Energy,* p. 63.

[8] *Report of the Mining Industries of the United States,* 1886, vol. 15, p. 666.

[9] *Commonwealth of Pennsylvania Department of Environmental Resources* (Harrisburg: State Printer, 1979), p. 84.

[10] William Fritz and Theodore A. Veenstra, *Regional Shifts in the Bituminous Coal Industry* (Pittsburgh: University of Pittsburgh Press, 1935), p. 19; *Pennsylvania Department of Mines,* Harrisburg, release, April 2, 1941.

[11] *The Pennsylvania Geological Review,* vol. 19, No. 5 (October 1988), p. 4; Donald M. Hoskins, "Celebrating a Century and a Half: The Geologic Survey," *Pennsylvania Heritage* (Summer, 1986); J. P. Lesley, *A Geological Hand Atlas of the Sixty Seven Counties of Pennsylvania* (Harrisburg: Board of Commissioners for the Second Geological Survey, 1885).

[12] Robert Stefano, *Coal Mining Technology: Theory and Practice* (New York: American Institute of Mining, Metallurgical, and Petroleum Engineers, Inc. 1983), p. 20.

[13] John C. Cassady, *The Somerset Outline* (Scottdale: Mennonite Publishing House, 1932), p. 47.

[14] Richard Quin, "Indiana County Inventory." Washington, D.C.: National Park Service, 1990, p. 6.

[15] *History of Pennsylvania Coal* (Harrisburg: Pennsylvania Department of Mines, n.d.), pp. 10-44.

[16] Howard N. Eavenson, *The First Century and a Quarter of American Coal Industry* (Pittsburgh: privately printed, 1942), p. 418.

[17] John W. Oliver, *History of American Technology* (New York: Ronald Press, 1956), p. 471; William Hogan, *Economic History of the Iron and Steel Industry in the United States,* vol. 2, part 3 (Washington, D.C.: Heath & Company), pp. 402-419.

[18] John Newton Boucher, *William Kelly: A True History of the So-Called Bessemer Process* (Greensburg: published by the author, 1924).

[19] Hogan, *Econonic History,* p. 363.

[20] Peter Temin, *Iron and Steel in Nineteenth Century America: An Economic Inquiry* (Cambridge: M.I.T. Press, 1954), pp. 268-269.

[21] William Sisson, Bruce Bomberger, and Diane Reed, "Iron and Steel Resources of Pennsylvania, 1716-1945" (Harrisburg: Pennsylvania Historical and Museum Commission, 1991), p. 60.

[22] Nichols, *The Story of American Coals,* p. 329.

[23] Long, *Where the Sun Never Shines,* p. 119.

[24] Hogan, *Economic History,* p. 395.

[25] Douglas Fisher, *Epic of Steel* (New York: Harper and Row, 1963), p. 110.

[26] John Enman, "The Relationship of Coal Mining and Coke Making to the Distribution of Population Agglomerations in the Connellsville (Pennsylvania) Beehive Coke Region" (Ph.D. diss., University of Pittsburgh, 1962), p. 65.

[27] E. C. Dixon, *Coke and By-Product Manufacture* (London: Charles Griffin & Company, 1939), p. 88.

[28] Nichols, *The Story of American Coals,* p. 332.

[29] John R. Lane, *Eliza: Remembering A Pittsburgh Mill* (Charlottesville, Va.: Howell Press, 1989), p. 102.

[30] Fisher, *Epic of Steel,* p. 110.

[31] Howard N. Eavenson, *The Pittsburgh Coal Bed: Its Early History and Development* (New York: American Institute of Mining and Metallurgical Engineers, 1938), p. 41; Nichols, *The Story of American Coals,* p. 332.

[32] *Pittsburgh: 50th Anniversary of the Engineers' Society of Western Pennsylvania* (Pittsburgh: Cramer Printing Company, 1930), p. 249.

[33] Kenneth Warren, *The American Steel Industry: A Geographical Interpretation 1850-1970* (Pittsburgh: University of Pittsburgh Press, 1973), p. 48.

[34] Franklin Ellis, *History of Fayette County, Pennsylvania with Biographical Sketches of Many of Its Pioneers and Prominent Men* (Philadelphia: Louis H. Everts and Company, 1882), pp. 246-247.

[35] George Littleton Davis, "Greater Pittsburgh's Commercial and Industrial Development (With Emphasis on the Contributions of Technology) 1850-1900" (Ph.D. diss., University of Pittsburgh, 1951), p. 346.

[36] *Pennsylvania Industrial Statistics,* vol. 15, 1887, pp. 2f.

[37] E. Willard Miller, *Pennsylvania: Keystone to Progress* (New York: Windsor Publication, 1986), p. 59.

[38] Enman, John A., *The Connellsville Coke Region* p. 152. A complete listing of coke works, mines and mining towns that opened in the Connellsville coke district during this decade can be found in this source.

[39] George Harvey, *Henry Clay Frick* (New York: Charles Scribner's Sons, 1924), p. 78; William Serrin, *Homestead: The Glory and Tragedy of an American Steel Town* (New York: Random House, 1992), pp. 46-47.

[40] Enman, "The Relationship of Coal Mining and Coke Making," pp. 434-439.

Coke Plants built by the H.C. Frick Coke Co. in the Connellsville Coke region

Frick (Novelty)	1871
Henry Clay	1871
Adelaide	1888
Brinkerton	1901
Chambers	1902
Bitner	1904
Shoaf	1904
Smiley	1904
York Run	1904
Hopwood	1906
Collier	1907
Phillips	1907

Mines	Year Closed
Star	1878
Ferguson	1880
Fountain	1881
Home	1881
Foundry	1888
Furnace	1888
Spurgeon	1882-1892
Washington	1882-1892
Hazlett	1893
Anchor	1896
Morrell	1898
Great Bluff	1900
Strickler	1900
Uniondale	1900
Sterling #1	1902
Diamond	1903
Mutual #2	1903
Rising Sun	1903
Wheeler	1908
Alverton #2 (Mayfield)	1910
Bridgeport	1910
Henry Clay	1910
Mullen	1910
Summit	1910
Hopwood	1911
Tip Top	1911
Franklin	1912
Southwest #4 (American, Warden)	1913
Chambers	1913
Monastery	1913
Painter	1913
Sterling #2 (Jimtown)	1914
Charlotte (Scottdale)	1910-1914
Clarissa	1910-1914
Empire (Bethany?)	1910-1914
Pennsville	1910-1914
Tyrone	1910-1914
Boyer	1917
Buckeye	1917
Byrne (Love)	1917
Eagle (Sherrick)	1917
Morgan	1917
Rist	1917
Veteran	1917
Marietta	1910-1918
Coalbrook	1918
Jackson	1918
Rainey (Fayette)	1918
Valley	1918
Spring Grove	1919
White (Globe)	1919
Enterprise	1920
Fort Hill	1920
Calumet	1922
Southwest #3 (Tarrs)	1922
Adelaide (Cupola)	1923
Baggaley	1923
Continental # 2 (Newcomer)	1923
Mutual #1	1923
Southwest #2 (Alice)	1923
Stewart # 1	1923
Acme	1925
Hecla #2 (Trauger)	1925
Juniata	1925
United #2 (Central)	1925
Continental #2	1926
Dorothy	1926
Lemont #1	1926
Oliphant	1926
Wynn	1926
Grace (including Moyer)	1927
Mammoth #1	1927
Redstone (Brownfield)	1927
Whitney	1927
Elm Grove	1928
Hecla #3	1928
Mayer	1928
Paul	1928
York Run	1928
Brinkerton	1929
Hecla #1 (Southwest)	1929
Marguerite (Klondike)	1929
Mount Pleasant (Carpentertown)	1929
Cora (Shannon)	1921-1931
Dexter	1921-1931
Mahoning (Mahaney)	1921-1931
Percy (Johnson?)	1921-1931
Mount Braddock	1930
Standard (Shaft and Slope)	1931
Hostetter (Lippincott, Jamison #21)	1931
United #1	1933
Revere	1934
Southwest #1 (Morewood)	1935
Bitner	1936
Saint Vincent	1936
Myers	1937
Lemont #1	1937
Trotter	1938
Davidson	1940
Crossland	1941
Leith	1942
Lemont #2	1942
Oliver #3	1944
Oliver #4	1944
Phillips	1944
Youngstown	1948
Clare (Jamison 20)	n.d.
Shoaf	1951
Smiley	1951
Continental #1	1954
Leisenring #1	1954
Kyle	1954
Leisenring #2	1957

[41] Serrin, *Homestead,* p. 47.

[42] Shyamal Majumdar and E. Willard Miller, eds., *Pennsylvania Coal: Resources, Technology, and Utilization* (University Park: Pennsylvania State University Press, 1983), p. 182.

[43] *Annual Report of the Secretary of Internal Affairs,* part 3, Industrial Statistics, vol. 15 (Harrisburg, 1885).

[44] Gray Fitzsimons, ed., *Blair County and Cambria County, Pennsylvania: An Inventory of Historic Engineering and Industrial Sites* (Historic American Building Survey / Historic American Engineering Record, National Park Service, 1990), pp. 86-87.

[45] Nancy Shedd, ed., *Huntingdon County: An Inventory of Historic and Engineering and Industrial Sites* (Washington, D.C.: National Park Service, 1991), p. 15.

[46] Fisher, *Epic of Steel,* p. 110.

[47] *Pittsburgh: 50th Anniversary,* p. 248.

[48] Ibid., p. 249.

[49] W. O. Hickok and F. T. Moyer, *Geology and Mineral Resources of Fayette County* (Harrisburg: Department of Internal Affairs, 1940), p. 92.

[50] Samuel S. Wyler, *The Smithsonian Institution Study of Natural Resources Applied to Pennsylvania's Resources* (Washington, D.C.: Smithsonian Institution, reprinted, 1923), p. 24.

[51] Hogan, *Economic History,* p. 380.

[52] Ibid., pp. 378-379.

[53] Wyler, *The Smithsonian Institution Study,* p. 30.

[54] Ibid., p. 24; Hogan, *Economic History,* p. 381.

[55] Sarah H. Heald, ed., *Fayette County, Pennsylvania: An Inventory of Historic Engineering and Industrial Sites* (Washington, D.C.: National Park Service, 1990), p. 19; T. E. Pierce, "Semet-Solvay Coke Plant at Cleveland," *Coal Age,* vol. 3, No. 23 (January 1913); Frederic Quivik, "Connellsville Coal and Coke." Washington, D.C.: National Park Service, 1991, pp. 46-47; Nichols, *The Story of American Coals,* p. 377.

⁵⁶ Quivik, "Connellsville Coal and Coke," p. 46; W. H. Blauvelt, "Development and Present Status of By-product Coking in the U.S.," *Coal Age,* vol. 6, No. 1 (July 1914); John L. Gans, "Beehive Oven Supremacy Passing—But Not Yet Passed," *Coal Age,* vol. 15, No. 2 (June 1919).

⁵⁷ "Beehive and By-Product Coke," *Coal Age,* vol. 15, No. 2 (1919).

⁵⁸ Elwood S. Moore, *Coal: Its Properties, Analysis, Classification, Geology, Extractions, Uses, and Distribution* (London: John Wiley and Company, 1940), p. 343.

⁵⁹ Oliver, *History of American Technology,* p. 471.

⁶⁰ Fisher, *Epic of Steel,* p. 201. The Koppers Company's corporate offices were located in the Union Trust Building in Pittsburgh circa 1930. Their laboratories were located in the Mellon Institute, Pittsburgh. The firm constructed Koppers and the newer Becker by-product ovens.

⁶¹ Ibid.; Sylvester K. Stevens, *Pennsylvania: Titan of Industry,* vol. 2 (New York: Lewis Historical Publishing Company, 1948), p. 382.

⁶² *Pittsburgh: 50th Anniversary,* pp. 237-238.

⁶³ Wyler, *The Smithsonian Institution Study,* p. 26.

⁶⁴ "By-products from Coal," *Coal Age,* vol. 8, No. 1 (1913); *Pennsylvania's Mineral Heritage* (Harrisburg: Pennsylvania State College School of Mineral Studies, 1944), p. 27.

⁶⁵ Warren, *The American Steel Industry,* p. 113.

⁶⁶ Quivik, "Connellsville Coal and Coke," p. 47.

⁶⁷ Hogan, *Economic History*, pp. 386-388. The study presents a complete listing of by-product coke oven installations in Pennsylvania and other states for the period between 1895 and 1920.

⁶⁸ Enman, "The Relationship of Coal Mining and Coke Making," p. 308

⁶⁹ *Pittsburgh: 50th Anniversary,* p. 32.

⁷⁰ Frank F. Marquard, "Mammoth Coke Plant," *The Coal Industry* (June 1919).

⁷¹ *Pittsburgh: 50th Anniversary,* pp. 238-239.

⁷² Hogan, *Economic History,* p. 523

⁷³ *Department of Mines—Commonwealth of Pennsylvania* (Harrisburg: Department of Mines, 1918), p. 7.

⁷⁴ Fisher, *Epic of Steel,* p. 201.

⁷⁵ *The Story of Pittsburgh* (Pittsburgh: First National Bank of Pittsburgh, 1920).

⁷⁶ *Pennsylvania's Mineral Heritage*, p. 28; John L. Gans, "Beehive Supremacy Passing-But Not Yet Passed," *Coal Age,* vol. 15, No. 4 (1919).

⁷⁷ Enman, "The Relationship of Coal Mining and Coke Making," pp. 313-314; Moore, *Coal,* p. 344.

⁷⁸ Quivik, "Connellsville Coal and Coke," p. 32.

⁷⁹ Hickok and Moyer, *Geology and Mineral Resources,* p. 355.

⁸⁰ Quivik, "Connellsville Coal and Coke," p. 88.

⁸¹ Enman, "The Relationship of Coal Mining and Coke Making," p. 95; Hickok and Moyer, *Geology and Mineral Resources,* p. 355.

⁸² Dever C. Ashmead, "Modern Rectangular Coke-Plant," *Coal Age,* vol. 13, No. 8 (February 1918); Quivik, "Connellsville Coal and Coke," pp. 87-88.

⁸³ Ibid., p. 60.

⁸⁴ Enman, "The Relationship of Coal Mining and Coke Making," p. 95.

⁸⁵ Quivik, "Connellsville Coal and Coke," p. 3.

⁸⁶ J. C. McLenthan, et al., *Centennial History of Connellsville, Pennsylvania, 1806-1906* (Connellsville: Reprinted by Connellsville Area Historical Society, Inc. 1974), p. 284.

⁸⁷ J. W. Barger, *Greene County Coal Book and Purchaser Official Guide* (n.p.: J. W. Barger, 1907), p. 4.

⁸⁸ Hickok and Moyer, *Geology and Mineral Resources,* p. 358.

⁸⁹ George W. Harris, "Changes in Beehive Coke Oven Construction Due to Mechanical Operation," *Coal Age*, vol. 15, No. 2 (January 1919); Newell G. Alford, "Improving Coal from Beehive Ovens," *Coal Age,* vol. 3, No. 23 (June 1913); E. C. Ricks, "Machinery for Beehive Coke Ovens," *Coal Age,* vol. 15, No. 2 (January 1919).

⁹⁰ Ricks, "Machinery for Beehive Coke Ovens"; McClenathan, *Connellsville,* p. 286.

⁹¹ Quivik, "Connellsville Coal and Coke," p. 67.

⁹² Louis Athey, *Kaymoor: A New River Community* (Washington, D.C.: National Park Service, 1986), p. 16; Quivik, "Connellsville Coal and Coke," p. 27.

⁹³ William Sisson and Bruce Bomberger, *Made in Pennsylvania: An Overview of the Major Historical Industries of the Commonwealth* (Harrisburg: Pennsylvania Historical and Museum Commission, 1991), pp. 5-6. The population of Pennsylvania according to the United States Census of 1900 was 6,302,150.

Native Born	5,316,685
Foreign Born	985,250
Native White—Native Parents	3,729,093
Native White—Foreign Parents	1,430,028
Foreign White	928,543
Non-White	160,451

Bruce T. Williams, *Coal Dust in Their Blood: The Work and Lives of Underground Coal Miners* (New York: AMS Press, Inc. 1991), p. 42.

⁹⁴ "The UMWA at 100 Looking Back, Looking Ahead," *United Mine Workers Journal* (August-September 1990), p. 8.

⁹⁵ Jeremiah W. Jenks and W. Jett Lauck, *The Immigration Problem* (New York, 1913).

⁹⁶ U.S. Immigration Commission, *Immigrants in Industries Part 1: Bituminous Coal Mining,* vol. 1 (Washington, D.C.: Governmmental Printing Office, 1911), pp. 21-22.

⁹⁷ Rochester, *Labor and Coal*, p. 75; *U.S. Immigration Commission,* p. 35. By 1909, only 15 percent of miners were native Americans or native born of native fathers, 9 percent were native born of foreign fathers, while 76 percent were foreign born.

[98] Edward T. Devine, *Coal: Economic Problems of the Mining, Marketing and Consumption of Anthracite and Soft Coal in the United States* (Bloomington, Illinois: American Review Service Press, 1925), p. 31.

[99] Rochester, *Labor and Coal*, p. 86.

[100] Ibid., p. 33.

[101] Long, *Where the Sun Never Shines*, p. 127.

[102] Quivik, "Connellsville Coal and Coke," p. 77.

[103] Long, *Where the Sun Never Shines*, p. 128.

[104] Mildred Biek, "The Miners of Windber: Class, Ethnicity, and the Labor Movement in a Pennsylvania Coal Town, 1880s-1930s" (Ph.D. diss. Northern Illinois University, 1989), p. 166; William Graebner, *Coal-Mining Safety in the Progressive Period: The Political Economy of Reform* (Lexington: University Press of Kentucky, 1976), p. 118.

[105] Tommy Ehraber, "King Coal," *Pitt Magazine*, vol. 5, No. 1 (February 1990).

[106] Long, *Where the Sun Never Shines*, pp. 127, 352; Paul Nyden, *Black Coal Miners in the United States* (New York: The American Institute for Marxist Studies, 1974).

[107] Rochester, *Labor and Coal*, p. 87.

[108] James B. Allen, *The Company Town of the American West* (Norman: University of Oklahoma Press, 1966), p. 10.

[109] Lola M. Bennett, *The Company Towns of the Rockhill Iron and Coal Company: Robertsdale and Woodvale, Pennsylvania* (Washington, D.C.: National Park Service, 1990), p. 19.

[110] Ibid.

[111] Curtis R. Lytle, *Landrus, PA: Pennsylvania Ghost Town and Electric Coal Mine.* (Mansfield: The Penny-Saver Printer, 1984); Leonard F. Parauchu, "Bitumen: All Gone With The Wind," *Pennsylvania Heritage* (Summer 1985).

[112] *Proceedings of the Coal Mining Institute of America* (Pittsburgh: C. Henry and Company, 1910), p. 111; Allen, *The Company Town,* p. 80.

[113] Barbara Ellen Smith, *Digging Our Own Graves: Coal Miners and the Struggle over Black Lung Disease* (Philadelphia: Temple University Press, 1987), pp. 37-40.

[114] Helene Smith, *Export: A Patch of Tapestry Out of Coal Country America* (Greensburg: McDonald/Seward Publishing Company, 1986), p. 107.

[115] John K. Gate, *The Beehive Coke Years: A Pictorial History of Those Times* (Uniontown: privately printed, 1991), pp. 79, 80, 82. These pages provide photographs with backyard baking ovens in use.

[116] Harold M. Watkins, *Coal and Men: An Economic and Social Study of the British and American Coal Fields* (London: George Allen & Unwin Ltd., 1934), p. 258. Readers interested in the social and economic condition of the wives and daughters of miners will find "Home and Environment Opportunities of Women in Coal-Mines Workers' Families, Bulletin Number 45," a useful document. "Houses for Mining Town, Bulletin 87" of the Department of the Interior (Bureau of Mines) provides details of houses, construction, and waste systems with plans and photographs.

[117] Long, *Where the Sun Never Shines,* p. 43; Dorothy Schwieder, "Italian Americans in Iowa's Coal Mining Industry," *Annals of Iowa,* vol. 46 (1982). Anne Marion Maclean, "Life in the Pennsylvania Coal Fields: With Particular Reference to Women," *American Journal of Sociology,* vol. 14 (1909).

[118] Michael Workman, *The Fairmont Coal Field: Historical Context* (Morgantown, W.V.: Institute for History of Technology & Industrial Archaelogy, 1992), pp. 118-119.

[119] *History of California, Pennsylvania—100 Years of Progress, 1849-1949.* (Waynesburg: Sutton Printing Company, 1949), p. 31.

[120] Biek, "The Miners of Windber," p. 136.

[121] Rochester, *Labor and Coal*, p. 87.

[122] Margaret Mulrooney, *A Legacy of Coal: The Company Towns of Southwestern Pennsylvania* (Washington, D.C.: National Park Service, 1991), p. 12; Lon Savage, *Thunder in the Mountains: The West Virginia Mine War, 1920-21* (Pittsburgh: University of Pittsburgh Press, 1990), p. xv.

[123] John L. Lewis, *The Miners' Fight for American Standards* (Indianapolis: Bell Publishing Company, 1925), p. 172.

[124] Mulrooney, *A Legacy of Coal,* p. 26; Devine, *Coal,* p. 252; Homer Lawrence Morris, *The Plight of the Bituminous Coal Miner* (Philadelphia: University of Pennsylvania Press, 1934), pp. 86-87.

[125] Devine, *Coal,* p. 253; Edward Eyre, Hunt Tyron, Fred G. Tyron, and J. H. Willits, eds., *What the Coal Commission Found* (Baltimore: Williams and Wilkins Company, 1925), pp. 143-144.

[126] Lytle, *Landrus, PA,* p. 54; George Korson, *Coal Dust on the Fiddle* (Hatboro: Folklore Associates, Inc, 1965), p. 53.

[127] Mulrooney, *A Legacy of Coal;* Shurick, *The Coal Industry,* pp. 313-315; A. F. Hauser, "Houses for Mine Villages," *Coal Age,* vol. 12, No. 17 (October 1917); Adam T. Shurick, "Colliery Dwelling Construction," *Coal Age,* vol. 6, No. 2 (October 1911); G. H. Prosser, "Housing Generalities and Some Particulars," *The Coal Industry* (June 1924); Leifur Mangusson, "Company Housing in the Bituminous Coal Fields," *Monthly Labor Review,* vol. 10 (1920).

[128] Mulrooney, *A Legacy of Coal,* p. 1.

[129] Ibid., p. 140.

[130] Ibid., p. 91.

[131] "Sociological Work Accomplished by the Consolidation Coal Company," *Coal Age,* vol. 15 (1919); Gwendolyn Wright, *Building the American Dream: A Social History of Housing in America* (New York: Pantheon Books, 1981), pp. 177-192.

[132] "Cambria Steel Co. Finds That the Good Housing Increases Output," *Coal Age,* vol. 23, No. 20 (May 1923); Donald J. Baker, "No. 1 Plant of the Mather Collieries," *Coal Age,* vol. 16, No. 20 (November 1919); J. O. Durkee, "Mather Mine in the Greene County Field," *The Coal Industry* (May 1920); Robert A. Korcheck, *Nemacolin: The Mine—The Community, 1917-1950* (Privately printed, 1980).

[133] Dennis Brestinsky, et al., *Patch /Work Voices: The Culture and Lore of a Mining People* (Pittsburgh: University of Pittsburgh Press, 1991), p. 42.

[134] *Pittsburgh-Buffalo Company* (Pittsburgh: Pittsburgh-Buffalo Coal Company, 1911); William Keyes, ed., *Historic Survey of the Greater Monongahela River Valley* (Harrisburg: National Park Service / Pennsylvania Historical and Museum Commission, 1991), p. 122; "Obituary for James Jones," *Coal Age* (March 1912); Lilian Potisek and Murchant Singadine, *A Bicentennial History of West Bethlehem Township and Marianna Borough: 1776-1976* (Marianna: privately printed, 1976).

[135] Thomas Coode, *Bug Dust and Black Damp and Work in the Old Patch Town* (Uniontown: Comart Press, 1986), p. 94. The village is six miles from Millsboro, five miles from Rices Landing, and a three-hour train ride from Pittsburgh. The Mather mine was the site of the second-worst mining disaster in the soft coal industry in Pennsylvania. The disaster occurred at 4:07 P.M. on May 19, 1928, when an explosion occurred as the second shift was reporting to work. The explosion killed 194 miners. Contemporary accounts were uncertain if the disaster was caused by gas or coal dust. J. O. Durkee, "Mather Mine in the Greene County Field," *The Coal Industry* (May 1920); Donald J. Baker, "No. 1 Plant of the Mather Collieries," *Coal Age*, vol. 16 (1919).

[136] Karen Anne Haley, *Indiana Township Preliminary History Book* (Indianola: Indiana Township Historical Commission, 1988), p. 159.

The list below identifies most of the ready-made prefabricated house companies in the nation:

Aladdin Redi-Cut Houses, Bay City, Michigan, 1906-1987
Armco, Hamilton, Ohio
Bennett Homes, North Tonawanda, New York, ca. 1930
Gordon-Van Tine, Davenport, Iowa, ca. 1910 1941.
Harris Home, Chicago, Illinois, ca. 1912-1930.
E. F. Hodgson Portable Homes, Dover, Massachusetts, ca. 1892-1970.
Lewis Built Homes Company, Chicago, Illinois.
Lewis / Liberty Manufacturing Company, Bay City, Michigan, 1914-1973.
Mershon and Morley, Saginaw, Michigan, 1899-1926.
Montgomery Ward and Company, Chicago, Illinois, 1912-1931.
Norwood Sash and Door Company, Wood, Ohio, ca. 1917.
Radford Architectural Company, Chicago, Illinois, ca. 1903-1920.
Sears, Roebuck and Company, Chicago, Illinois, ca. 1908-1940.
Standard Homes Company, Washington, D.C.
Sterling System Homes, Bay City, Michigan, 1915-1971.

[137] A. F. Brosky, "Indianola Pumps Water from the Allegheny River Wells and Treats its Sewage by Bacteria and Chlorination," *Coal Age*, vol. 20, No. 1 (September 1921).

[138] Athey, *Kaymoor*, p. 32.

[139] Mulrooney, *A Legacy of Coal*, p. 139.

[140] Ibid., p. 67.

[141] Mack H. Gillenwater, "Cultural and Historical Geography of Mining Settlements in the Pocahontas Coal Fields of Southern West Virginia, 1880-1930" (Ph.D. diss. University of Tennessee, 1972), p. 91.

[142] Eavenson, *The First Century*, p. 418.

[143] Long, *Where the Sun Never Shines*, p. 57.

[144] Devine, *Coal*, p. 50.

[145] Long, *Where the Sun Never Shines*, p. 121.

[146] Ibid., p. 120.

[147] *Commonwealth of Pennsylvania Report of the Department of Mines of Pennsylvania* (Harrisburg: C. E. Aughinbaugh, State Printer, 1912), p. 29.

[148] Rochester, *Labor and Coal*, p. 123; Mary Ann Kleeck, *Miners and Management* (New York: Russell Sage Foundation, 1934), p. 179; George S. Rice, "Should New Mines Be Opened?" *Coal Age*, vol. 13, No. 16 (1918); 1900, 234 days worked; 1905, 211 days worked; 1910, 217 days worked; 1915, 203 days worked.

[149] Long, *Where the Sun Never Shines*, p. 123.

[150] Richard Bissell, *The Monongahela* (New York: Rinehart and Company, 1952), p. 153; *The Story of Pittsburgh and Vicinity* (Pittsburgh: Pittsburgh Gazette Times, 1908), p. 229.

[151] Stevens, *Pennsylvania: Titan of Industry*, vol. 3, pp. 613-614.

[152] Ibid., p. 614.

[153] *The Story of Pittsburgh*.

[154] Mary Ann Kleeck, *Miners and Management* (New York: Russell Sage Foundation, 1934), p. 181.

[155] John H. Jones, "Government Regulation of Coal Prices," *Coal Age* (October 1914).

[156] Bruce T. Williams, and Michael D. Yates, *Upward Struggle: A Bicentennial Tribute to Labor in Cambria and Somerset Counties* (Johnstown: University of Pittsburgh, 1976).

[157] Devine, *Coal*, p. 176.

[158] *Thirteenth Census of the United States: Abstract* (Washington: U.S. Department of Commerce and Labor, 1913), p. 553.

[159] *Fourteenth Census of the United States: Mines and Quarries—Statistics for Pennsylvania*, vol. 9 (Washington, D.C.: Bureau of the Census, 1919), p. 270.

[160] Ehraber, "King Coal."

[161] Moore, *Coal*, p. 117.

[162] Long, *Where the Sun Never Shines*, p. 138.

[163] J. W. Koster, "Gases Commonly Met with in Coal Mines," *The Coal Industry* (February 1919).

[164] Moon, "The Standard Wolf Safety Lamp."

[165] *Commonwealth of Pennsylvania Report of the Department of Mines of Pennsylvania*, p. x.

[166] Rochester, *Labor and Coal*, pp. 75-77.

[167] George Miller, *A Pennsylvania Album* (University Park: Keystone Books, 1986), p. 71; C. K. McFarland, "Crusade for Child Laborers: Mother Jones and the March of the Mill Children," *Pennsylvania History*, vol. 38 (1971).

[168] Louis Poliniak, *When Coal Was King: Mining Pennsylvania's Anthracite* (Lebanon: Applied Arts Publishers, 1977), p. 21.

[169] L. B. Smith, "Handling Mine Waters," *Coal Age*, vol. 10, No. 27 (December 1916); "The Development of Modern Types of Mine Pumps," *Coal Age* (April 1914).

[170] Carter Goodrich, *The Miners' Freedom* (Boston: Marshall Jones Company, 1925), pp. 15-55. See Chapter 2, "The Jobs of the Mine Workers"; Harold W. Aurand, *From the Molly Maguires to the United Mine Workers: The Social Ecology of an Industrial Union, 1869-1897* (Philadelphia: Temple University Press, 1961), pp. 184-190. This study identifies the principal mining occupations and their duties in the anthracite region of Pennsylvania.

[171] Devine, *Coal,* p. 214.

[172] Heald, *Fayette County,* pp. 89-97.

[173] Quivik, "Connellsville Coal and Coke," pp. 37-39.

[174] Baird Halberstadt, *Mineral Map of the Bituminous Coal Fields of Pennsylvania.* (Pottsville: Halberstadt Publisher, 1903, 1907); John N. Boucher, *History of Westmoreland County* (New York: Lewis Publishing Company, 1906); Enman, "The Relationship of Coal Mining and Coke Making"; *Topographic Maps of the Connellsville Coke Region from Surveys by the H. C. Frick Company* (J. R. Paddock, Chief Engineer, Kenneth Allen, Engineer-in-Charge), 1892; *A Town That Grew at the Crossroad* (Scottdale: Laurel Group Press, 1978); *Keystone Consolidated Publishing Company—The Catalog, Including Directory of Mines* (Pittsburgh: Keystone Consolidated Company, 1914, 1935, 1938); *Department of Mines of Pennsylvania* (Harrisburg: Board of Commissioner Publisher, 1929, 1931).

[175] Fred C. Keighley, "The Connellsville Coke Region," *Engineering Magazine,* vol. 20 (1901).

[176] *U.S. Thirteenth Census: Mines and Quarries* (Washington, D.C., 1909).

[177] Donna M. Ware, *Green Glades and Sooty Gob Piles* (Crownsville: Maryland Historic Trust, 1991), p.122.

[178] *The Black Diamond's Year Book and Directory* (New York: Black Diamond Company, Publisher, 1910); *Keystone Consolidated Publishing Company: The Catalog;* John N. Boucher, *Old and New Westmoreland* (New York: American Historical Society, 1918). The Carnegie Coal Company also operated the Carnegie, Oakdale, and Primrose mines,Washington County, in 1910. The production of these three drift-entry mines was 350,000 tons.

[179] Eavenson, *The Pittsburgh Coal Bed,* p. 40.

Preparation Plants in Western Pennsylvania 1890s

County	Plants	Washers	Type
Allegheny	8	5	Diescher
		4	Stutz
		3	Endres
		1	Slush
Fayette	1	1	Stutz
		1	Waverly Coal Co.

[180] *Coal Age,* June 1, 1912.

[181] Charles Gersha, *From the Furrows to the Pits* (Van Voorhis: McClain Publishing Company, 1987), p. 145.

[182] This area was the site of underground mining prior to 1925. Mining today is limited to surface coal mining. Jimmy Hoffa (1913-1975?), Teamster president between 1957 and 1971, was born in Brazil, Indiana, on February 13, 1913. His father was a coal miner.

[183] Eavenson, *The First Century,* p. 377.

[184] Long, *Where the Sun Never Shines,* p. 134; Andrew Roy, *A History of the Coal Miners of the United States* (Westport, Ct: Greenwood Press, 1970), pp. 149-151.

[185] Shurick, *The Coal Industry,* p. 127.

[186] Moore, *Coal,* p. 295.

[187] Lewis, *The Miners' Fight,* pp. 94-95.

[188] Smith, *Digging Our Own Graves,* p. 48.

[189] Moore, *Coal,* p. 204; Roy, *A History of the Coal Miners,* p. 152; Thomas Fry, "Development of Coal Cutting Machines," *Coal Age,* vol. 4, No. 26 (January 1913); C. E. Warbom, "A New Type of Coal Cutter," *Coal Age,* vol. 4, No. 25 (December 1913).

[190] Rochester, *Labor and Coal,* p. 108; Long, *Where the Sun Never Shines,* p. 136.

[191] *Mechanization, Employment, and Output per Man in Bituminous Coal Mining,* vol. 1 (Washington, D.C.: U.S. Bureau of Mines, 1939), pp. 24-29; David R. Shearer, "Electricity in Coal Mining," *Coal Age,* vol. 6, No. 2 (May 1914); A. C. Callen, "Electric Locomotives for Coal Mines," *The Coal Industry* (March 1918); C. L. Packard, "New Methods of Handling Coal Electrically," *Coal Age,* vol. 13, No. 20 (May 1918); William Van C. Brandt, "Storage Batteries for Mine Locomotives," *Coal Age,* vol. 4, No. 23 (December 1913).

[192] "The UMWA at 100: Looking Back, Looking Ahead," *United Mine Workers Journal* No. 7 (August-Septmber 1990), p. 4.

[193] McAlister Coleman, *Men and Coal* (New York: Farrar & Rinehart, Inc., 1943), p. 121.

[194] John Brophy, *A Miner's Life* (Madison: University of Wisconsin Press, 1964), p. 45.

[195] Long, *Where the Sun Never Shines*, p. 50.

[196] Korson, *Coal Dust on the Fiddle,* p. 229.

[197] Ibid., p. 47.

[198] Ibid.

[199] Long, *Where the Sun Never Shines,* p. 49.

[200] William A. Pavlik, *Johnstown Rolling Mill Mine: Disaster in Johnstown: The Rolling Mill Mine Exposion, 1902, and Its Immigrant Victims* (Johnstown: privately printed, 1979), p. 11.

[201] Graebner, *Coal-Mining Safety*, p. 118.

[202] *Bituminous Coal Data—1960* (Washington, D.C.: National Coal Association, 1961), p. 98; Alexander Trachtenberg, *The History of Legislation for the Protection of Coal Miners in Pennsylvania* (New York: International Press, 1942), p. 228; Pennsylvania Department of Mines, Harrisburg, release, April 2, 1941.

[203] Samuel M. Cassidy, ed., *Elements of Practical Coal Mining* (Baltimore: American Institute of Mining, Metallurgical, and Petroleum Engineers, Inc. 1973), p. 243.

[204] *The Engineering and Mining Journal* (February 1900); *Coal Age,* April 1, 1916, October 31, 1918.

[205] Poliniak, *When Coal Was King,* p. 22; Moon, "Carbide Safety Lamp"; Edwin M. Chance, "Portable Miners' Lamp," *Coal Age,* vol. 11, No. 17 (April 1917); H. O. Swoboda, " Self-Contained Portable Electric Mine Lamp," *Coal Age,* vol. 15, No. 20 (May 1914); "Improvement in Carbide Lamp," *Coal Age,* vol. 10, No. 15 (June 1916).

[206] *Annual Report of Mining Activities, 1979* (Harrisburg: Commonwealth of Pennsylvania; Department of Environmental Resources, 1978), pp. 94-95.

207 Devine, *Coal,* p. 232. The worst mining disaster in the Connellsville coke district occurred at the H. C. Frick Coke Company's Mammoth Mine. On January 21, 1891, 109 miners were killed at this mine at Mount Pleasant, Westmoreland County. The accident was caused by an explosion of gas and dust.

George H. Ashley, *Bituminous Coal Fields of Pennsylvania* (Harrisburg: Department of Forests and Water, 1928), p. 195. Some accidents and deaths in the mines were probably caused by carelessness on the part of miners. The principal causes of fatal mine accidents in 1928 were roof falling (60%), carelessness in the moving of coal from the rooms to the surface tipple (20%), and gas and dust explosions (10%), while the remaining 10% were such miscellaneous accidents as falls down shafts or slopes and electrocution.

208 Graebner, *Coal-Mining Safety,* p. 118.

209 Ibid., p. 139.

210 Aurand, *From the Molly Maguires,* p. 43.

211 Frank Hall, "Pennsylvania Bituminous Mine Inspectors Given Life Tenure of Offices," *Coal Age,* vol. 12 (1915). The first district comprised parts of Washington, Westmoreland, Fayette, and Allegheny Counties; the second district comprised the counties of Beaver, Warren, Mercer, Crawford, Erie, Lawrence, Forest, Venango, Clarion, Jefferson, Indiana, Armstrong, Butler, and part of Allegheny. The third district comprised the counties of Cambria, Blair, Huntingdon, Centre, Clearfield, Elk, Cameron, Mckean, Potter, Clinton, Lycoming, Tioga, and Bradford.

212 Harold W. Aurand, "Mine Safety and Social Control in the Anthracite Industry," *Pennsylvania History,* vol. 54, No. 4 (October 1985); Trachtenberg, *The History of Legislation*, p. 200.

213 Michael Workman, *The Fairmont Coal Field: Historical Context* (Morgantown, W.Va.: Institute for History of Technology & Industrial Archaelogy, 1992), pp. 97-98; *Annual Report of Mining Activities, 1979* (Harrisburg: Department of Environmental Resources, 1978), pp. 94-95. Two major accidents occurred in Pennsylvania in 1907. The first at the Naomi Mine of the United Coal Company at Fayette City, Fayette County, on December 1 killed 34 miners. The tragedy was caused by an explosion, although few miners died from the blast itself. Most died from the rapid spread of black damp, a mixture of lethal gas which causes choking and death by suffocation. The second disaster happened at the Darr Mine of the Pittsburgh Coal Company near Jacobs Creek, Westmoreland County, which claimed 239 lives on December 19.

214 Williams, *Coal Dust in Their Blood,* p. 109; Trachtenberg, *The History of Legislation.* Trachtenburg's study, first published in 1917, is the most complete study covering nearly a century-long struggle by miners for protective legislation in the coal industry. Aurand, "Mine Safety and Social Control in the Anthracite Industry." The first federal coal mine inspection law was adopted by Congress in 1941.

215 *Pittsburgh: 50th Anniversary,* pp. 229-230. Charles Enzian, "Rock-dusting at Berwind-White Mines Costs Less than One Cent per Ton," *Coal Age,* vol. 30, No. 2 (July 1926); John E. Miller, "Permissible Explosives," *Coal Age,* vol. 16, No. 6, August 7, 1916.

216 Walton H. Hamilton and Helen R. Wright, *The Case of Bituminous Coal* (New York: Macmillan Company, 1926), p. 76.

217 Muriel Earley Sheppard, *Cloud by Day: The Story of Coal and Coke and People* (Pittsburgh: University of Pittsburgh Press, 1991), p. 110.

218 David Allen Corbin, *Life, Work, and Rebellion in the Coal Fields: The Southern West Virginia Miners, 1880-1920* (Urbana: University of Illinois Press, 1981), p. 10.

219 Mulrooney, *A Legacy of Coal,* p. 9.

220 Ibid., p. 26.

221 Bennett, *The Company Towns,* p. 1.

222 Long, *Where the Sun Never Shines*, p. 82.

223 Ibid., p. 80.

224 John Enman, "Coal Company Store Prices Questioned: A Case Study of the Union Supply Company, 1905-1906," *Pennsylvania History,* vol. 41, No. 1 (1974).

225 *Annual Report of the Secretary of Internal Affairs* (Harrisburg: Lane S. Hart, 1884), pp. 337-338.

226 *Ibid.*

227 Long, *Where the Sun Never Shines,* p. 123; Trachtenberg, *The History of Legislation,* p. 182.

228 Ibid., pp. 183-184.

Donohoe Coke Co.
Donohue $153,636.33   85,625.33   55.7
Latrobe - Connellsville C.& C. Co.
Saxman $105,511.28   18,108.60   17.2
Superior No. 1 $111,678.65   21,148.69   18.1
Superior No. 2 $10,625.86   2,794.13   26.3

229 Donald O. Edkins, comp., *Edkins' Catalogue of Coal Company Scrip* (New Kensington: The Catalogue Committee of the National Scrip Collectors Association, nd.), pp. 138-151.

230 Gordon Dodrill, *20,000 Coal Company Stores in the United States, Mexico and Canada* (Pittsburgh: privately printed, 1971); John K. Gate, *The Beehive Coke Years: A Pictorial History of Those Times* (Uniontown, PA: privately printed, 1991).

231 Linda Atkinson, *Mother Jones: The Most Dangerous Woman in America* (New York: Crown Publishers, Inc. 1978); Dale Fetherling, *Mother Jones: The Miners' Angel* (Carbondale: Southern Illinois University Press, 1974).

232 Ibid., p. 16.

233 Long, *Where the Sun Never Shines,* p. 335.

234 Jim Steiner, "The Noble Order of the Knights of Labor," *Mining Artifact Collector* (Fall 1989).

235 David J. McDonald and Edward A. Lynch, *Coal and Unionism: A History of the American Coal Miner's Union* (Silver Spring, Md.: Cornelius Printing Company, 1939), p. 23.

236 Watkins, *Coal and Men,* p. 166.

237 Roy, *A History of the Coal Miners,* pp. 271-272.

238 Biek, "The Miners of Windber," p. 293; Roy, *A History of the Coal Miners,* p. 278.

[239] McDonald and Lynch, *Coal and Unionism,* p. 23.

[240] Ibid., p. 25.

[241] Keith Dix, *What's a Coal Miner to Do? The Mechanization of Coal Mining* (Pittsburgh: University of Pittsburgh Press, 1988), p. 116; Roy, *A History of the Coal Miners,* pp. 279-280; Williams and Yates, *Upward Struggle.*

[242] Morton S. Baratz, *The Union and the Coal Industry* (New Haven: Yale University Press, 1955), p. 52.

[243] Long, *Where the Sun Never Shines,* p. 154.

[244] Ellis W. Roberts, *The Breaker Whistle Blows* (Scranton: Anthracite Museum Press, 1984), pp. 73-75.

[245] Irving Bernstein, *The Lean Years: A History of the American Worker, 1920-1933* (Boston: Houghton Mifflin Company, 1960), p. 126; Roberts, *The Breaker Whistle Blows,* p. 79.

[246] Long, *Where the Sun Never Shines,* p. 163.

[247] Smith, *Digging Our Own Graves,* p. 46.

[248] Coleman, *Men and Coal,* p. 63; McDonald and Lynch, *Coal and Unionism,* p. 48.

[249] Dix, *What's a Coal Miner to Do?* p. 150; Coleman, *Men and Coal,* pp. 58-80; McDonald and Lynch, *Coal and Unionism,* pp. 48-79.

[250] "Public History—Workers' History: "For Our Own Goods." *Pennsylvania Heritage* vol. 58, No. 4 (October 1991).

[251] George S. Rice, "Should New Mines Be Opened?" *Coal Age,* vol. 13, No. 16 (April 1918).

[252] Rochester, *Labor and Coal,* p. 15.

[253] Charles Reitell, *The Shift in Soft Coal Shipments* (Pittsburgh: University of Pittsburgh Press, 1927), p. 10. Pennsylvania produced 31 percent of the nation's bituminous coal in the five-year period from 1922 to 1926.

[254] Sisson, Bomberger, and Reed, "Iron and Steel Resources," p. 19.

[255] McDonald and Lynch, *Coal and Unionism,* p. 130.

[256] Robert A. Korcheck, *Nemacolin: The Mine—The Community 1917-1950* (Privately printed, 1980), p. 167.

[257] Martin L. Primack and James F. Willis, *An Economic History of the United States* (Menlo Park, Calif.: Benjamin/Cumming Publishing Company, 1980), p. 319; Patrick Renshaw, *The Wobblies: The Story of Syndicalism in the United States* (New York: Anchor Books, 1967).

[258] David Brody, *Steelworkers in America: The Nonunion Era* (New York: Harper & Row, Publishers, 1960), pp. 231-262. Chapter 12 is an excellant narrative of the "The Great Steel Strike of 1919." Sisson, Bomberger, and Reed, "Iron and Steel Resources," pp. 105-107.

[259] Heald, *Fayette County,* pp. 84-87, 216.

[260] *Blossburg Centennial, 1871-1971* (Blossburg, 1971), pp. 6-7.

[261] Coleman, *Men and Coal,* p. 98.

[262] Jeremy Brecher, *Strike* (Boston: South End Press, 1972), p. 136. Keeny was head of District 17, West Virginia district of the UMWA. His stronghold was located in the Kanawha and Fairmont Fields in the northern part of the state.

[263] Berton Braley, "The Miner," *Coal Age,* vol. 13, No. 7 (February 1918).

[264] Leo Wolman, *Ebb and Flow in Trade Unionism* (New York: National Bureau of Economic Research, 1936), p. 217.

[265] James P. Johnson, *The Politics of Soft Coal* (Urbana: University of Illinois Press, 1979), p. 26.

Retrenchment, Decline, and the Mechanized Mine, 1920-1945

Introduction

The 1920s were characterized by most contemporary American economists as a period of robust expansion for the domestic economy. A leading economist confidently asserted at the end of the decade that the United States was entering into an era of "permanent prosperity." However, the prosperity decade predicted by many economists did not come true for some industries, including the bituminous coal industry. It was recognized by many as "a mighty sick industry" in an otherwise expanding and robust economy. The 1920s were an economic and social catastrophe for both union and nonunion miners, the United Mine Workers of America, and the nation's coal companies. The demand for coal during and immediately following World War I, and the attendant high price of coal, led to a great expansion in production capacity.[1] The industry's economic prospects soured during the 1920s, as demand which had been temporarily increased during the war years began to level off. The American bituminous coal industry, which had experienced nearly three decades of phenomenal growth, was entering into a prolonged crisis of overproduction. Government and private coal commissions identified this "over development" and "over expansion" of the industry as the principal causes of the long-term coal crisis of the 1920s and 1930s. Coal output nationally in each successive decade since the 1850s had practically doubled. The United States Bituminous Coal Commission (1920) and the Brookings Institution (1926) noted the productive capacity of the American bituminous coal industry at over 800 million tons per year, although annual production never exceeded 579 million tons (1918). By 1926, the capacity of American mines stood at one billion tons, although the demand for coal was half that amount. Soft-coal sales reached nearly 569 million short tons in 1920. But except for 1926, sales fell below that throughout the 1920s. Employment in the coalfields peaked in 1923, when the industry employed over 862,000 men and boys. A popular story told in the many mining communities during the period addressed the fundamental problems plaguing the industry. A miner's son asked his mother, "Why don't you light the fire? It's so cold." "Because we have no coal. Your father is out of work, and we have no money to buy coal." "But why is he out of work, Mother?" "Because there's too much coal."[2]

The bituminous coal industry had developed over a century and a half without a coherent, long-term economic plan, and according to historian James P. Johnson, had become "a splintered and competitive industry that could not find a means to save itself through industrial self-government."[3] The industry was characterized by great seasonal and annual irregularity, recurrent and violent labor difficulties, and unrestrained competition with uncontrolled prices and speculation. These problems were collectively called the "Chaos of Coal" within the industry. The severe economic decline in the industry during the 1920s and 1930s was also exacerbated by new competition from alternative fuels like oil, natural gas, and hydroelectricity. Operation of both union and nonunion mines created a wide wage differentiation within the industry. As long as there was the potential for an oversupply of coal, some form of regulation of price and production seemed necessary to effect stability in the industry. Many leading coal companies decried these conditions and the lack of a stable pricing system, although the industry as a whole was unwilling or simply unable to find adequate remedies for these problems.

These apparent structural weaknesses within the industry were inadequately addressed by the divided and highly competitive coal companies, the UMWA, and

Coal Production in the United States

	Bituminous	Anthracite	Total	Percent Increase/Decline	Mines
1920	568,667	89,598	658,256	1910-1920, 31.2%	8,921
1925	520,053	61,817	581,870		7,144
1930	467,526	69,385	536,911	1920-1930, 18.45	5,891
1935	372,373	52,159	423,532		6,315
1940	460,772	51,485	512,257	1930-1940, 4.6	6,234
1945	577,617	54,934	623,551		7,033[4]

the federal government during the 1920s; consequently the bituminous coal industry witnessed increased interregional competition between union and nonunionized operators in a shrinking market. These harsh economic realities fundamentally transformed existing relationships between management and workers in the bituminous coal industry. The Great Depression of the 1930s simply worsened the basic situation of the industry, the depression not having created the inherent structural shortcomings, all of which had been left unresolved during the 1920s. It was only government intervention and regulation of the industry that permitted a temporary revival of the industry between 1933 and 1945 and the rebirth of the moribund miners' union.

The bituminous industry of the United States entered a period of decline following World War II that continued during the 1950s and 1960s. Coal was soon replaced as the nation's principal source of energy, first by petroleum and then by natural gas. Output and employment declined in Pennsylvania's soft coal industry as many of the traditional markets disappeared. The large coal companies, in an attempt to remain competitive with oil and natural gas, survived by embracing mechanized mining practices following the Second World War. Tens of thousands of small and medium-size coal companies were forced into bankruptcy or were acquired by large firms.

The Challenge to "King Coal" from Alternative Sources of Energy

Coal had replaced wood as the nation's principal energy source by the 1880s and its dominance among competing fuels in the United States was unchallenged until the 1920s. From 1901 to 1905 coal (bituminous 70.7 percent and anthracite 18 percent) supplied 88.7 percent of the nation's energy needs, compared to domestic oil 6 percent, natural gas 3.2 percent, and water power 2.1 percent.[5] Oil's commercial history began with the gusher brought in by "Colonel" Edwin Drake (1819-1880) at Titusville, Pennsylvania, in August 1859. Oil markets grew as crude oil production increased sevenfold between 1900 and 1920. Its principal use was gasoline for automobiles and kerosene for illumination, cooking, and home heating. Gasoline had replaced kerosene as the oil industry's principal product by 1920. The natural gas era dates from the 1880s, when it was used for heating furnaces, making steam at iron and steel works, and heating kilns at glass factories. The natural gas industry was limited to factories in Ohio, West Virginia, and Pennsylvania.

This new competition was a subtle and long-term trend as competing fossil fuels began to erode incrementally coal's market share in the 1880s. Although competition from natural gas and oil was beginning to be felt in the industry, coal still provided the nation's principal source of energy. Coal produced 68 percent of the fuel consumed in 1924 (bituminous 56.9 percent, anthracite 11.1 percent), while oil provided 21.6 percent (domestic 19.4 percent and imported oil 2.2 percent,; natural gas provided 5.2 percent and water power 5.2 percent.[6] Bituminous coal's share of the energy industry declined to 55 percent in 1930, 48 percent in 1940, 39 percent in 1950, and 26 percent in 1960.[7] This challenge to coal from these alternative fossil fuels accelerated following World War II. The proportion of energy supplied by bituminous coal dropped from nearly 60 percent to approximately 45 percent between 1940 and 1950. Petroleum's market share rose from about 17 percent to 23 percent during this decade, while natural gas generated 23 percent of production, up from 12 percent in 1940.[8] Petroleum (1952) and later natural gas (1962) eclipsed bituminous coal as the nation's premier energy source.

UMWA leadership and the rank and file during the 1920s still believed in "King Coal." They accepted explicitly the 1890 preamble to the constitution of the United Mine Workers of America that stated, "There is no truth more evident than that without coal there could not have been such marvelous social and industrial progress as makes present-day civilization." John L. Lewis was elected the ninth

UMWA president on New Year's Day 1920 and declared repeatedly throughout the 1920s to his coal miners that "when we control the production of coal we hold the vitals of our society in our hands."[9] Most coal miners still saw themselves as "the creators of all wealth and power." These competitive fuels might chip away at the market share of the coal market; they would not replace coal as the dominant fuel, at least in the immediate future.

Decline in Demand and Use for Bituminous Coal

A more immediate cause of concern to the bituminous coal industry during this period was the slackened demand for coal and coke from traditional markets. High coal prices and new technologies encouraged a trend toward economy in fuel consumption by consumers of coal and coke. Ironically, the consumption of bituminous coal in the mines of Pennsylvania declined. Mines consumed 1.9 percent of their production in the generation of power and heat in 1915, while their coal use declined to .07 percent in 1929. This reduction was caused by mine electrification and energy conservation programs after 1900. Electrically powered machinery, used in undercutting, blasting, and hauling coal, had replaced coal steam-powered machinery in the larger mines.[10]

The periodic coal shortages caused by strike activity, wildcat work stoppages, and increasing coal prices, especially between 1918 and 1923, stimulated a long-term energy-saving program by the principal coal users. Bituminous coal sold for only $1.12 per ton as late as 1910, but between 1916 and 1922 the price per ton more than doubled.[11] Railroads, electric power plants, iron and steel companies, manufacturing industries, and domestic users were the major consumers of bituminous coal. The industry produced 443,492,000 tons in 1915, and the principal coal consumers, in order of significance, were industrial establishments (including electric utilities) 36 percent, railroads 27.5 percent, coke manufacturers 14 percent, households 12.4 percent, the export market 4.2 percent, steam-driven ocean liners (bunker coal) 2.7 percent, used at the mines 2.2 percent, and gas manufacturers 1 percent.[12]

Railroad locomotives were major consumers of coal. They consumed about 135 million tons of bituminous coal, nearly one-quarter of the nation's total consumption at its peak just after World War I. Many large railroad companies operated their own "captive" mines to supply their huge demand for steam coal for their locomotives. The volume of rail transportation continued to grow from 1918 through 1929, but more-efficient designs in steam locomotives reduced their use of coal by almost 30 percent during this period. An average locomotive consumed 174 pounds of coal per thousand freight-ton miles in 1920, 125 pounds in 1929, and 121 pounds in 1933. The conversion to diesel locomotives after World War II ended the reign of coal-fired steam locomotives and with this conversion a 135-million-ton consumer was gone.

Coal use by utilities in the production of electrical energy was reduced by 48 percent per kilowatt-hour of electricity generated between 1919 and 1928. By 1900 electricity had been applied to railroading and was used to power coal-mining machinery. Utilities used on average 3.2 pounds of coal for each kilowatt-hour of electricity generated in 1919, 1.76 pounds in 1928, 1.42 pounds of coal in 1937, and less than .08 pound in 1952.[13] The production of electric power, between 1920 and 1940, increased about threefold.

This trend in fuel efficiency continued in the production of iron and steel, which increased from 1904 to 1929 although manufacturers reduced their consumption of both coal and coke. The number of tons of coal used per gross ton of iron and steel was reduced from 2.01 tons in 1904 to 1.85 tons in 1914 and 1.41 tons in 1927. Coke has been the principal mineral fuel in iron and steel production since the 1880s, and the amount of coke required to make a ton of pig iron was reduced from 1.6 tons in 1860 to 1.2 tons in 1890, 1.0 in 1920, and 0.9 in 1940.[14] The continual shift of coke production from the beehive coke oven to the by-product coke oven was a major source in the reduction of coal and coke by the steel industry. This transition was most pronounced in the United States with the start of World War I. By-product ovens, introduced during the 1890s, were more efficient users of fuel. Pennsylvania produced 61.1 percent of the nation's coke production as late as 1915, and beehive coke ovens in western Pennsylvania produced 54.2 percent.[15] A majority of coal used in beehive coke production came from the famous Pittsburgh seam of western Pennsylvania and northern West Virginia.[16]

The growth of the beehive coke industry in southwestern Pennsylvania was spectacular, but its decline was even more striking. Beehive ovens produced 40 percent of the nation's coke in 1920, 6 percent in 1930, and 5 percent in 1945. The beehive coke industry of the famous Connellsville district hit its production peak in 1918 when 38,986 ovens produced 18,135,000 tons of coke valued at $117,000,000.[17] The number of beehive ovens in this coke district declined after this peak period from 15,333 in 1925 to 7,393 in 1930, and 4,355 in 1935. There were fewer beehive coke ovens in operation in 1927 than had operated in the district during the 1880s. The amount of coal used for coke production in the antiquated beehive ovens of Pennsylvania between 1915 to 1929 decreased from 32,498,000 to 7,308,000 tons.

The H. C. Frick Coke Company, the largest beehive coke producer in the United States, reported only twenty-five beehive coke plants in operation on December 31, 1929. The plants, located in Fayette, Washington, Greene, and Westmoreland Counties, had only 9,324 ovens, including 2,261 inactive.[18] There were only seventy-one beehive coke plants in the nation, with 13,012 ovens, in 1936 and beehive coke production was usually restricted to boom periods.[19] A quarter of the annual production of coke in the United States was produced in by-product ovens in 1912, 60 percent in 1920, 94 percent in 1930, and 95 percent by 1940. Raw coal, not coke, was increasingly shipped from the coalfields by river barges directly to the steel plant and converted to coke in the batteries of by-product coke ovens. The hauling of coal by railroads between 1915 and 1929 was reduced as river shipments on huge barges almost trebled during this brief fourteen-year period.

	Beehive and By-Product Coke Production in the United States				
	By-Product Ovens (net tons)	**Beehive Ovens** (net tons)	**Total**	**By-Product Ovens** (percent)	**Beehive Ovens** (percent)
1920	30,833,951	20,511,092	51,345,043	60 percent	40 percent
1925	39,912,159	11,354,784	51,266,943	78 percent	22 percent
1930	45,195,705	2,776,316	47,972,021	94 percent	6 percent
1935	34,224,053	917,208	35,141,261	97 percent	3 percent
1940	54,014,300	3,057,800	57,072,100	95 percent	5 percent
1944	66,627,381	6,990,308	73,617,689	90 percent	10 percent[20]

The increasing acceptance of the by-product coke oven by steel corporations diminished the need for beehive ovens and high-quality Connellsville coal. The trend in the coke industry from beehive coke to by-product coke is clearly shown by the following figures:

	1893	1923	1937
Beehive coke produced, net tons	9,464,730	19,379,870	3,156,300
By-product coke produced, net tons	12,850	37,957,664	49,205,798[21]

The change in coke production from beehive to by-product ovens resulted in a complete change in the location of the coke industry. Beehive coke was produced near the mine, where the ovens were located. The acceptance of the by-product ovens, located at the steel plant and not at the coalfield, ended the competitive advantages that the Connellsville district had enjoyed for generations in three distinct ways:

(1) By shifting most of the coke industry out of the Connellsville area to steel-producing centers all over the country.
(2) By further encouraging the substitution of non-Connellsville coal in coking through provision of greater latitude in the coking process.
(3) Indirectly, by exposing the steel industry to the geographical attraction of supplies of coal and coking other than those of the Connellsville area. Shifts in the steel industry have been important in terms of the market situation for the region's coal.[22]

The Department of Mines of Pennsylvania wrote an obituary for the production of coke in the beehive oven in its 1929 annual report:

> Bee-hive coke production is rapidly becoming a lost art as the advantages of by-product coke manufacture are so apparent that it is only a question of a short time until the banks of abandoned bee-hive ovens scattered throughout the coke region will be the only symbol remaining of what was once the leading industry of southern Pennsylvania.[23]

Many beehive coke plants were simply abandoned as demand for coke decreased, and the ovens remained unused except for brief periods of unusually high demand. Homeless families moved into the deserted ovens during the Depression and there established legal residences so they might draw their relief checks. The increased demand for coke in the steel industry during World War II brought a temporary revival of the moribund beehive coke industry as abandoned ovens were repaired or entirely rebuilt and placed in operation. The annual beehive coke capacity in 1944 was 22 percent higher than in 1940, although it accounted for only 10 percent of the nation's coke production. The expansion of by-product and beehive coking capacity enabled the steel industry to meet essential wartime requirements for metallurgical fuel. This economic revival was essentially the "last hurrah" for the venerable beehive oven. Coke production was negligible nationally after the conclusion of the war, although Jones and Laughlin Steel Corporation's Aliquippa Works, Beaver County, made coke in beehive ovens as late as 1948. New coke ovens constructed in the United States after 1920 were almost exclusively by-product ovens, which produced more efficiently. They produced high-quality metallurgical coke and recovered gas and other valuable chemicals lost in the earlier process. A major exception to this long-term trend was the construction of a battery of beehive coke ovens at Lucerne Mines, near Homer City, by the Rochester and Pittsburgh Coal Company of Indiana County. The 264 ovens were constructed at a cost of $2 million in 1952.[24] High-quality coke produced at this plant was sold and used in local blast furnaces. The coking plant was operated by the company until 1957, when the facility was sold to Shenango, Inc. The ovens operated continuously until the Pennsylvania Department of Environmental Resources (D.E.R.) forced their permanent closure in October 1972. The owners were simply unable to meet strict air-quality standards established by the state. Similar beehive coke operations at Shoaf, Fayette County, and at Alverton, Westmoreland County, were closed during the 1970s by D.E.R. The beehive coke plant at Shoaf was one of the last major producers of beehive coke in the United States. The H. C. Frick Coke Company constructed this beehive coke plant in 1902. The Shoaf facility, located about seven miles southwest of Uniontown, was later owned by the Menallen Coke Company, which shipped its last railroad cars of beehive coke to market in March 1972. The owners of these ovens attempted unsuccessfully to modify their design to comply with the stringent air-quality standards established by D.E.R.

The economies of Westmoreland and Fayette Counties, based principally since the 1880s on coke production from beehive ovens, saw the closing of mining and coke plants after 1945. This permanent economic downturn was as devastating to the region's economy as the Great Depression. Fayette County had a higher percentage of unemployment and persons on welfare than the state or nation as a whole. The population of the county declined by 25 percent between 1950 and 1970, from 200,909 to 154,667, as former miners left the county in search of employment.

The Bituminous Coal Industry of Pennsylvania during the 1920s and 1930s

Pennsylvania employed more miners than any other coal-producing state in 1920. There were 184,168 bituminous workers and 149,117 anthracite workers, accounting for 43 percent of all coal miners in the United States. West Virginia was second with 105,000 workers while Kentucky and Illinois ranked third and fourth.[25] The number of coal miners in the Commonwealth had shrunk by 1945 to 174,989 while production had dropped to about 185 million net tons (the 1945 figures are combined anthracite and bituminous production totals). Pennsylvania's soft coal industry had employed more than a hundred thousand workers continuously from 1903 to 1943, but the industry employed fewer than a hundred thousand workers (99,942 employees) for the first time in 1944.[26] In 1931, for the first time since 1904, bituminous coal output in Pennsylvania fell below 100 million tons.

Pennsylvania producers, as a group, had not held their own against rival producers in competing coalfields. The Commonwealth averaged more than 35 percent of the bituminous coal market from 1890 to 1920, but its average market share nationally shrank to about 25 percent during the 1920s. The center of the American bituminous coal industry was the Appalachian fields extending from northern Pennsylvania, across Ohio, West Virginia, Kentucky, and Tennessee into Alabama. Coal resources in the interior and western coalfields were of poorer quality. Pennsylvania, West Virginia, Illinois, Kentucky, and Ohio mined about four-fifths of the nation's production in 1923, while the coal states of Pennsylvania and West Virginia extracted almost one-half of all bituminous coal. Pennsylvania's production experienced a fairly gentle decline after 1921, but during the 1930s the downward trend became more pronounced.[27] Pennsylvania's national share declined from 36 percent in 1912 to 26 percent in 1931 as output dropped from approximately 173 million tons in 1923 to 96.8 million in 1931. Twenty-four counties in Pennsylvania mined coal between 1918 and 1939, although Allegheny, Cambria, Clearfield, Fayette, Greene, Indiana, Somerset, Washington, and Westmoreland Counties were the principal coal counties. Five adjacent counties in southwestern Pennsylvania, Fayette, Washington, Allegheny, Cambria, and Westmoreland, were producing about 70 percent of the state's total by 1920. Pennsylvania production was 19.6 percent less in 1929 than in 1918 (the production peak), and 17.4 percent less than in 1913.[28] Seven of the eight leading counties produced less coal during this period. Fayette, Washington, Cambria, and Allegheny Counties suffered declines of about 12 percent between 1917 and 1929 while Westmoreland County's decline of 38.7 percent was the largest. Greene County, which had the largest coal reserves in the state during this period, was the sole exception to this trend as output increased from about nine hundred thousand tons in 1917 to 6.2 million in 1929. The original coal deposits in the county were estimated at more than nine billion tons. Greene County was part of the Pittsburgh district, but annual production in this rural county was modest until the 1920s. The coal deposits of eastern Greene and eastern Washington Counties were nearly identical geologically with those of the Connellsville district. This coal has sufficient carbon content to make good coke. The rich Pittsburgh seam that underlies Greene County from the Ruff Creek district east to the Monongahela River, and south through Jefferson to Muddy Creek, and continues south toward the state line had been opened for coal exploration since 1900. Steel companies, including U.S. Steel, Crucible Steel, Inland Steel Corporation, and Picklands, Mather and Company, purchased coal lands in the county around this period and constructed a number of large "captive" mines with accompanying mining communities. The county produced more than a million tons annually between 1918 and 1945.[29]

Shifts in American Bituminous Coal Production Million net tons			
State	1912-1916	1917-1921	1922-1926
PA	811	788	703
IL	304	394	342
OH	146	200	155
WV	373	418	559
VA	42	48	59
KY	103	156	250
TN	31	29	27
MD	22	18	12
IN	86	127	113[30]

Bituminous coal production in the nonunion coal states of West Virginia and Kentucky made remarkable gains in contrast to Pennsylvania's declining industry. These two southern Appalachian mining states produced less than 11 percent of the nation's coal in 1895 in contrast to Pennsylvania's 37.1 percent. West Virginia's (23.7 percent) and Kentucky's (10.1 percent) proportions of the nation's soft coal production rose to 33.8 percent while Pennsylvania's percentage declined to 26.7 percent in 1925.[31] West Virginia permanently displaced Pennsylvania as the leading soft coal producer during the 1930s. The coal-producing states, in descending order of importance, mining more than 10 million tons in 1940 were West Virginia, Pennsylvania, Illinois, Kentucky, Ohio, Indiana, Alabama, and Virginia.

Some of the principal causes put forth by mining experts to explain the decline of the once flourishing coal industry in Pennsylvania during this period were (1) the competition of soft coal from southern coalfields which, encouraged by preferential rates, had been able to ship coal beyond the boundaries of their natural markets; (2) the competition of other sources of energy such as fuel oil, natural gas, and electric power; (3) increasing efficiency in coal burning and in the use of fuel; (4) previous overdevelopment in mine capacity; (5) curtailment of demand due to the Depression; and (6) lower production costs in southern fields.[32] A variety of remedies were proposed to stop this erosion. However, all these programs were generally ineffective in reversing this downward spiral. All indices of activity in the industry, with the exception of capital investment—value of products, number of mines, total employees, total compensation, and tons produced—were lower in 1930 than in 1923.

Decline of the Bituminous Coal Industry in Pennsylvania		
	1923	1930
Value of Products	$453,003,200	$209,274,300
Capital Invested	443,516,800	473,695,000
Number of Mines	1,617	785
Total Employees	189,226	128,905
Total Compensation	314,807,100	140,982,700
Tons Produced	172,158,436	121,384,040[33]

Coal Companies' Response to Their Declining Industry

Both union and nonunion coal operators employed a variety of strategies to maintain their market share in an ever-shrinking and increasingly competitive market. Mergers of coal companies and informal pools were formed to control and limit production and establish consistent and long-term tonnage rates. These methods had been used since the 1890s by coal operators but were generally ineffective. The price per ton of coal plunged from a historic high of $3.75 per ton in 1920 to $1.78 in 1929. A number of large coal companies were firmly established through consolidation by the 1920s. Coal production was increasingly concentrated in larger mines. Class 1 mines, those producing two hundred thousand tons or more annually, accounted for 31.5 percent of total coal production in 1922 and 65.2 percent of the nation's output in 1929. But no single company produced more than 5 percent of coal production nationally. The industry, in sum, was increasingly concentrated into fewer and larger companies and in larger mines; but no one firm or group of firms was dominant. Nationally 12 percent or 553 bituminous coal operators produced 79.7 percent of all bituminous coal in 1929; the seventeen largest operators produced 19.9 percent of the total tonnage. There were 218 coal producers nationwide who produced half a million tons or more in 1929. These large producers mined 59.8 percent of the total bituminous coal production of about 534 million tons in 1929.[34] The seventeen largest coal operators, who mined nearly 20 percent of the nation's bituminous coal, each mined more than three million tons of coal annually. Ten of these operators sold their coal and coke on the commercial market while the remaining seven companies operated "captive" or "consumer" mines. Chicago, Wilmington, and Franklin Coal Company, Consolidation Coal Company, Island Creek Coal Company, New River and Pocahontas Consolidated Coal Company, Old Ben Coal Company, Peabody Coal Company, Pittsburgh Coal Company, Pocahontas Fuel Company, W. J. Rainey Company, and Stonega Coal and Coke Company were the ten largest in 1929. Five of the seven captive coal operators were steel corporations: the H. C. Frick Company, the Tennessee Coal, Iron and Railroad Company, and the United States Coal and Coke Company—U.S. Steel, parent company; Bethlehem Mines Corporation—Bethlehem Steel, parent company; and Vesta Coal Company—Jones and Laughlin Steel, parent company. The Superior Coal Company, whose parent company was Chicago and Northwestern Railroad, and the Union Pacific Coal Company, whose parent company was Union Pacific Railroad, were the largest operators of captive mines owned by corporations.

Six of the largest seventeen coal operators in the nation operated mines and coke plants in the four bituminous coalfields of Pennsylvania. They were the Consolidation Coal Company, Pittsburgh Coal Company, W. J. Rainey Company, Bethlehem Mines Corporation, H. C. Frick Company, and Vesta Coal Company (subsidiary of Jones and Laughlin Steel Corporation).[35] There were thirteen hundred coal operators in Pennsylvania in 1930. One hundred operations produced more than two hundred thousand tons annually, nearly 80 percent of the state's bituminous coal production. There were twenty-five corporations with annual production exceeding more than a million tons, while the ten largest corporations had annual production over two million tons and each employed more than two thousand workers. Pittsburgh Coal Company was the largest employer (11,485), while the H. C. Frick Coke Company was the leading coal producer in the state (12,878,579 tons). The ten largest bituminous coal companies and counties with mining operations in Pennsylvania in 1930 were as follows:

Company	Counties with Mines
H. C. Frick Coke Co.	Fayette/Greene/Westmoreland
Pittsburgh Coal Co.	Allegheny/Fayette/Washington/Westmoreland
Vesta Coal Co.	Washington
W. J. Rainey, Inc.	Fayette/Greene/Washington/Westmoreland
Westmoreland Coal Co.	Westmoreland
Pittsburgh Terminal Coal Corp.	Allegheny/Washington
Rochester and Pittsburgh Coal and Coke Co.	Clearfield/Indiana/Jefferson
Bethlehem Mines Corp.	Cambria/Indiana/Westmoreland
Hillman Coal and Coke Company	Allegheny/Fayette/Somerset/Westmoreland
Berwind-White Mining Company	Somerset/Cambria[36]

This concentration of bituminous coal production nationally and within Pennsylvania was neither a monopoly nor an oligopoly in the classic business sense. The bituminous coal industry remained fiercely competitive, unlike the anthracite industry, which had established a coal monopoly. The anthracite region of northeastern Pennsylvania was controlled by a small number of railroad companies, known simply as the "Companies": the Delaware, Lackawanna and Western; Delaware and Hudson; Lehigh Valley; Reading; Erie; New York, Ontario and Western; Lehigh and New England; Central of New Jersey, and Pennsylvania. These railroad companies owned and controlled about 75 to 80 percent of all anthracite mined and 90 percent of future supplies in 1923. The bulk of anthracite extracted from these mines was sold directly to retailers or consumers through their own selling departments, affiliated selling companies, or contract commissioners. The remaining anthracite was produced by over one hundred relatively small producers, known as the "independents in the trade."[37]

Large corporations produced a majority of the bituminous coal and coke, and employed a majority of workers, but they could not dictate production tonnage or maintain or "fix" the price of coal throughout the industry. Small or medium-size companies did not disappear; in fact, their numbers continued to increase throughout this period. There were 4,612 companies operating 6,057 mines producing 534 million tons in the United States by 1929. These small and medium-size mines posed a constant threat to price stability in the industry. C. E. Lesher, editor of *Coal Age*, in 1921 addressed why he believed the bituminous coal industry was unable to form a monopoly:

> So easy is the coal of access and so easy is the initial work of opening a mine that every period of unusual demand, in which prices rise more than a few cents above the cost of production, find many entering the business. Having your own coal mine is almost as simple as having a war garden. Raising hogs, cotton, and corn are no more competitive than bituminous coal.[38]

Coal mining, unlike other extractive industries, remained essentially a labor- and not a capital-intensive industry. A new coal operator required very little capital to establish a mining operation. The drift-entry mine required no machinery to open and its maintenance cost was low. There was also a surplus of unskilled immigrant labor to pay a low piece-rate, by the ton, to mine coal. The mine could be opened quickly if demand and price rose, and easily closed if demand and tonnage rates were small. High-quality bituminous coal was still dispersed over a wide area in six principal coalfields in the United States. High-quality bituminous coal occurred in Pennsylvania and was mined in more than two dozen counties. None of the large-scale coal companies operated in more than four counties at one time. The means to impose some form of monopolistic control or government regulation to establish minimum coal prices and regulate production were not forthcoming during the 1920s. Controls were finally imposed on the splintered industry by the federal government during the first New Deal.

A second distinction that made it such a cut-throat industry was its operation simultaneously with union and nonunion mines. Journals as diverse as *The New Republic* and *Coal Age* asked, "Can coal continue two-thirds free (unionized) and one-third unfree (unorganized)."[39] The unorganized southern coalfields acted "like a pistol pointed at the heart of the union" during this entire period.[40] Unionism prevailed in the Pittsburgh district and northern portion of Pennsylvania, in the Broad Top Coal Field, and in Illinois, Indiana, and Ohio. The union prevailed in parts of West Virginia—Fairmont Field until 1925, Kanawaha Field until 1924, and New River Field until 1922. These states had been strongholds of unionism since the formation of the Central Competitive Field by northern coal operators after the 1897 strike. A second, smaller region of unionism was the South Western Interstate Field, lying between Iowa and northern Texas and consisting of Arkansas, Kansas, Missouri, and Oklahoma. North of the South Western Interstate was the organized Iowa Coal Field and in the south the partially organized state of Texas. Union fields were also located in the Rocky Mountains in Wyoming and Montana.[41]

The predominant nonunion areas in western Pennsylvania were the Connellsville coke district, Irwin gas region, the Greensburg, Latrobe, and Ligonier area of Westmoreland County, and the Somerset-Meyersdale Field in Somerset County. The coalfields of southern Appalachia that had developed since the 1880s had increased the number of nonunion mines. Coal production had been insignificant in these fields in comparison to the northern coalfields until this period. The high demand for coal and the high price per ton during the Great War prompted local operators to expand their production. The war had brought on overproduction and these nonunion southern operators had not reduced their output as demand for coal diminished in the 1920s.

The recent coal boom of World War I and 1919 and 1920 was soon to become a memory. The continual decline of coal prices from the historic high of $3.75 a ton in 1920 to $1.78 a ton in 1929 prompted coal operators to slash wages and increase hours of work to remain competitive and solvent. A number of variables were used in calculating the operating cost of mining coal, including royalties, operator association dues, compensation insurance, taxes, power used at the mine, and labor. Labor costs represented from 60 to 70 percent of the cost of producing each ton of coal. They accounted in 1918 for 69.3 percent of the operating cost in the union-operated mines in the Pittsburgh district; 65.6 percent in the nonunion mines in Pocahontas, West Virginia; and 67.7 percent in the union mines of central Illinois. These new economic realities made it increasingly difficult for the older northern unionized mines in the Central Competitive Field, which paid higher wages, to compete effectively and to maintain their market share against nonunion operators in a shrinking industry. The union mines reduced production and wages and laid off workers in an attempt to remain solvent. Paradoxically, as union operators reduced their production, nonunion Appalachian operators increased production. Their labor costs were less and, therefore, they had an economic advantage even in a shrinking market.

The nonunion operators of western Pennsylvania and southern Appalachia, according to union-mine operators, had an unfair competitive edge in the production and sale of coal throughout the period. They paid lower wages, produced superior-quality coal, and received favorable freight rates from the railroads. The southern Appalachia coalfields, unlike Pennsylvania's sagging coal industry, increased their production throughout the 1920s. Coal production increases from this region were so large that they more than offset the declining production in Pennsylvania, Ohio, Indiana, and Illinois. The ratio between coal mined by union and nonunion companies was about two-thirds to one-third in 1922-1923. West Virginia produced 23.7 percent of the nation's bituminous coal production while Kentucky produced 10.1 percent in 1925. Some 42 percent of all bituminous coal workers and 32 percent of all coal miners were working in nonunion southern mines by 1929.[42]

Percentage of Bituminous Coal Mined by Principal States

Year	PA	WV	KY	OH	IN	IL
1895	37.1	8.4	2.4	9.8	2.9	13.1
1905	37.5	11.9	2.6	8.1	3.7	12.2
1915	35.6	17.4	4.83	5.0	3.8	13.2
1925	26.7	23.7	10.1	4.9	4.2	13.1[43]

Coal companies employed a variety of strategies to remain competitive and profitable in a declining and overdeveloped industry. The collapse of the war boom and the crisis of overproduction and declining coal prices pushed coal operators into a fever of competitive cost-cutting strategies. Overcapacity brought on cutthroat competition that forced many companies into receivership and bankruptcy. Mass unemployment of miners began with the collapse of the coal market in 1923, when 1,745 mines closed throughout the nation, squeezing at least eighty-five thousand miners out of the industry between 1920 and 1923.[44]

The reduction of labor costs became the coal operators' point of attack to remain competitive. They attempted to decrease wages and increase the daily hours of work throughout the 1920s. A number of union-organized mines began experimentation with mechanical loaders (led by the Pittsburgh Coal Company) in an attempt to reduce their labor costs, but mechanizing coal loading was not a major factor in solving the problems of high labor costs during the 1920s. By 1920, 60.7 percent of all coal being mined was mechanically undercut by electric machines but less than 1 percent (.3 percent) was mechanically loaded nationally.[45]

Practically all bituminous coal was shipped as run-of-mine coal throughout the nineteenth century, unlike anthracite coal. "Rashed coal" is unpure and unmarketable coal, mixed with clay, slate, shale, or other foreign substances. Anthracite coal, on the other hand, was sorted into at least nine different sizes at the breaker before delivery to market. Most anthracite was used for home heating, stores, factories, and the like. In contrast, the largest consumers of bituminous coal were railroads, factories, coke ovens, and electric and gas-producing plants.[46] Some larger firms attempted to improve the quality of the coal that they shipped to market during the 1920s. Coal operators, facing keen competition in a shrinking market, believed that improving the quality of their coal would make it easier to sell. Cleaning coal upgraded its quality and heating value by removing or reducing the amount of pyretic sulfur, rock, clay, and other ash-producing material. Coal-washing equipment was installed at many mines to enhance the marketability of coal. Consumers were increasingly insistent upon the best quality of coal. The demand

Preparation plant at Mather Colliery, Mather, Greene County, c1980. Carmen DiCiccio.

for most types of coal was decreasing overall, but consumer demand for certain sizes was increasing. Slack (fine) size coal was in demand as fuel with the development of new steel furnaces after World War I. These furnaces used small pieces of coal that could not be hand cleaned, so coal operators were required to construct cleaning plants to market high-quality slack coal.

Companies constructed elaborate and expensive preparation or cleaning plants during the 1920s, hoping to attract a broader market and offset the declining demand for coal. Before the construction of the large preparation plants, raw run-of-the-mine coal was only roughly screened and sized in tipples. Coal was then dropped into railroad cars below for shipment to market by railroad or by river. Coal screening in the tipple was inefficient. There were no loading booms and no facilities for removing impurities in the coal. The treatment of coal was resumed by larger coal companies in the Pittsburgh district and throughout the bituminous fields during this decade. Several cleaning facilities were also constructed in eastern Ohio by the Hanna Coal Company, and in northern Virginia at the Pursglove Coal Mining Company. There were twenty-five cleaning plants operating in the Pittsburgh District in 1938.[47]

Company	Mine	Year started	Capacity Ton per Hour
Humphrys C. & C. Co.	Greensburg	1929	50
Clinton Block Co.	Imperial	1928	299
J. & L. Steel Corp.	Hazelwood	1930	300
J. & L. Steel Corp.	Aliquippa	1912	80
Keystone C. & C. Co.	Salem	?	150
Saxman C. & C. Co.	Latrobe	1927	50
Buckeye Coal Co.	Nemacolin	1929	700
Carnegie-Illinois Steel Co.	Clairton	1931	800
Jamison C. & C. Co.	Hannastown	1931	350
Pgh. Terminal Co.	Coverdale No. 8	1931	500
Acme Coal Cleaning Co.	Avella	1933	300
Deep Vein Connellsville Co.	Brier Hill	1929	200
Hillman C. & C. Co.	Naomi	1933	100
Lincoln Gas Coal Co.	Lincoln No. 1	1928	250
Panhandle Mining Co.	Midway	?	?
Washington Coal Co.	Tyler	?	?
Pittsburgh Coal Co.	Champion No. 1	1928	700
	Champion No. 2	1936	100
	Champion No. 3	1927	300
	Champion No. 4	1929	450
	Champion No. 5	1929	450
	Champion No. 6	1933	400
Westmoreland Coal Co.	Magee	1938	275
Westmoreland Coal Co.	Hutchinson	1931	300

All these preparation facilities, except the Jones and Laughlin plant at Aliquippa, Beaver County, were constructed during the 1920s and 1930s. Raw run-of-the-mine coal was cleaned, screened, washed, and dried in the preparation plant. Cleaned coal of different sizes and properties could be blended by the coal company to meet specific consumer requirements at the preparation plant. The preparation plants of this era had the capacity to clean and wash from fifty to seven hundred tons of coal per hour. Since the construction of the plant was costly, nearby coal operators often washed and cleaned coal from their competitors' nearby mines for a fee. The Acme Preparation Plant, near Avella, Washington County, and the Champion Preparation Plant Number One of the Pittsburgh Coal Company, near Imperial, Allegheny County, cleaned and washed coal from more than one mine. Coal from Pittsburgh Coal Company's Westland, Montour Number 4, and Margerun Mines was cleaned and wetted at this Champion plant.[48]

Coal was transported to the surface in coal cars that stopped at the rotary dumpers, designed to empty each car with minimum coal breakage. The dumped coal was fed into the preparation plant where it passed over and through screens which accurately separated it into several sizes. The larger coal passed over combination picking tables and shaker screens where slate and impurities were removed, and it was then ready to be loaded for market. The small coal passed through another mechanical cleaning, where the rock, slate, bone coal, and pyrites were removed. At some preparation plants coal was also cleaned. Coal cleaning consisted of the following steps: (1) crushing or breaking the coal to prepare it for washing; (2) sizing to separate coal into different dimensions, both to match the specifications for the various washing devises and to meet market requirements; (3) washing to remove impurities from coal; (4) drying to remove excess moisture prior to shipment. There were no universal standards in the bituminous coal industry as to the degree of cleanliness, trade names, and sizes as late as 1940, such as were found in the anthracite industry.

Coal preparation consisted of two interrelated processes—sizing and cleaning. Coal was sorted according to size through a series of different-size screens at the preparation plant. Coal cleaning involved the separation of impurities by the miners hand picking underground as they loaded the coal or by mechanical cleaning at the surface. The underground mine foreman checked the loader's wagon to see that these impurities were removed. The introduction of mechanical loading machines, during the 1920s, made necessary more surface preparation of coal to separate impurities. Impurities were hand-picked from the coal in the tipples and it was shipped as run-of-the-mine coal. Picking and shaking screens were found in a more-elaborate tipple from which the coal was shipped to market in different or mixed sizes. The more elaborate and expensive preparation plant screened coal from rock and other impurities, including a mixture of rocks, slate, shale, boney coal, and coal dust. There were four general sizes of coal available in the bituminous industry as of 1925, although size and name had not yet been standardized: (1) slack, everything less than 3/4 inch; (2) nut, 3/4 to 1 1/4 inches; (3) egg, 1 1/2 to 4 inches; (4) lump, everything over 4 inches. The measurements indicated the diameter of the screen openings used to separate the coal. Steam coals, especially those used as bunker coal in ships, in locomotives, and in stationary boilers, were usually cleaned and sized.

General view of tipple and screening plant. Coal Age c1910.

Coal used in the production of metallurgical coke that had high ash and sulfur content was washed to reduce its high content.[49] Effective washing could reduce sulfur content by as much as 50 percent, depending upon the efficiency of the washer and the chemical properties of the coal. Coal was treated with oil, calcium chloride, or other chemicals to minimize dust and prevent it from freezing while in transit to market by rail or river barge.

The Berwind-White Coal Company of New York constructed two cleaning plants in western Pennsylvania during the 1920s, the first at Eureka Mine Number 37 in 1926. Another was contructed at Eureka Mine Number 40, Windber, by Roberts & Schaefer Company of Chicago in 1928. The three-story building was a reinforced-concrete tipple and cleaning plant costing about a half million dollars to construct.

Charles Enzian, a graduate of Lehigh University, served as president and general manager of the Liberty Coal Corporation of Kentucky before his appointment as chief engineer of the Berwind-White Mining Company. Enzian noted,

The economic conditions of the bituminous coal industry which have existed in the past and no doubt will exist in the future, require exceptional alertness on the part of the operator to create additional demand through the improvement of the product from the mines so that he may be insured of retaining the market already supplied and a reasonable hope of gaining new markets.[50]

The United States Bureau of Mines noted that 7.8 percent of Pennsylvania's coal output was mechanically cleaned and that the state ranked behind Alabama, Washington, and Colorado in percentage of mechanically cleaned coal in 1929. The tonnage of coal cleaned in the United States in 1936 (13.9 percent) was double the amount cleaned in 1932.[51] The growth of the mechanical cleaning of coal in Pennsylvania paralleled that in the nation as a whole.

The Percentage of Coal Mechanically Cleaned			
	1927	1936	1940
U.S.A.	5.3	13.9	20.9
PA	4.2	19.6	23.6[52]

Workers' Responses and the National Coal Strikes of the 1920s

The UMWA, under the leadership of their fiery President John L. Lewis, had become the most powerful union in the United States by 1920. Lewis (1880-1969) was the most important leader in the history of the union, and for the next forty years would lead the miners' union until his voluntary retirement in January 1960. His parents had migrated from the coal-mining district of southern Wales, and settled and married in Lucas, Iowa, in 1878. Lewis was born in this small southern Iowa mining town on February 12, 1880. He attended the public schools in Lucas and at the age of seventeen entered a local mine and worked with his father and brother. He was a mule driver and worked ten hours a day, receiving $1.60 daily. Lewis had a varied career in the coal, copper, and silver mining industry throughout the West before working as a coal miner in the coalfields of Illinois, where he began his rise in the miners' union bureaucracy. He served as pit committee boss in 1909 and was chosen by Illinois miners in 1910 as their state legislative representative for District 12 of the UMWA. He secured the passage of the Workmen's Compensation Act and a series of mine safety laws in Illinois. Samuel Gomper, the AFL president, who was impressed by Lewis's organizational abilities, appointed him a field and legislative representative for the union in 1911. UMWA President White appointed Lewis chairman of the resolution committee at the union's 1916 annual convention. Lewis served as a union statistician and as a manager of *The United Mine Workers Journal* before his election in 1919 as the ninth president of the United Mine Workers.[53]

The union, like coal operators, was becoming an unwilling victim of an ailing industry during the 1920s. President Lewis and the UMWA were becoming prisoners of events, rather than the shapers of economic events, throughout the decade. Lewis wanted to raise wages to force operators to mechanize the industry. He believed the more efficient, high-wage union mines would force the less productive and less mechanized southern operators out of business. The union was unable to solve the intense interregional and interfirm competition, and this inability to find solutions to these complex issues made the union increasingly impotent in stopping the slide in the average price of a ton of coal and subsequent wage reductions. The nation's most viable and powerful union had become a shattered and decimated labor organization by the end of the 1920s.

The national coal strike of 1922 became inevitable when coal operators and the union could not agree on a scale of wages to replace the expiring 1920 Washington Contract. Coal operators had campaigned vigorously since 1920 to maintain their profits and reduce the economic gains made by the union during the 1919 strike, and had rolled back wages and hours to the lower wage standard established in 1917. Some union operators expressed the view that the Central Competitive Field interstate bargaining structure should be immediately dissolved, but most executives of the coal companies were content simply to reduce wages and hours to the lower, 1917 wage structure.

John L. Lewis asserted, "We have tried to bring the mine owners into wage conference and now we are going to strike," and with that declaration the first UMWA strike of the 1920s was called on April 1, 1922.[54] Lewis called this strike "the Verdun of Organized Labor." The union was optimistic that its strike action would succeed because of its victory in the 1919 strike. Some confident union leaders boldly asserted to their membership that "before the snow flies, we will win."[55]

The strike was the first defensive struggle of the decade by the UMWA to maintain the $7.50 a day wage it had won for its membership in the 1919 strike. The strike followed the expiration of the 1919 contract on March 31, 1922. Both union and nonunion mines of the anthracite and bituminous fields of Pennsylvania were effectively shut down for the first time in the state's history. Some six thousand mines nationally were closed and the U.S. Department of Labor estimated 610,000 of the nation's 795,000 coal workers were on strike and manning the picket line by the first week of August 1922.

Union organizers actively enlisted support from nonorganized miners, especially in District 2, Pennsylvania, including the counties of Somerset, Cambria, Clearfield, and Centre and the Broad Top semibituminous coal region. Union operators in District 2 had produced 72 percent of the region's coal in 1920, but in 1921 they mined only 62 percent of the total.[56] In 1922-1923 the division of coal production was about two-thirds union and one-third nonunion and by the summer of 1925 coal production was fifty-fifty.[57] The rapid growth of coal tonnage in the nonunion fields made it increasingly difficult for the union to maintain higher wages for its membership. For example, production in nonunion mines increased by 48.6 percent in Kentucky and 8.3 percent in Virginia.[58]

John Brophy (1883-1963), District Two president, was convinced that if the union-led strike of 1922 was to succeed in western Pennsylvania, the support of nonunion miners was essential during the strike. Brophy was the son of a Lancashire miner, born in St. Helens, England, in 1883. He had a limited formal education since he had entered the mines when he was barely twelve and had worked in a number of mines at Carolltown, Greenwich, and Nanty Glo, Cambria County. He was elected a union checkweighman at the South Fork mine and was subsequently elected local secretary. Brophy was elected District 2 president in 1917 and served in this position until 1928. A centerpiece of his tenure as district president was his five-point miners' program, a detailed program of what the union needed to do to insure the future prosperity of its membership. He played a prominent role with President Lewis in the formation of the Committee for Industrial Organization (CIO) in 1935.[59]

Union leadership of District 2 issued a strike call to all nonunion men from their strike headquarters at Cresson, Somerset County. Organizers distributed twenty thousand copies of the union-sponsored paper *Penn Central News* in the nonunion coalfields of Somerset County, parts of Cambria County around Johnstown, and the Black Lick Field north and west of Nanty Glo, Cambria County. They circulated pamphlets with the following call: "To All Non-Union Mine workers: STRIKE! QUIT WORK! THIS IS YOUR FIGHT! LEAVE THE MINE! BRING OUT YOUR BUDDIES!"[60] They circulated cards throughout the nonunion "patch" towns declaring "Non-Union Towns are Towns of Fear!" because "Nonunion miners have lived in Fear! Fear of the Boss—Fear of the Spies and Spotters—Fear of Gunmen and Coal and Iron Police—Fear of the Blacklist—Fear of Evictions."[61] This appeal to "Join the Union and Quit Being Afraid of the Boss" met with success as nonunion workers deserted the mines and manned the picket lines. The strike spread quickly into the heart of the nonunion counties of Somerset and Cambria, and the Connellsville coke district of Pennsylvania. Workers joined the strike to protest working and living conditions around Windber, Somerset County. Miners from the St. Michael Shaft mine of the Berwind-White Mining Company, Cambria County, were the first nonunion mine in the district to strike, as more than two hundred miners walked to South Fork, the nearest town, to sign up with the union. This was the deepest bituminous shaft-entry mine in the Commonwealth during this time. The mine, located in Adams Township, was built circa 1908 and operated continuously until it ceased production in 1958.[62]

The Berwind-White Coal Mining Company and the Rockefeller-controlled Consolidation Coal Company were the principal nonunion mining companies in Cambria and Somerset Counties. The Consolidation Coal Company was incorporated on March 9, 1860, by an act of Maryland's legislature, when the Frostburg Mining Company, the Ocean Steam Coal Company, and the Mount Savage Iron Company in the Georges Creek Region of Maryland combined their interests to form the new corporation. The company was not incorporated until April 19, 1864. The delay was due to the Civil War. Consolidation Coal Company opened its first mines in the semibituminous Georges Creek coalfield of Maryland in 1864. The firm developed a close business relationship with the Baltimore and Ohio Railroad between 1876 and 1906, during which time the two companies monopolized the production and distribution of the soft coal industry in Maryland. The Interstate Commerce Comission held an investigation of this business relationship and concluded in 1906 that it was a monopoly. The Baltimore and Ohio Railroad was ordered to sell back its 52 percent interest to the Consolidation Coal Company. The company expanded its regional coal operations beyond Maryland during the first decade of the twentieth century. The company acquired interests in the Fairmont Coal Company, West Virginia, and the Somerset Coal Company, Somerset County, in January 1903. The Fairmont Coal Company was itself created by the consolidation of twenty smaller companies. The company owned thirty-seven mines, 1,060 beehive coke ovens, stores, and company houses. The company had produced about 3.8 million tons in 1902 when acquired by the Consolidation Coal Company. The company opened a number of mines and constructed the company towns of Gray, Acosta, Jenners, and Bell in Somerset County. Consolidation Coal entered the eastern Kentucky coalfield in 1909 by acquiring thirty thousand acres of coal and surface lands in Johnson, Martin, and Lawrence Counties. The Rockefeller family purchased 38 percent of the firm's securities and acquired a controlling interest in the firm in 1915. John D. Rockefeller Jr. controlled 72 percent of preferred stock and 28 percent of common stock by 1928. The Consolidation Coal Company was one of the largest coal operators in the nation during the 1920s, having mining facilities and coal reserves in Pennsylvania, West Virginia, Maryland, and Kentucky. The company operated simultaneously both union and nonunion operations, and owned three hundred thousand acres of coal land with reserves of more than 1.8 billion tons. The Baltimore-based corporation completed its first seventy years in 1934, operating twenty-two mines in four states.

The Berwind-White Coal Mining Company was formed in 1874 when the four Berwind brothers of Philadelphia, Charles, Edward, John, and Henry, and retired Judge Allison White formed a partnership. White was a Philadelphia businessman and a former partner of the White and Linge Coal Company. Charles F. Berwind, the principal leader of the firm, entered the coal trade in 1861 at the age of fifteen as an office boy for Robert Hare Powell, a Philadelphia coal merchant. In 1869, Berwind founded his own company named Berwind and Bradley. The company dissolved, however, and in 1874 Charles Berwind and his brothers formed a partnership with Judge Allison White. The Philadelphia-based company opened its first mine, Eureka Number 1, in Houtzdale, Clearfield County, in 1874. The company was operating eleven mines and had acquired coal lands in Clearfield and Jefferson Counties by 1885. The following year the partners incorporated as the Berwind-White Company, with Charles F. Berwind president, Edward J. Berwind vice-president, Fred Owen secretary, and Judge White treasurer. The company's annual production had exceeded three million tons by 1892. The firm had established a steamship bunker coal business with the U.S. Navy valued at two million tons annually by 1903. The firm had contracts with the principal steamship lines of New York, giving the firm a virtual monopoly of the trans-Atlantic steamboat bunker coal market.[63] The company owned and operated vessels for coal transportation to Panama and other ports. They operated docks at Superior, Minnesota, for the Great Lakes trade and at Newport News, Virginia, and South Amboy, New Jersey, for overseas coal shipments.

The firm expanded its mining into rural Somerset County by acquiring sixty thousand acres of coal lands from local farmers during the 1890s, and it opened Eureka mine 30 in September 1897. The firm founded Windber, Somerset County—the name being a transposition of the syllables in Berwind—as a model industrial community to serve as Berwind's headquarters for its extensive coal-mining in the region. Windber was laid out in 1897 and incorporated as a borough on July 2, 1900. Berwind-White erected thirteen Eureka mines, Numbers 30 through 42, around the new company town and acquired two shaft mines named Maryland 1 and 2 near St. Michael and Wilmore, Cambria County, respectively. Eureka Numbers 30 to 42 extracted coal in the "B" or Lower Kittanning seam (mines 30, 31, 32, 33, 34, 35), or the "C" Prime or Upper Kittanning coal seam (mines 36, 37, 38, 39, 40, 41, 42). Eureka mines 35, 36, 37, 40, and 42 all operated into the 1950s. Between 1897 and 1962 the thirteen drift-entry mines of the Windber region produced over 150 million tons of coal. The company shipped coal to market on the South Bend Branch of the Pennsylvania Railroad. The company's expansion continued during the next two decades and by the advent of the First World War, Berwind-White Mining Company had acquired coal lands and opened new mines in Westmoreland and Cambria Counties in Pennsylvania, and in Kentucky, Virginia, and West Virginia. It was one of the largest independent coal companies in the nation by the 1920s, and maintained corporate offices in Philadelphia; sales offices in New York, Philadelphia, Chicago, Baltimore, and Boston; and shipping piers at Newport News and Norfolk, Virginia, Philadelphia, Baltimore, New York harbor, Duluth, Minnesota, and Superior, Wisconsin. Most of the firm's coal was sold to ocean-going steamships as bunker coal and it had a lucrative contract to supply coal to the Interborough Rapid Transit Company in New York City. This coal was an excellent fuel for manufacturing, steamship, and railroad use. The firm called its coal Berwind's Eureka Bituminous, Berwind's Standard New River, and Berwind's Standard Pocahontas. The Berwinds integrated their diverse and widespread coal businesses by forming a variety of subsidiary companies to mine, supply its miners with goods and services, and ship its coal to market. The New York-based company was operating the New River Consolidated Coal Company, the Porto Rico Coal Company, the Kentland Coal and Coke Company, the Windber Electric Corporation, the Atlantic Coal Company of Massachusetts, the Herminie Land Company, and the Eureka Supply Company by 1945.

Union and nonunion coal companies immediately mobilized and fought back, attempting to break the strike and curtail the power and influence of the aggressive and powerful miners' union. The strike was both bitter and violent from the start. There were violent confrontations throughout the coalfields of Pennsylvania, West Virginia, and Utah between militant nonunion and union miners on the one hand and hired police forces employed by the coal companies. Nineteen strikebreakers were killed in retaliation for the murder of two strikers by mine guards at Herrin, Illinois, on June 1922.

Coal companies possessed an arsenal of resources at their disposal to break the strike. Spies and paid informants were used to infiltrate union meetings to obtain information on union strategy. Strikebreakers (called "scabs" or "rats" by miners), who included local farmers, transient workers, and African Americans, were immediately imported to work the idled mines. The companies ordered local county deputies, commissioned and paid by the county, to be stationed at the mine to protect scab workers and prevent vandalism of mine property by striking miners. The eviction of miners and their families from their rented company houses was a common practice. The lease was terminated automatically whenever a worker ceased to work for the coal company. Coal and Iron Police were used to remove the striking miners and their families from seventeen mining towns in Cambria and Somerset Counties by the first week of June 1922—Conemaugh, Bituman, Hollsopple, Ralphton, Acosta, Jerome, Ankeny, Macdonaldton, Kiel Run, Pretoria, Blough, Gray, Bell, Husband, Revloc, Twin Creek, and Windber. Miners and their families spent the harsh winter of 1922-1923 living in tarpaper shacks, tents, or crudely constructed barracks.[64] A striking nonunion miner at Maryland No. 1 Shaft of the Berwind-White Mining Company reported his eviction: "They made us move out. They threw our furniture out onto the street if you tried to organize. They had what you would call pussyfoots [Coal and Iron Police] riding horses, and if you would get four or five men in a group and try to organize, they would come and break it up."[65]

Coal companies of Pennsylvania had their own police force, known as the Coal and Iron Police, legally recognized by the state of Pennsylvania. This industrial police force was created by two statutes passed by the General Assembly of Pennsylvania in 1865 and 1866. This police force was the primary law-enforcement agency used by coal and railroad companies in the anthracite coal region. Its presence permitted the anthracite coal and railroad companies to exercise both social and military control of the area. The police had been employed to suppress the militant Irish secret society, the Molly Maguires, during the 1860s and 1870s. These policemen were recruited from the ranks of the unemployed or were former policemen and were hired by strike-breaking detective agencies for their ability to use rifles and blackjacks. No special training was required and a fee of one dollar per commission payable to the state of Pennsylvania qualified them to carry the Coal and Iron Police badge and a variety of weapons—rifles, 38 Colt pistols, black jacks, and billy clubs.

These police had been used regularly in the bituminous coalfields of western Pennsylvania since at least the 1922 strike. Miners contemptuously called them "Cossacks," "Pussy-footers," "Company Men," and "Yellow Dogs."[66] Cossack was a generic term used by most industrial workers to describe state troopers or paramilitary police organizations used by management to suppress strikes or to spy on their union activities. Coal miners, like steel and railroad workers, saw them as simply common gunmen, hoodlums, and adventurers. This private police force provided a variety of essential services for their employers. They patrolled the company town, evicted striking miners from company housing, and enforced their private laws in the "patch." They recruited, imported, and protected strikebreakers, and provided security for mining property against violent attacks of strikers. They were hated and feared for their excessive brutality and lawlessness in their vigorous attempts to suppress strikes and maintain control in the company towns. Their methods often provoked violence from discontented miners.

Deputy sheriffs and Coal and Iron Police were conspicuously active during the 1922 strike. The coal companies of Fayette County alone hired twenty-five hundred deputy sheriffs to protect their property and evict miners. There was one deputy for every twelve striking miners in the county.[67] These private guards each cost the coal companies about $5.60 to $9.00 per day in Fayette and Greene Counties. *Coal Age,* the principal coal mining weekly trade journal, reported the strike with weekly articles published between April and October 1922. It reported on April 13, 1922, that "in the Uniontown area many plants are being enclosed by barbed wire and hundreds of special deputies have been sworn in to guard mine property." Violence between deputy sheriffs and nonunion miners broke out at numerous mines in Fayette and Westmoreland Counties as striking miners destroyed company property at the mines. The *Penn-Central News* reported on the numerous violent confrontations between striking nonunion miners and hired company police in District 2 in 1922. The following excerpts report the typical repressive tactics employed by coal companies to suppress the strike in the coal region:

> Charles Dias, after helping five miners evicted from the Vinton Collieries Company at Claghorn, Indiana County on May 8, was beaten up by two Coal and Iron police while on the station platform. He was hammered over

the arms and head with a club and blackjack, denied a doctor, was driven down the railroad track at a revolver point [sic].

Tony Lilko, at Berwind No. 38 mine, while helping a comrade evicted from a company house, was arrested for "trespass;" two guards beat him up and added the charge of resisting officer; he was taken by the company store boss to a neighboring town where a justice was found to hold Lilko on the charges.

At Kelso on June 7 three women and one man were arrested by a parade of fourteen deputies and guards, armed with guns and clubs, fined $25.50 to $39.00, each and jailed in Somerset on refusing to pay; charges "disorderly;" they had been picketing.[68]

The Harding administration attempted to mediate a solution to the strike as it spread throughout the coalfields, and acts of violence between labor and management escalated. The Republican administration was fearful of a coal shortage during the upcoming winter. President Harding ordered miners and the operators "to resume mining operations at once and then adjust their differences in joint conferences." President Harding had threatened to use federal troops if the strike was not immediately resolved. Coal was still the nation's main energy provider, and a coal shortage from a prolonged strike would cripple transportation, communication, home heating, and industry throughout the nation. The strike had already cut weekly coal production between four to six million tons. Coal reserves had fallen from over 63 million to approximately 22 million net tons, the second lowest reserve on record. Nonunion operators in the South, West Virginia, and Pennsylvania were unable to produce sufficient coal to meet the national demand.

Union leaders, coal operators, and government officials met in Cleveland throughout the summer and on August 15, 1922, a new labor contract was negotiated and signed. The contract ended the great strike of 1922 after four and a half months. The agreement maintained the economic gains acquired by the UMWA since the 1920 agreement. Some forty-five thousand union mine workers in District 2 and five hundred thousand miners in other parts of the country immediately returned to the mines. John L. Lewis proclaimed the strike a complete victory for the miners, although the nearly seventy thousand nonunion miners who had so passionately supported and participated in the strike were excluded from the labor settlement. Brophy and other union organizers harshly criticized Lewis and their union for their abandonment of the new union recruits from Cambria and Somerset Counties following the Cleveland settlement.

The management of Berwind-White Coal Company was, traditionally, virulently antiunion and refused to sign a union contract with its workers. The company resorted to numerous methods to suppress their workers' drive for union representation during the strike. Some miners, accompanied by their families, went to New York City to picket the company's office. The firm had a lucrative contract to supply coal to the city's subway system. John F. Hylan, mayor of New York City, concerned by these noisy demonstrations, appointed a five-member committee, chaired by David Hirshfield, to investigate the mining and living conditions of Berwind-White coal towns. The committee traveled to Windber and called a number of meetings to investigate conditions experienced by striking miners and their families but the firm boycotted all the committee's hearings. Committee members visited the firm's numerous mines and company towns around Windber and recorded the inhuman and depressing conditions. The company had evicted the strikers and their families, who now resided in makeshift "tent towns" that sprang up near the company towns. Their diets were poor and hygienic conditions in these makeshift tent towns were simply appalling. The committee released its report in 1923, declaring that miners worked and lived in settings "worse than the conditions of the slaves prior to the Civil War."[69]

W. J. Lauck, of the United States Immigration Commission, attributed the following characteristics to southern and eastern European workers in 1911:

> ... when aroused the Slavic races have demonstrated their inclination to follow their leaders to any length, but in the normal life of the coal fields the miners of recent immigration are usually tractable and easily managed, and are imposed upon without protest. This temperament would give the impression of subserviency to their employers and to older workers.[70]

His views were shared by most coal operators of the period and some leaders within the UMWA. The immigrant miners' militant and unified action during the strike forever destroyed this pervasive belief that they were passive and malleable workers. A majority of the workers in the Eureka and Maryland mines of the Berwind-White Mining Company were immigrants from southern and eastern Europe. There were only 411 American workers in a labor force at these mines numbering 2,894 in January 1922. These foreign-born nonunion miners from District 2 felt betrayed by union officials after they had so passionately supported the union's strike call. They continued the strike for union recognition with little overt union support until miners' delegates from Somerset County met in Johnstown on August 14, 1923, and called off their seventeen-month-long strike.[71] They were forced back to work under conditions similar to those that had prompted them to strike. They remained demoralized and unorganized miners until the next major organization drive by the UMWA in 1933.[72]

Brophy spoke on behalf of these unorganized immigrant miners before the United States Coal Commission in Washington, D.C., in 1923. He identified six demands of the striking nonunion men as follows: (1) collective bargaining and the right to affiliate with the union; (2) a fair wage; (3) accurate weight of the coal they mined; (4) adequate pay for "dead work"; (5) a system by which grievances could be settled in a peaceful and conciliatory spirit by the mine committee representing the miners and a representative of the operator; (6) their rights as free Americans against the state of fear, suspicion, and espionage prevailing in nonunion towns and against a small group of operators controlling the life, liberty, and pursuit of happiness of a large number of miners. They wanted to put an end to the absolute feudal control of the coal operators.[73] Most of these demands were guaranteed to union miners, but did not apply to the nonunion miners of District 2 or other miners employed by nonunion coal companies.

The U.S. Coal Commission, before which Brophy had testified, was also known as the Harding Commission. This commission was yet another in a series of government commissions established to examine the ailing American coal industry.[74] The federal government proposed some sixty laws, commissions, or investigations of the bituminous coal industry between World War I and the New Deal. Public concern with the violence engendered by the recent strike had prompted the commission's establishment on October 10, 1922, "for the purpose of securing information in connection with questions relative to interstate commerce in coal, and for other purposes."[75] The president appointed a seven-man commission to undertake a comprehensive investigation of the industry and to recommend procedures for the industry's reorganization. The commission was essentially a fact-finding body and during the next eleven months researchers scrutinized all phases of the bituminous coal industry in the major coalfields in the East and Midwest. This voluminous study was followed by a second and third extensive government survey investigating conditions in the coalfields in 1947 and 1980 respectively.[76]

The Harding Commission issued a massive four-volume study comprising 800,000 words on September 22, 1923. The study consisted of a series of reports on various phases of the mining industry and provided a detailed portrait of all facets of the industry during the 1920s.[77] The commission conclusions, adopted by unanimous vote of the commissioners, rested on two broad principles: (1) coal is a public utility, a public necessity, affected by a public interest, and its mining, transportation, and distribution are therefore subject to supervision and regulation by the federal government; and (2) the main responsibility for solving the problem of the industry must rest on the industry itself, "not through governmental coercion but through the enlightened self-interest of producers and consumers. The real remedy is to be sought; and the coal industry can reform itself from within."[78] John Hays Hammond, chairman of the Harding Commission, suggested four areas of improvement to make the industry a more rational and profitable business: (1) development of machinery "to replace the irksome and solitary operation of hand loading," (2) improved control of underground operation, (3) improvement in the work of the individual, and (4) standardization of equipment.[79] The commission's recommendations on what ailed the industry and possible solutions were generally ignored and summarily dismissed by the union, the coal operators, and the federal government. Many of their recommendations were later implemented during Roosevelt's first New Deal.

The two-year Cleveland Agreement expired on April 1, 1924, and the international officers of the UMWA, coal operators, and government officials met in Jacksonville, Florida. Both management and the UMWA were aware of the two salient facts now confronting the industry: it was capable of producing twice as much coal as the nation could consume and there were twice as many miners as were needed in the industry. Nevertheless the UMWA insisted on maintaining the wage scale of $7.50 established in 1920 and maintained in the 1922 strike. The Jacksonville Interstate Joint Agreement signed in New York by the Central Competitive Field operators maintained the 1920 wage scale and served as the basis for coal agreements signed in other bituminous coalfields, including those fields in the southwestern states. The contract stated simply in two paragraphs that the Cleveland Agreement, signed on August 20, 1922, and expiring on April 1, 1924, would be carried forward and that district and subdistricts then in effect would be continued without further negotiations. The three-year Jacksonville Agreement was signed in February 1924 and was effective until its expiration on April 1, 1927. The union believed that high wages would stabilize the industry without government intervention because the high wage scale of $7.50 a day would close down high-cost and unprofitable mines.

The UMWA Journal and most union miners hailed the agreement as the best contract ever negotiated with the coal operators of the Central Competitive Field. They were elated and after its ratification, by a vote of 164,858 to 26,253, chanted "we got a three-year contract; next time we'll make it five."[80] Herbert Hoover, secretary of commerce, who helped to mediate successfully the Jacksonville Agreement, called the agreement "the most constructive development in the bituminous industry for years" and later declared:

> The coal industry is now on the road to stabilization. The benefits lie not only in the position of coal to consumer at lower prices than have been attained at any time since the beginning of the war. The gradual elimination of high-cost, fly-by-night mines is bringing about a greater concentration of labor upon a smaller number of mines, the increase in days of employment per annum, and thus a larger return to the workers. The inherent risk in the industry will be decreased because the efficient and stable operators will no longer be subjected to the type of competition that comes when those mines exist only to take advantage of profiteering periods.[81]

Coal operators, under the new contract, agreed to keep wages at $7.50 a day base pay for company men and a $1.80 per ton piece-rate. This rate was considerably higher than the prevailing wage scale paid workers in the nonunion coalfields. The contract, in essence, simply maintained the status quo of the previous miners' contract, which was a "small miracle" considering the rapidly deteriorating state of the soft coal industry since 1923. Coal prices had declined by 40 percent since 1922 while the number of unemployed mine workers had risen by eighty thousand. Some union operators followed suit and renewed contracts according to the 1920 scale, while the union operators in the Kanawha region of West Virginia refused to sign the union contract and attempted to operate their mines as open shops.[82] They refused to accept the union agreement, contending they could not compete with the nonunion Mingo, Logan, Williamson, Pocahontas, New River, and eastern Kentucky regions. The ten coalfields of West Virginia were Panhandle, Kanawha, Fairmont, Upper Potomac, Elkins, Gauley-Greenbrier, New River-Winding Gulf, Pochahontas, Williamson, and Logan. The Kanawha Field is located in Boone, Logan, Kanawha, Putnam, and Mason Counties while the Fairmont Field is located in Barbour, Preston, Taylor, Harrison, Marion, and Monongalia Counties. Coal companies in Colorado, Utah, Texas, Maryland, and Virginia were already operating 100 percent nonunion mines at this time. The southwest mining region, once a stronghold of unionism, had begun to break up by 1924. The Jacksonville Agreement was signed by one-fourth of the mine owners in Oklahoma although the union was not represented in the larger mines in other southwestern states.[83] This breakdown of collective bargaining happened in spite of strong resistance by the UMWA. The inability of the union to organize these areas and establish collective bargaining covering both northern and southern operations made the future success of the union precarious.

Some Central Competitive Field operators, who had voluntarily and in good faith signed the three-year Jacksonville Agreement, asked Lewis and the UMWA for an immediate wage cut before the first year of the contract was fulfilled. Union operators faced with overproduction and falling coal prices during the postwar years sought to reduce operational costs, especially those of labor. They blamed their economic difficulties on the relatively high wages paid union miners in their mines. They declared that their signing of the agreement had been a grave economic mistake on their part and demanded that the union accept a six dollars a day wage for its members. They asserted that without these concessions by the union more union mines would close. The UMWA power had been centered in the Central Competitive Field of western Pennsylvania, Ohio, Indiana, and Illinois since 1897, when John Mitchell signed the union's first collective bargaining contract. The UMWA had successfully organized most mine workers in these states by 1924. All miners of western and central Pennsylvania were unionized except for the following districts: Clearfield, Cambria, Somerset, Myersdale, Indiana, Latrobe, Connellsville, Youghiogheny, and Westmoreland. All of northern West Virginia, except the Upper Potomac Field, as well as the Georges Creek Field in Maryland were unionized. Virginia was unorganized and only a small section in northern Kentucky had union representation. The balance of the Appalachian region through Kentucky, Tennessee, and Alabama was nonunion.[84] The nonunion operators were paying lower wages as a rule, some as low as the 1917 level. The table to the left compares the different wage scales for underground and surface workers and clearly demonstrates that union operators were economically handicapped in competing with nonunion operations.

Labor accounted for as much as two-thirds of the cost of mining a ton of coal. Nonunion coal companies could produce coal at lower costs, sell it at a lower price per ton, and open more new mines while union operators were closing some of theirs permanently and dismissing miners. Competition from nonunion southern coal eventually forced union operators to break their union contracts and slash wages.

President Lewis, the union's national leadership, and the rank and file were all politically and philosophically committed to maintaining the Jacksonville Agreement. Although the efforts of the union to organize miners in the South had failed dismally in 1921, Lewis embarked on his "no backward step" program. Lewis told a newspaper reporter in May 1923 that "the union miners cannot agree to the acceptance of a wage principle which will permit his annual earnings and his living standards to be determined by the hungriest unfortunate whom the nonunion operators can employ."[86] Lewis's firm policy of "no backward step" regarding demands for wage concessions was strongly endorsed by union miners at their annual convention

Average Wages of Union and Non-Union Miners in 1920

	Union	Non-Union
Blacksmith	$7.57	$6.36
Carpenter	$7.14	$5.04
Engineer	$7.29	$5.97
Fireman	$6.95	$5.41
Stableman	$6.07	$5.07
Laborer	$6.55	$4.58
Miscellaneous	$6.74	$5.37
Bratticeman	$7.42	$5.96
Doortender	$4.48	$4.26
Driver	$7.47	$6.22
Pumpman	$7.14	$5.53
Timberman	$7.52	$6.26
Tracklayer	$7.35	$6.09
Laborer	$7.16	$5.08
Miscellaneous	$7.46	$5.99[85]

in 1924. He insisted upon the retention of the $7.50 base rate in the Central Competitive Field and noted that even if a wage reduction was granted by the union to the coal companies, "the nonunion crowd would simply make another cut in the wages of their miserable workers and the same relative condition would continue."[87] The miners' union president refused and informed the coal companies that his union would not renegotiate the contract until its expiration in 1927. He was contemptuous of coal companies and their managerial skills, accusing them of "being incompetent, inefficient, backward, lazy and disunited."[88] Lewis was a firm believer in the natural law of survival of the fittest and argued that the future economic prosperity and stability of the coal industry was possible only if the industry operated with fewer mines and fewer miners. Strict adherence to the higher wages of the Jacksonville Agreement would force inefficient coal operators to put the management of their mines on a more efficient basis, to consolidate their operations and close unprofitable mines, and to install labor-saving machinery, especially the new mechanical coal-loaders. The union believed that high union wages would drive inefficient coal companies into bankruptcy. The uncompromising and intransigent position taken by Lewis and unanimously supported by the rank and file of the UMWA was fraught with danger in this period of severe economic decline. Lewis expressed his own sense of deep apprehension to a newspaper reporter when he stated that "we expect losses, perhaps heavy losses, but we are confident of victory in the end."[89] "In light of Lewis' refusal to revise the Jacksonville Agreement," historian Irving Bernstein noted, "the operators set about the systematic destruction of the union."[90]

The Jacksonville Agreement was first abrogated by Rockefeller's Consolidation Coal Company and the Bethlehem Steel Company in the coalfields of northern West Virginia in July 1925; daily wages were reduced to $6.00 a day. The successful "open shop" drive by these companies in West Virginia was immediately followed by the announcement of the Mellon-controlled Pittsburgh Coal Company the following month that it would operate on a strictly nonunion basis. This Pittsburgh-based company, founded in 1899, was the largest coal company in the world at this time. The firm operated over one hundred mines employing some seventeen thousand union workers in Pennsylvania, Ohio, and Kentucky.[91] Management of the company, as a member of the Central Competitive Field, had honored all union contracts it had signed with the UMWA since its formation. The prominent Mellon banking family of Pittsburgh, led by Treasury Secretary Andrew W. Mellon and his brother Richard, had acquired 25 percent of the company's stock in 1924 and controlled the board of directors of the firm.[92] The new Mellon management team believed the firm could not continue to effectively compete with rival nonunion coal operators, especially those from West Virginia, Kentucky, Tennessee, Virginia, and Alabama, without wage concessions. Their solution was simply to crush the miners' union. Management renounced the Jacksonville Agreement unilaterally and on August 1925 posted notices at their mines that wages were to be reduced immediately to $6.00 a day and that all mines would operate on a nonunion basis.

James D. A. Morrow, president of the Pittsburgh Coal Company, in an open letter to striking employees, clearly stated the producer's intent to operate its mines on the "open shop" principle: "Don't believe any story that this company is going to sign up with the union on April 1 or any other time. This is not true and has been put out to scare you and make you unhappy. We will never sign a scale with any union again. We will always have open-shop mines. We will never run any mine any way but shop open." Morrow was a former executive of the Joy Manufacturing Company and had left to become vice president of sales for the Pittsburgh Coal Company. He was soon promoted to president of the company, a position he held for fourteen years until his return to Joy Manufacturing in 1940.[93] William G. Warden, chairman of the board of Pittsburgh Coal Company, testifying before a Senate committee in 1925, was asked why the firm locked out its striking workers. He answered simply that "we were losing money. The firm had operated under a deficit for seven years in the decade after World War One."[94]

While Consolidation Coal Company and Bethlehem Steel Company were able to break the union at their West Virginia mines easily, Pittsburgh Coal Company spent three long years fighting the union. The company's losses in 1924 approximated ten cents per ton on all coal mined.[95] The struggle changed the financial status of the firm, and the company went out of business, on paper, to avoid being sued for breach of contract by the union, and reorganized. This was a terribly expensive and destructive fight and started the company on a twenty-year period of no dividend payments to stockholders. Pittsburgh Coal Company prepared for a violent and protracted conflict with its union miners by doubling the number of its Coal and Iron Police, importing African American workers from the South and white strikebreakers to mine coal, and obtaining a federal injunction restricting the number of pickets at its mines. Strikebreakers were regarded by striking union miners as "bums and thugs." African American miners were employed chiefly in three southern coal-producing states during the 1920s: more than twenty-five thousand in West Virginia (most worked in the southern counties), some seventeen thousand in the Birmingham district of Alabama, and nearly ten thousand in Kentucky (most in the western Kentucky field). About thirty-five hundred worked in the coalfields of Virginia and Tennessee.[96] Striking white miners called the "scab" African American labor force "Roanoke niggers," because many were recruited by agents employed by the Pittsburgh Coal Company from this Virginia city and the surrounding rural communities.

Coal companies had imported thousands of African Americans from the South as strikebreakers throughout the strife-ridden 1920s. They were recruited by the promise of high wages. "Scabs" or "blacklegs" were derisive terms used by union workers to describe workers who refused to join the union and instead replaced them in the mines. Racism, combined with the hatred of scabs, frequently led to violence in the coalfields. A popular union song of striking miners during the 1920s clearly expressed their contempt for "scab" labor: "Just like a mule, A goddam fool, Will scab until he dies."[97] African Americans constituted half or more of the nonunion miners in a number of Pittsburgh Coal Company mines; by the mid-1920s they composed up to one-third of the nonunion labor at the mines of Pittsburgh Terminal Coal Company and other mining companies in southwestern Pennsylvania. Strikebreakers did not produce much coal because many were unskilled at mining. Some miners were racists and called them "cotton pickers," "blackbirds," and other names. Veteran miners believed that management brought them to the mines not to mine coal but to humiliate and "to break their hearts" by showing them that anyone could mine coal, "even niggers." Some companies built barracks and tents to house their strikebreakers and segregated them according to race. This area of the "patch" was referred to after the strike as "scab hill."

Pittsburgh Coal Company mines were reopened on an open-shop basis three months later with wages reduced by one-third. Management slashed miners' wages an additional 20 percent in 1927. The destruction of the union by the Pittsburgh Coal Company did not bring economic prosperity; the company lost even more money in 1925 and its deficit rose to $2,175,000 in 1926.[98] The inability of the company to make a profit was shared by other union mines. Income and tax returns from coal companies during the 1920s showed more firms reporting no net income with each passing year between 1920 and 1925.

Returning former union miners of the Pittsburgh Coal Company were forced to sign a "Pocahontas" labor contract, better known as the "yellow-dog contract," as a condition of their employment. They were called yellow-dog contracts because, supposedly, only a "yellow-dog" would sign one. The contracts signed

Income and Tax Returns of Bituminous Coal Companies

	Returns	Firms Reporting Net Income	Firms Reporting No Net Income
1920	1234	1152	82
1921	1234	503	731
1925	3650	1065	2585
1928	2705	863	1842
1929	2469	934	1535[99]

by miners as a condition of their employment stated, "I am not now a member of the United Mine Workers and I enter this employment with the understanding that the policy of the company is to operate a nonunion mine and would not give me employment under any other condition."[100] Their use was criticized and despised by all miners and the UMWA. One miner clearly voiced this sentiment stating: "This Yellow Dog Contract, knocks a man out of his citizenship rights. A man hain't got no protection under it. He can be kicked out any time like a yellow dog. Under the contract which I signed if I get fired I can be forced to move out of my house inside of an hour."[101] Labor injunctions were also used by coal companies during the 1920s, especially in the 1925 and 1927 strikes.[102] Millions of dollars were spent by the UMWA to have them struck down as unconstitutional. Miners who refused to work in the nonunion mine found themselves fired and their families forcibly evicted from their rented houses. The UMWA spent nearly $8 million to support these evicted miners and families by building temporary housing, including wooden barracks and tents, near the "patch" towns. The barracks were constructed very crudely with tongue-and-groove board and were not insulated.[103]

A primary cause for the decline of the UMWA during the 1920s was the union's inability to organize the expanding southern coalfields. Northern union operators had been urging the union to organize these regions ever since the 1890s, but unionization attempts in the expanding southern fields were sporadic and generally ineffective. The wage scale of the Jacksonville Agreement of 1924 was well above the prevailing rates in the nonunionized coalfields. Southern coal operators paid their workers lower wages, and these savings permitted them to undersell their coal in the shrinking commercial coal market and to acquire more coal lands. The capacity of coal stood at one billion tons in 1926; the nation's demand was half that. The nonunion mines had seized almost one-half the nation's coal market. Lewis accused these nonunion operators of perpetuating a system of feast or famine in the coal industry, noting that the coal operators of Logan County, West Virginia, were "practicing economic feudalism as the operators governed by terror and peonage of workers, a regime of private government backed by an army of mercenaries."[104] The shift of coal production from northern union fields to southern nonunion fields continued at a steady pace throughout the 1920s. The demands for wage and hourly concessions by union coal companies were motivated by genuine economic distress and not simply by corporate avarice.

The Pittsburgh Coal Company strategy to transform itself from a union to a nonunion producer was copied by rival union coal companies in Pennsylvania, Ohio, and West Virginia. Company after company followed Pittsburgh Coal's lead and by Christmas of 1925 almost 100 coal companies, some 110 union mines in Pennsylvania, and 50 mines in West Virginia had rid themselves of union representation. The Inland Collieries Corporation, a subsidiary of Bethlehem Steel Corporation, the Youghiogheny and Ohio Coal Company, the Paisley Coal Company of Ohio, and the Rochester and Pittsburgh Coal Company of Indiana County all decided to run their mines as nonunion operations. These companies represented some of the largest coal operators in Pennsylvania and neighboring states. The principal objective of these companies was not simply to extract temporary wage concessions from the union but to operate their mines on a nonunion basis. Some Central Competitive Field producers honored the Jacksonville Agreement until its formal expiration on April 1, 1927. Frank E. Herriman, president of the Clearfield Bituminous Coal Corporation, regarded the 1925 wage agreement with the union as "legally and morally binding" and honored the contract until its expiration.[105] This company operated a number of mines at the company towns of Rossiter, Barr, and Commodore, Indiana County. Unionized companies which honored the Jacksonville Agreement until its expiration thereafter refused to recognize the UMWA as a bargaining agent for their miners and demanded significant wage and hour concessions from them.

There had been closures of numerous unprofitable mines and companies ever since 1923, although not in sufficient numbers to decrease the capacity for overproduction of coal significantly. Union mines began to close in great numbers during 1925 and 1926 because operators could not mine coal profitably at $2.04 per ton and compete with nonunion southern coal operators. Unemployed union miners were forced to find work at nearby nonunion mines or move south to find work at nonunion mines operating at the lower 1917 wage scale. Individual UMWA locals held selective strikes between 1925 and 1927 in a futile attempt to stop the erosion of union-operated mines in western Pennsylvania. The UMWA spent more than seven million dollars during 1925 and 1926, largely to conduct selective strikes at mines where the Jacksonville Agreement had been abrogated. The UMWA scale committee authorized local districts to make local agreements for its members with individual companies. The union scale of $7.50 was a memory by 1926, this daily rate being maintained only in Illinois, although locals there accepted day rates as low as $5.00 with the expiration of the Jacksonville Agreement.

The union accused Andrew W. Mellon of Pittsburgh Coal Company, Charles M. Schwab of Bethlehem Steel Company, and John D. Rockefeller Jr. of Consolidation Coal Company of conspiring to destroy their union mines in Pennsylvania, West Virginia, and Ohio. They called upon Herbert Hoover, secretary of commerce, to intervene and stop this apparent union-busting, reminding him, "you sanctioned the Jacksonville Agreement, now back it up."[106] The government did not intervene and by this inaction tacitly condoned the union-busting activities of the coal operators.

With the formal expiration of the three-year Jacksonville Agreement, the UMWA leadership called for the obligatory general strike on April 1, 1927, under the slogan "no backward step."[107] The strike call initiated one of the most bitter and prolonged labor disputes of the twentieth century, which continued until October 1928 when the strike was officially declared over. Despite the Jacksonville Agreement of 1924, which guaranteed miners a wage of $7.50 a day, few union mines were still operating under this agreement in Pennsylvania, Ohio, Indiana, and Illinois when the union called the strike of 1927. The union had been driven out of New River, West Virginia, after 1922; Colorado in 1922; Kanawha, West Virginia, and western Kentucky in 1924; and northern West Virginia in 1924-1925. Most operators in western Pennsylvania, Ohio, Oklahoma, Arkansas, Maryland, and Iowa had discarded their union contracts and operated nonunion mines since 1925.[108] This strike, unlike the 1922 strike, was doomed because no coal shortages in the nation would determine the course of the strike. The nation's coal needs could be easily filled by production from the nonunion southern coalfields. The production ratio was about two-thirds union production and one-third nonunion in 1922-1923.[109] The share of the nation's coal output from nonunion mines increased from about 28 percent in 1919 to about 60 percent in the summer of 1925. The traditionally unionized northern fields of Pennsylvania, Ohio, Indiana, and Illinois lost about forty-four million tons of yearly production from 1924 to 1927. By 1926, for the first time in about thirty years, nonunion mines in the Pittsburgh district were producing more coal than union mines. The nonunion fields of southern Appalachia increased their production by fifty-seven million tons during this same period.[110] A southern Ohio coal operator reminded his striking union miners that "the mines of West Virginia are today working full time filling your orders while you have no work. These mines could take care of their trade and also all the trade formerly held by the Ohio mines."[111]

The strike from the very outset was especially violent, some two hundred thousand union miners having left the pit on April 1, 1927. The strike involved as many as one hundred thousand miners in western Pennsylvania, who effectively closed down all mining activity in Pennsylvania. The nonunion miners south of the northern West Virginia fields were unaffected by the strike. The union-busting coal operators were victorious by employing strategies similar to those they had used during the 1922 strike. The operators used dismissal, blacklisting, "yellow-dog" contracts, sheriffs' deputies, injunctions, importation of "scab" labor, private security police, and state and federal troops to suppress the strike. Pittsburgh Coal Company, H. C. Frick Coke Company, Bethlehem Mines Corporation, and Pittsburgh Terminal Railroad Company, for example, employed the notorious and hated Coal and Iron Police to protect their properties and intimidate striking miners and their families. These police wielded all the powers of publicly employed police officers. Governor John S. Fisher (1927-1931), a former executive of the Clearfield Bituminous Coal Corporation, Indiana County, and a close personal friend of the Mellons of Pittsburgh Coal Company, issued so many Coal and Iron Police commissions during the strike period that angry union leaders complained that "they were running amuck in the coal mining regions of western Pennsylvania." Governor Fisher issued four thousand commissions and permitted hundreds of uncommissioned men to wear the Coal and Iron badge during his term.[112] The coal company paid one dollar to the Commonwealth for each commission between 1871 and 1929. There is no record of a single commission having been revoked during this entire period. In hearings in 1928 before the Committee on Interstate Commerce, documented in *The Conditions in the Coal Fields of Pennsylvania, West Virginia, and Ohio,* it was observed:

> [The coal and iron police] are all very large men; most of them weighing from 200 to 250 pounds. They are all heavily armed and carry clubs, usually designated as "blackjacks." Everywhere the committee visited they found victims of the coal and iron police who have been beaten up and still carrying scars on their faces and rough treatment they had received . . .[113]

The use of a private police agency by the coal operators of Pennsylvania was not unique. The Baldwin-Felts Detective Agency played a similar role for the coal operators of Virginia and West Virginia. William G. Baldwin and Thomas L. Felts established the Baldwin-Felts Detectives Agency in Virginia and the agency was subsequently incorporated during the 1890s. Mine guards were hired by coal companies ostensibly to break strikes, spy on miners, maintain order in the company towns, and evict striking workers from company-owned housing. Baldwin-Felts gunmen were involved in the infamous 1921 Massacre at Matewan, West Virginia. The West Virginia State Police was not established until 1919. A West Virginia law, enacted in 1933, prohibited the employment of private police by coal companies, and the law helped bring about the inevitable demise of the Baldwin-Felts Detective Agency in 1935. The mine guard system was not formally abolished in West Virginia until the administration of Governor Holt in 1937.

The United States Senate Committee on Interstate Commerce established an ad hoc committee in 1928 to investigate charges that the John D. Rockefeller-controlled Consolidation Coal Company, General W. W. Atterbury of the Pennsylvania Railroad, and the Mellon-controlled Pittsburgh Coal Company, were "responsible for hunger and radicalism of the reddest kind in the mine fields."[114] This subcommittee (U.S. Senate, *Subcommittee of the Committee on Interstate Commerce Conditions in the Coal Fields of Pennsylvania, West Virginia and Ohio*, 70th Congress, 1st session, 1928) was one of the few congressional bodies to leave Washington and conduct its inquiry in the coalfields of the Pittsburgh district, Pennsylvania, West Virginia, and Ohio. Committee agents visited numerous isolated mining communities and immediately described "the squalor, suffering, misery, and distress as a blotch on American civilization."[115] Senator Frank Goodling of Idaho visited a number of Pittsburgh Coal Company mine towns in western Pennsylvania, and reported "women and children living in hovels which are more unsanitary than a modern swine pen. They are breeding places of sickness and crime."[116]

The Committee's findings filled two immense volumes entitled *Conditions in the Coal Fields of Pennsylvania, West Virginia, and Ohio,* which carefully documented the worsening conditions in the strike-torn mining communities of these regions. The dismal quality of daily life of miners and their families in western

Pennsylvania was chronicled by newspaper reporters. Journalist Lowell Limpus of the New York *Daily News* visited the strike areas and reported in explicit terms the conditions he encountered:

> I have just returned from a visit to "Hell in Pennsylvania." I have seen horrible things there; things which I almost hesitate to enumerate and describe.... I went into the coal camps of western and central Pennsylvania and saw for myself. We saw thousands of woman and children, literally starving to death. We found hundreds of destitute families living in crudely constructed bareboard shacks. They had been evicted from their homes by the coal companies. We unearthed a system of despotic tyranny reminiscent of czar-ridden Siberia at its worst. We found police brutality and industrial slavery.... we unearthed evidence of terrorism and counterterrorism; of mob beatings and near lynching; of dishonesty, graft, and heartlessness.... The mine fields are a bubbling cauldron of trouble. If it boils over—and it threatens to do so—blood must flow freely and many lives pay the forfeit.[117]

UMWA membership fell from nearly four hundred thousand members to less than one hundred thousand during the first decade of Lewis's presidency. In 1928 President Lewis told the Senate subcommittee that "the bituminous industry is today in the worst state of demoralization it has ever known. In some regions the work day had been increased from eight to ten hours while wages were slashed to $2, $3, and $4 dollars a day."[118] Wages were lower after the strike than they had been in 1924. Coal sold below actual production costs and miners' wages plummeted to as little as $2.50 a day. The strike proved to be a complete disaster for the union, and its consequences were much worse than anything Lewis could have imagined. The union was simply decimated! The UMWA membership in the bituminous fields dropped from 386,000 members in 1920—nearly two-thirds of all mine workers then employed—to about 84,000 or barely one-sixth of all miners employed in 1929. The Canadian membership in the UMWA fell from 20,600 to 12,900.[119] Seventy-one union mines between Connellsville and Brownsville, employing about 20,000 workers, had become nonunion operations by February 1928.

Lewis and the UMWA executive leadership abandoned any attempt to maintain a national centralized wage policy for its membership when the strike finally collapsed in October 1928. Individual districts were instructed by the UMWA to negotiate with individual coal companies the best possible contract for their membership. The UMWA bargained collectively for about eighty-four thousand dues-paying members or less than 20 percent of the nation's bituminous miners. Agreements in Indiana and Illinois resulted in 18.7 percent wage cuts, while Ohio miners saw their wages reduced by one-third. The UMWA's only remaining strength lay in the dissension-ridden Illinois and Indiana coalfields and in the anthracite districts of northeastern Pennsylvania. The Central Competitive Field as a bargaining organization was gone, never to return. Josephine Roche, president, manager, and the majority stockholder of the Colorado Coal Company, accepted union representation and signed a two-year wage agreement in 1928 with the UMWA. Most operators in Colorado opposed her policy of conciliation with the miners' union. Roche signed the contract with the UMWA covering about six hundred workers for the purpose of "stabilizing employment, production and marketing through co-operative endeavor and the aid of science."[120]

Lewis's leadership was seriously challenged by both moderate and radical factions within the UMWA, although he maintained control of the decimated union. He was leader of a miners' union that lacked members, financial resources, and a coherent plan for rebuilding its power.

UMWA Membership and Coal Miners			
	1910	1920	1930
Number of employees	613,924	733,936	621,661
UMWA Membership	226,000	373,800	205,100
Percentage Organized	36.8%	50.9%	33.0%[121]

Union membership was less in 1930 than at any period since 1910, 168,800 members fewer than 1920.

There was general dissatisfaction expressed with Lewis's leadership by some union members after the 1922 strike and the exclusion of nonunion miners from the contract that ended the strike. Events in the coal industry after 1925 made Lewis a vulnerable target for radical insurgents within the union. A coalition of Communists, Socialists, and moderate reformers all challenged Lewis's leadership in the UMWA presidential election of 1926. John Brophy posed the chief threat to Lewis's control of the union and discontented miners supported his candidacy for the presidency in 1926. The heart of Brophy's campaign was the nationalization of the mines. Lewis defeated Brophy's "Save the Union" campaign by attacking him as a dual unionist and a Communist sympathizer. Opponents of Lewis charged the election results were fraudulent due to numerous irregularities in the vote count. This serious dispute eventually led to a split in the union with the creation of a rival miners' union. Miners in the important Illinois coalfields formed the Progressive Miners of America. Competing union leaders, including John Brophy, developed a variety of competing strategies to resurrect their decimated union after the strike. They demanded that the new Hoover administration provide legislation to stabilize the industry, while other reformers advocated nationalization of the coal industry as a feasible solution. Still others proposed the creation of a labor political party and closer cooperation between anthracite and bituminous coal districts. All these reform proposals, however, met with little success as the industry continued its economic decline.

The anti-Lewis coalition split within the UMWA over internal differences, and disenchanted miners met in Pittsburgh at a convention in September 1928. This Communist-led splinter group of radical miners formed the National Miners' Union (NMU) with delegates from eleven states. The NMU made numerous unsuccessful attempts to remove Lewis as president, accusing him of both corrupt and ineffectual leadership. He was accused of squandering union money and providing high-paying patronage jobs for his friends, who rarely entered nonunion territory to organize miners, but instead were content to spend their time in the union halls. The NMU succeeded in organizing a few locals in Pennsylvania, West Virginia, and Ohio. The organization was not interested in negotiating wage agreements because its sole purpose was revolution. The union was not willing to work within the framework of the UMWA. The leadership believed miners were ready for a policy of violent confrontation and class struggle with the operators. The union's constitution asserts emphatically the class-struggle basis of the union: ". . . our organization declares that the interests of the employers and those of the workers have nothing in common. . . . The history of coal miners . . . is that of an incessant struggle between these two classes—the class struggle. . . . Our organization shall ever remain truly class-conscious[ness] . . . and proceed as an organization of the class struggle for the abolition of capitalistic exploitation."[122] The NMU, in accordance with its principles of international class conflict, was affiliated with the Red International of Labor Unions. This Marxist-controlled miners' union was responsible for numerous violent acts, including shootings of coal officials and dynamiting of their mining properties in Allegheny and Washington Counties. The leaders of the NMU were aware of the causes of the deplorable conditions miners were experiencing, but the organization had no concrete solutions, except the use of violence and the overthrow of capitalism. The NMU could claim two major victories during its brief existence. Its actions led to the passage of the Anti-Injunction Act (Norris-LaGuardia Act) of March 23, 1933. Senator George Norris of Nebraska toured the coalfields and spoke with miners and their families. He was touched by the sheer scale of the despair and poverty that he encountered and he was determined to improve the plight of the miners and their families. Norris sponsored federal legislation that made the "yellow-dog" contracts unenforceable in the courts, limited the power of courts to issue injunctions against labor, and asserted the right of labor to organize. The Anti-Injunction Act served as a background to Section 7a of the National Industrial Recovery Act of 1933. It declared that "the worker shall have full freedom of association, self-organization, and designation of representation of his own choosing to negotiate the terms and conditions of his employment, free from employers interference in these or other concerted activities for mutual aid or protection."[123] The NMU also lobbied successfully for the elimination of Pennsylvania's Coal and Iron Police. The union had faded away by the end of the 1930s.

FDR, The Great Depression, and the Revitalization of the UMWA

The fragile bituminous coal industry virtually collapsed during the Great Depression of the 1930s.[124] The industry had been devastated during the 1920s; but the nadir of the industry was reached in 1932 when prices, wages, output, and profits reached their lowest levels since the first decade of the twentieth century. Output declined nationally by 47 percent and employment by 61 percent between 1918 and the low point in 1932.[125] This economic decline caused severe hardship to coal operators and miners alike. Bituminous coal output was reduced from 573,366,985 tons in 1926 to only 309,709,872 tons in 1932, the lowest annual production since 1904. This reduction in output witnessed the decline in the number of workers employed in the bituminous coal industry from 593,647 to 406,380 during the same period. The average number of days worked annually declined from 215 to 146 days. Average hourly earnings of miners fell from seventy-six cents to fifty cents. Coal's share of the American energy market stood at 44.7 percent in 1932 while its share had been 66.6 percent in 1919. The average price per ton of coal fell from $1.78 to $1.31 between 1929 to 1932.

Coal operators, like coal miners, were victimized by the price wars of the 1920s and 1930s. The number of operating independent mines in the bituminous coalfields declined by twenty-five hundred between 1920 and 1935. Cutthroat competition and decline in the average cost of a ton of coal drove thousands of coal companies into receivership or bankruptcy. There were 6,070 coal mines producing 534,989 tons in 1929; in 1932 there were 5,427 mines producing only 309,710 tons.[126] Only 16 percent of nineteen hundred coal companies filing federal income taxes showed net income after taxes in 1932.

Bituminous coal production in Pennsylvania had fallen to 142 million tons by 1929—a decline of 35 million tons from its peak 1918 production of 177 million tons. Employment in the industry was 135,272, a reduction of 46,000 workers since 1918. The surviving mining firms operated on a part-time basis, open only three or four days a week.

The Great Depression hit the many isolated coal communities of western Pennsylvania especially hard and made life even more bleak. Unemployment and underemployment drastically reduced income and created severe hardships for the miner, his wife, and their children. The economic collapse of the soft coal industry can be easily measured, but its long-term hardship and misery is more difficult to measure. Economic collapse deepened poverty and hunger in the coalfields as miners suffered a drastic curtailment of their wages, if they were still employed at all. The daily life of miners and their families in western Pennsylvania was never easy, but life was increasingly harsh as unemployment led to poverty and hunger in many company towns. "You didn't live then," one contemporary miner noted, "you just existed." Some years later another miner asked to describe the period observed: "Well, there's one thing you can say about that time. We were all equal. Nobody had nothing."[128] "The fare of the workers and their dependents," Lewis wrote President Hoover in 1932, "is actually below domestic animal standards."[129] Governor Gifford Pinchot of Pennsylvania, in a letter to President Roosevelt in February 1933, clearly expressed the miners' economic

Bituminous Coal Mines			
	Number of Mines	Percentage of Change	Production (thousands of tons)	Percentage of Change
1929	6,057	-27.4	534,989	
1930	5,891	-42.2	457,526	-12.6
1931	5,642	-38.1	382,089	-18.2
1932	5,427	-2.3	309,710	-18.9[127]

despair in Pennsylvania: "[M]en were working there six and seven days a week who cannot earn enough to feed their families—working full time and not yet on relief."[130]

By 1933, over 35 percent of Pennsylvania's working population was out of work.[131] Unemployment in the coal-producing counties was even worse. Estimates of unemployment in Armstrong, Blair, Cambria, Cameron, Clearfield, Somerset, Washington, and Westmoreland Counties neared 40 percent while unemployment in Dunbar, Everson, and Fayette City in Fayette County soared beyond 40 percent in 1934.[132] Most miners still lived in isolated and often drab private mining camps owned and ruthlessly controlled by the coal company; its "feudal rule" was strictly enforced by private security police. Evicted miners and their families lived in makeshift tents, coke ovens, and other forms of accommodation that afforded them minimal shelter. Fannie Hurst, a New York labor journalist, visited many coal towns of America during the 1930s and in her writings painted a disturbing portrait of absolute despair and ugliness. According to Hurst's writings, "every aspect of their lives is ugly and anesthetic." These people "are living in shambles." "Children were reared under conditions that are shocking beyond description . . . where human beings are living under conditions that generate hate, you can see that seeds of revolt are being sown."[133] Many unemployed miners took to the road in search of work during the 1930s and when they could not find it returned home to the "patch." The following writings of contemporary coal miners give a poignant vision of daily life and the sense of hopelessness and resignation that had overwhelmed them:

> I've tried all up and down the river and can't get work no place. At one time I was gone nine days but couldn't find anything. I got so tired being turned down I got sick and disgusted and came home. There ain't nothing a man can do these times.
> I've travelled 3,000 miles in the past ten weeks trying to find a job. But it ain't no use. There ain't no job any place.
> We find we are starving even at our work, as we can't get any food or money that we have sweated so hard to earn. . . . We have no doctor. We have no hospital for our sick, no graveyards for our dead. We have gotten nothing to eat at this job for ten days.[134]

Coal companies had reduced hourly wages and the number of days their mines were open. Many companies maintained unchanged the rents on company houses that were at best dismal with few material amenities. Working miners and their families still payed exorbitant prices for food and necessary mining equipment at the "pluck-me store." Studies undertaken by private relief agencies during the 1930s to compare prices charged at company and noncompany stores showed that prices charged for goods at the company store were consistently much higher than at neighboring independent stores.[135] Miners were paid twice monthly, and from their meager wages had to pay rent for their company-owned house, buy household coal, pay the company doctor, pay school taxes, buy mining supplies, and purchase food and clothing. The company checkweighman who weighed their coal was employed by the operators in nonunion mines, and he alone decided if the scales were accurate. Cheating or "shortweighting" the miners at the tipple's weigh station was a well-established custom that had been employed by many nonunion companies since before the Civil War.

Bread was the principal food consumed daily by miners and during the Depression their lunch was often bread and coffee, called "coffee soup." The dinner menu for many mining families consisted of "bulldog gravy" made of flour, water, and some grease. They also ate "miner's strawberries"—beans—and for variety white beans one day and red beans the next. A miners' "water sandwich" for the lunch was stale bread soaked in lard and water.[136] A miner described his family's menu: "We have been eating wild green . . . Such as Polk salad. Violet tops, wild onions,

Coal companies sponsored competitions for the best vegetable garden in the mining community. Pennsylvania State Archives.

forget me not, wild lettuce and such weeds as cows eat as a cow won't eat a poison weeds [sic]."[137] Families with gardens enjoyed fresh vegetables. The sole aid in their economic plight was local charity, which was generally limited, if available. The economic plight of the coal miners was not regarded as a problem for the nation as a whole. It would take the Great Depression to convince society that society as a whole was responsible for the care of those members unable to find adequate employment. The relief programs of the federal government under New Deal legislation were the concrete expression of this new idea. Lewis told UMWA delegates in the lightly attended 1932 annual convention, "Fear and hesitancy dominate American thought in industry and finance."[138] The coal industry was mired in the worsening depression and prospects for improvement in the immediate future were dim. The extent of the decline, the chaos of the union's fortunes, and conditions of American miners were clearly addressed in the government report of the international officers to the 33rd Constitutional Convention in 1934:

> The United Mine Workers of America were driven from one field after another by the law of injunctions and the rule of gunmen; the right to collectively bargain for their wages was denied the mine workers and there was substituted the individual system of employment, in which the worker foreswore his right to belong to a union; wage rates were arbitrarily posted at the tipple; the right of mine workers to check-weigh their own coal was denied; wages were cut time and again and further sweated by the rent of company houses and prices charged at the company stores. The free hand which the corporation thus exercised in labor relations was the chief cause for the increasing demoralization of the industry.[139]

Lewis delivered a radio speech on September 11, 1932, identifying the principal causes of the Depression:

> Labor protests against further over-expansion of production facilities and asserts its opposition to employees being forced to carry the burden of fixed charges on the unnecessary investment involved in over-expanded industrial plants, labor seeks definitively to eliminate the manufacturer whose sole ability to remain in business is geared to pauperize wage rates and cutthroat sales prices.[140]

John L. Lewis believed prosperity would return to the United States when the following aims of organized labor were introduced: a shorter working day and week, the creation of a national economic council, and the enactment of a national industrial code. He was convinced that legislation that would stabilize the coal industry and require coal operators to recognize and bargain with all miners collectively was the only way to save the UMWA and save the coal industry. Lewis was re-elected president of the UMWA on January 31, 1933, and with his new mandate decided it was time to reorganize and revive the decimated union. He summoned his closest lieutenants to Washington, D.C., to plot a long-term strategy to recoup the union's devastating losses since the ill-fated 1927 strike. President Lewis planned the UMWA resurgence before Roosevelt had taken office and his administration had drafted the National Industrial Recovery Act. The reasons for his sudden decision to mount a major union-organizing drive at this time are still unclear, but perhaps he was emboldened by the recent election of Roosevelt in November 1932.

The Great Depression had hit America and the rest of the capitalist world in 1929. From 1929 to 1933 economic conditions worsened. Unemployment reached a peak in the winter of 1932-1933 at around 25 percent. The federal government did not systematically collect unemployment statistics. The practice of keeping accurate unemployment records was not introduced until 1940. The Bureau of Labor Statistics later estimated that 12,830,000 persons were out of work in March 1933, at the bottom of the depression, about one-quarter of the civilian labor force of over fifty million. Approximately twenty-eight million Americans were supported by some type of public or private relief.

Roosevelt, who swept the 1932 presidential election against incumbent President Herbert Hoover by carrying forty-two states took office on March 4, 1933, and observed in his inaugural address that "a host of unemployed citizens face the grim problem of existence, and an equally great number toil with little return. Only a foolish optimist can deny the dark realities of the moment." Roosevelt, working closely with the Congress, launched the first New Deal, a series of federal programs aimed at lifting the deteriorating American economy out of the Depression by dealing with high unemployment and farm relief. The Roosevelt administration sent fifteen major legislative proposals to the seventy-third Congress between March and June of 1933. All these proposals were adopted during this legislative session later known as the "First Hundred Days." The first New Deal legislation greatly extended the influence of the federal government in regulating many areas of the slumping economy. Government spending in federal social welfare rose by 500 percent between 1933 and 1936; the federal deficit nearly doubled to $1.8 billion during this period. Congress established scores of new federal regulatory entities. Notable among them were the Agricultural Adjustment Administration (1933-1942), Civil Works Administration (1933-1934), Civilian Conservation Corps (1933-1942), Farm Security Administration (1937-1944), Federal Emergency Relief Administration (1933-1937), Public Works Administration (1933-1939), and the National Recovery Administration (1933-1935).

The National Industrial Recovery Act, called the NIRA, passed by Congress on June 16, 1933, was one of the most significant pieces of labor legislation passed during the first New Deal. Roosevelt noted that "history will probably record the National Industrial Recovery Act as the most important and far reaching legislation ever enacted."[141] The NIRA was administered by the National Recovery Administration (NRA), directed by General Hugh S. Johnson under the "Blue Eagle" symbol. Its essential purpose was to stimulate industrial and business activity by a series of agreements or codes, drawn up to govern each industry and to reduce unemployment. The NRA was based on the principle of self-regulation, operating under government supervision, through a system of fair-competition codes concerning working conditions, wages, and business, by dividing markets, boosting prices, and stopping wage cuts and layoffs. The NIRA represented "the first serious and far-reaching attempt in peace time on the part of the federal government to regulate an industry that had been so committed to laissez-faire."[142] Under the NRA, 765 codes were drawn up by the federal government to regulate output, fix prices, reduce working hours, and increase wages in various industries, including the coal and steel industries.

The coal industry in 1932 was ripe for the unionization which came with the NIRA. Both coal companies and miners sought relief in any form. Production had declined by some 40 percent and many coal operators were bankrupt or on the verge of bankruptcy during the exceedingly competitive 1930s. Enormous capital values had been lost in the coal industry as many companies failed to survive. The

Rockefeller family, which owned 38 percent of the Consolidation Coal Company, took a complete loss on its investment and retired from the coal business in 1932.

Lewis's union drive was initiated in the northern coalfields on "Mitchell Day," April 1, 1933, at Blythedale, Fayette County, when nearly two thousand miners gathered to hear a call for unionization and solidarity among the rank and file. Union miners had been granted the eight-hour workday in 1897 under the leadership of John Mitchell, who was called "the father of the eight-hour day." Mitchell Day, now called Miner's Day, is celebrated every April 1 as a contractual miners' holiday. Philip Murray, vice-president of the UMWA, was appointed by Lewis to lead the union drive in Pennsylvania and the northern coalfields. Murray, a Catholic Irishman, was born in Lanarkshire, Scotland, the son of an Irish immigrant miner. The family emigrated to Westmoreland County on Christmas Day 1902. He worked as his father's helper in a number of coal mines, first in Scotland and later in western Pennsylvania. Murray was later employed at the Keystone Shaft Mine of the Keystone Coal and Coke Company, Westmoreland County. The lanky eighteen-year-old Murray smacked a weigh boss suspected of cheating him and other miners on weighing coal at the tipple. The company evicted the Murray family from their company house and they were forced to live in a tent on the outskirts of the town. Angry and sympathetic miners left the pits and refused to mine coal. The strike fizzled after a month when hunger forced them back to work. Murray was personally escorted to the county border and told not to return by company officials. This action made him immensely popular with the rank and file and he rose rapidly in District 5 of the UMWA. Murray was an International Board member by 1912 and four years later was elected president of District 5. Lewis made him a vice-president in 1920 and during the next decade he was an invaluable ally and confidant to Lewis.

On May 1933, Philip Murray spoke to four thousand miners in Clymer, Indiana County, and recruited 693 new union members who later organized mines in the region owned by the Delano side of President Roosevelt's family. (Roosevelt's grandfather Warren Delano and James Roosevelt, his father, had both been members of the board of directors of the Consolidation Coal Company. Delano had served as a company director from its founding until 1874 while James Roosevelt was a director from 1868 to 1875.) Murray and other union activists had attracted 150,000 miners to the swelling union ranks by the end of 1933. Union organizers entered such former citadels of nonunionism as the Connellsville coke district and Somerset and Cambria Counties, Pennsylvania. The Keystone State became the backbone of the new, emerging UMWA.

This burst of union activity by the UMWA in early 1933 was both dramatic and unprecedented in the history of the American labor movement. Historian Leo Wolman asserted that "there was nothing in the annals of labor . . . comparable to the dynamic burgeoning of the UMWA under the NIRA."[143] A contemporary song sung throughout the "patch towns" of western Pennsylvania reflects the miners' despair over the collapse of their union and the subsequent hard times, "We will have a good local in heaven, Up there where the password is rest, Where the business is praising our Father, And no scabs ever mar or molest."[144] Hunger and misery had stalked the dismal and often oppressive mining camps since the horrific union defeat in 1927, and these years of economic deprivation had intensified the miners' desire and need to band together. James A. Wechsler, a labor journalist of the newspaper *PM,* observed that "to the miners the union was more than a collective bargaining association, it is the pillar of their hopes. As long as the union was preserved they are not serfs, they retain a glimpse of freedom and an awareness of potential power. The fortunes of the union are completely entwined with their own personal histories."[145]

The federal government, under Section 7a of the National Industrial Recovery Act, also legalized and guaranteed that employees have "the right to organize and bargain collectively through representatives of their own choosing, and shall be free from interference, restraint, or coercion of employees in the designation of such representation."[146] The union succeeded in getting a bill introduced in the U.S. Senate in 1928 that would guarantee the rights of workers to organize and bargain collectively. The bill drafted by the UMWA's general counsel, Henry Wehrum, never made it out of the first congressional committee to consider the bill. Section 7a gave official approval to collective bargaining between employers and their workers by the federal government. UMWA organizers, like other industrial labor organizers, saw its passage as the critical ingredient for the resurgence of their impotent union. Membership had declined sharply in all unions in the 1920s. American workers called the NIRA "Labor's Bill of Rights." "From the standpoint of human welfare and economic freedom," said President Lewis, "we are convinced that there has been no legal document comparable with it since President Lincoln's Emancipation Proclamation."[147]

The repressed and moribund American labor movement of the 1920s stirred to action under the impetus of the NIRA and other pro-labor New Deal legislation. American trade union membership soared as 775,000 workers flocked into numerous labor organizations—500,000 to the AFL, 150,000 to independent unions, and 125,000 in the Trade Union Unity League (TUUL).[148] American trade union membership, averaging only 3.3 million in 1930, rose continuously during this period, to 3.9 million in 1933, 4.7 million in 1936, 8.2 million in 1939, 12 million in 1942, and 13.6 million in 1945.[149]

The Bituminous Coal Code, under the NRA, represented "the first serious and far-reaching attempt in peace-time on the part of the federal government to regulate an industry that had been so committed to laissez-faire."[150] Bituminous coal operators were required under the code to abide by a Code of Fair Competition that called for (1) the maintenance of minimum prices and (2) elimination of unfair competitive practices. The code had the force of law and was binding on all coal operators. A minority of coal companies refused to comply immediately with the NIRA and Section 7a because they doubted the law's constitutionality. The Code of Fair Competition was initially successful in the bituminous coal industry, but code violations were increasingly frequent and complaints of cutthroat competition and unfair price fixing became more numerous. The code was in force from October 1933 to May 1935. The NIRA was abruptly struck down by the Supreme Court as unconstitutional on May, 27, 1935, in Schechter Poultry Corp. v. U.S. (known as the sick chicken case). The Supreme Court ruled the NIRA unconstitutional on the ground that Congress had delegated too much authority to the agency. The court noted that the act was encouraging monopoly and creating cartels to the detriment of small businesses. The voiding of the NIRA ended the coal Code of Fair Competition.

An angry Lewis stated that "we are living in a state of continuous crisis under the negative autocracy of five former corporation lawyers on the Supreme Court bench. Only industrial democracy can save America from a position of permanent social and economic disequilibrium."[151] Lewis, with Congressman John Buell Snyder of Fayette County and Senator Joseph F. Guffey of Pennsylvania, wrote a new bill providing for the regulation of the bituminous industry, similar to the recently voided coal code under the NIRA. The Bituminous Coal Conservation Act, commonly known as the Guffey-Snyder Coal Act, was passed by Congress in August 25, 1935. The new law, called coal's "little NRA," was yet another attempt by the federal government to establish a balance between production and consumption in the coal industry. The Bituminous Coal Labor Board and the National Bituminous Coal Commission were created to administer production quotas, price-fixing, and labor regulations based on the NRA soft coal code. A 15 percent tax was levied on producers based on the market value of coal, with some 90 percent of the tax to be remitted to coal producers complying with the code. The Guffey-Snyder Coal Act was jubilantly received by the union miners and by a majority of northern coal operators who vividly remembered the ruinous competition in their formerly unregulated industry. The new act was strongly resisted by southern operators who unsuccessfully lobbied against it. James Walter Carter, president of the Carter Coal Company, filed suit immediately to test the constitutionality of the new law. The United States Supreme Court ruled the act unconstitutional in Carter v. Carter Coal Company et al. in May 18, 1936. The majority decision asserted Congress had no power to regulate wages, hours of labor, and working condition in an industry not directly engaged in interstate commerce. The Court ruled the coal industry was not interstate commerce but, like manufacturing, a local business.[152] Lewis responded angrily again to the Court's decision: "[I]t is a sad commentary on our form of government when every decision of the Supreme Court seems designed to fatten capital and starve and destroy labor."[153]

The Guffey-Vinson Bituminous Coal Act was passed by Congress in 1937. The act re-enacted all the principal provisions of the unconstitutional Guffey-Snyder Act of 1935 with the exception of the wages-and-hours clause. The act created a new code of fair competition for the bituminous coal industry. New federal regulations laid a revenue tax of one cent a ton on soft coal, and imposed on noncode coal producers a 19.5 percent penalty tax of the sales price. The Supreme Court upheld the act as constitutional. Associate Justice William O. Douglas, who wrote the court's majority opinion, made the following astute observation regarding the recent tragic history of the bituminous coal industry:

> Labor and capital alike were the victims. Financial distress among the operators and acute poverty among the miners prevailed during periods of general prosperity. This history of the bituminous coal industry is written in blood as well as ink.[154]

Government intervention to regulate coal prices was successful in ending the destructive cutthroat, competitive wars of the 1920s. One of the major causes of instability in the bituminous coal industry was overdevelopment, made possible by the abundance of coal located over a wide geographic area. The inability of coal companies to regulate production and marketing and to control coal costs had made some form of external regulation inevitable. It is ironic that federal control over the bituminous coal industry came not because the industry was monopolistic but rather because the industry was not. The splintered coal industry was simply unable to find means to solve its problem of overproduction through industrial self-government.

The American Federation of Labor, formed on December 8, 1886, in Columbus, Ohio, was guided for years by conservative and cautious leadership. President Samuel Gompers was the union's president until his death in 1924. William Green, UMWA secretary-treasurer, was elected AFL president following his death. President Green, like his predecessor, was also a conservative and cautious leader. *Fortune* magazine, in December 1933, noted that the AFL "has been suffering from pernicious anemia, sociological myopia, and hardening of the arteries for many years."[155] The heart of the AFL was still craft unions, such as the carpenters and the machinists. The AFL leadership favored the continuation of craft or "horizontal" unions. Union membership in the United States in 1933 had reached its lowest point since 1920 and some labor leaders within the AFL favored organizing workers in industrial or "vertical" unions—unions which took in all workers in an industry. At the November 1935 AFL convention in Atlantic City, David Dubinsky of the International Ladies' Garment Workers and Sidney Hillman of the Amalgamated Clothing Workers joined Lewis, John Brophy, and other UMWA officials to organize American labor on an industry-wide basis. They believed that the time was ripe for an all-out effort to organize mass-production industries. Proponents of the new industrial union met immediately following the AFL convention to form the Committee on Industrial Organization (CIO). The CIO was a conscious attempt, led by the miners' union, to organize workers employed in the mass-production industries. The new union was headed by Lewis and included the leaders of the Amalgamated Association of Iron, Steel and Tin Workers, Mill and Smelter Workers, Federation of Flat Glass Workers, United Textile Workers, and other large unions joining the CIO throughout 1935. Lewis was vice-president of the AFL but resigned in 1935. Conservative AFL leadership was opposed and felt threatened by the growth of industrial unionism and on August 3, 1936, the executive council of the American Federation of Labor suspended all CIO unions over a jurisdictional dispute. The expelled unions established a rival organization, the Congress of Industrial Organizations. The CIO held its founding convention on November 14-18, 1937, in Pittsburgh. John L. Lewis presided over the four-day convention. The bitter conflict between the leaders of the AFL and CIO was essentially craft versus industrial unionism. The CIO's subsequent organizing victories were seen as personal triumphs for the UMWA's John L. Lewis. The AFL and the CIO were competitive unions until their merger at a New York City annual convention on December 5, 1955.

Lewis was elected the first CIO president and supported Franklin D. Roosevelt in his 1932 and 1936 presidential campaigns against Herbert Hoover and Alf Landon. Lewis and the United Mine Workers contributed a half-million dollars to Roosevelt's 1936 presidential campaign and the same to Labor's Non-Partisan League (LNPL). The LNPL was an independent political-action organization created by the leader of the UMWA and other unions as an independent political organization. The organization gave the miners' union and other participating unions the means to maintain their own political identity. Lewis, however, refused to support Roosevelt in his bid for a third term in 1940. Lewis believed Roosevelt had not delivered on all the union's demands to the government. He was also concerned that organized labor was becoming too closely wedded to an increasingly conservative FDR and Democratic Party, and he backed Republican Wendell Willkie's presidential bid in 1940. Lewis resigned his presidency of the CIO following the 1940 presidential election, having failed to divert labor's support from Roosevelt to Willkie. Philip Murray was elected the new CIO president and served from 1940 to 1952.

The cry to "ORGANIZE" spread across the desolate mining villages from Pennsylvania to the coalfields of Alabama where southern miners sang,

> In nineteen hundred and thirty-three
> When Mr. Roosevelt took his seat
> He said to President John L. Lewis
> In the Union we must be.[156]

UMWA organizers constantly reminded West Virginia, Pennsylvania, Illinois, and Kentucky miners that "the President wants you to join the union" and "the law is on our side." "The old union is coming back, by God" was yet another rallying cry of union organizers.[157] Miners responded immediately and almost unanimously. Organizers went into such antiunion bastions as Mingo County, West Virginia, and the Connellsville coke region of Pennsylvania. Van A. Bittner, a West Virginia mine organizer, reporting on the speed with which miners were returning to the union just a week after the passage of the NIRA, observed, "We expect to be practically through every mine in the state and have every miner under the jurisdiction of our union by the first of the week."[158] John Brophy, in his autobiography, *A Miner's Life*, noted that "within ninety days the industry was organized." There was no need to campaign; an organizer had only to see that he had a good supply of application blanks and a place to file them, and the rank and file did the rest. The fact that workers organized so quickly after 1933 was an indication of the general breakdown of the coal operators. They couldn't marshal sufficient strength to resist this crusade of organization that was sweeping the nation.[159] Membership in the UMWA quadrupled in a few short months of the passage of the NIRA. Miners joined the union in record numbers, membership increasing by some three hundred thousand in only one year. Philip Taft, in *Organized Labor in American History,* wrote that "it can truly be said the revival of the United Mine Workers of America was the greatest labor event in the short history of the National Recovery Administration, an event which was to make possible the forthcoming labor revolution in the late 1930s."[160] The UMWA became the biggest and strongest of American labor unions virtually overnight. Much of the membership drive was completed by August 1, 1933, in every coal-producing region except the steel companies' "captive" mines.

The reorganized and vibrant UMWA won significant gains for its members in terms of wages, hours, and working conditions after 1933. No basic agreement had existed from April 1, 1927, until September 1933 in the bituminous coal industry, although some district agreements were signed in Indiana and all the states to the west. Seventy-two percent of bituminous coal production was nonunion, with wages from $1.25 to $2.84 for a nine- to ten-hour workday.[161] The union concluded a new collective wage agreement in the former northern Central Competitive Field (CCF) region and in the nonunion coalfields of West Virginia, Kentucky, and Tennessee on October 2, 1933. A single contract that would apply uniform wages and working conditions to all miners and employers in the industry was unheard of until the 1930s. Commercial coal operators signed five collective wage agreements with the UMWA between 1933 and 1939 that were applicable to all union miners and employers.[162] The first contract established minimum coal prices at the established 1929 level of $1.78 a ton. The object of the contract was to prevent the price-cutting and wage reductions which had characterized the industry in the 1920s. Most commercial coal mining companies accepted the NIRA because they felt that the new contract with the UMWA could act to stabilize and maintain a minimum price for their coal. This first agreement and subsequent contracts were all collectively called Appalachian Agreements, and unlike the pre-Depression era contracts signed between the UMWA and CCF operators, were very explicit contracts, detailing in specific terms the conditions of employment.

The first Appalachian Agreement, effective between October 1, 1933, and March 31, 1934, formed new arrangements for collective bargaining in the bituminous coalfields, and included coal operators from both the defunct Central Competitive Field and the nonunion southern Appalachian coalfields. This contract was a landmark agreement in the soft-coal industry, covering as it did 340,000 miners—the largest number of workers ever included under one labor agreement in American history. The agreement was the beginning of a new era in the task of stabilizing and modernizing the economic process of this basic industry.[163] Wage differentiation in the industry had existed between union and nonunion mines since the 1890s. The new contract narrowed wage differences among the various coalfields, and at last the UMWA could bargain collectively with both southern and northern coal operators. Daily wages were stabilized by establishing a minimum national wage of $3.40 a day: Northeast, $4.60; South, $4.20; Midwest, $4.57 to $5.00; Southwest, $3.75; Northwest, $4.00 to $5.63; Deep South, $3.40 to $3.84.[164] The agreement reintroduced the eight-hour day, the five-day forty-hour workweek, the right of miners to choose their own checkweighmen, a dues checkoff, the abolition of child labor in the mines, and policies for handling labor disputes. Some nonunion operators were recalcitrant and refused to operate their mines with union workers. The UMWA had led a strike against nonunion operators in Harlan County, Kentucky, but by May 1931, admitting defeat, had withdrawn its strike action. Harlan County was called "that little ugly running sore" by union organizers because the union had fought one of the longest and bloodiest organization drives in this county. The Harlan County Coal Operators' Association was composed of staunch antiunion operators who ruled their mines and workers like feudal barons. They were successful in thwarting unionization attempts until July 19, 1939, when they finally signed the national UMWA contract.

The UMWA was once again a potent force in the coalfields and by 1934 the ranks of the union stood at about four hundred thousand dues-paying members. The union was reestablished as the collective bargaining agent for coal miners under the first Appalachian Agreement. Each successive Appalachian contract signed with management by the UMWA between 1933 and 1941 improved wages and maintained the forty-hour workweek. The forty-hour weekly schedule, eight hours per day and five days a week, was established and remained in force except for a few years during World War II. The average bituminous coal miner in the United States had worked sixty hours weekly before 1898, reduced to fifty-two hours weekly during the period from 1898 through 1916, to forty-eight hours weekly between 1917 and 1932, and to forty hours after 1933.

Coal miners were generally the lowest paid of all industrial workers in the United States. The ratification of the Appalachian contracts increased their wages, and miners' income became competitive with workers in the other extractive industries. It is difficult to calculate with any certainty how much these wage increases helped workers' real income. Some coal companies simply offset their rising wages by increasing rents on company houses and the cost of foodstuffs and mining supplies in their company stores. The passage of the first Appalachian Agreement was, at least for the short term, a festive time for overworked and underpaid miners who had endured a decade of decline in their income. Some eighty-five thousand more miners nationally were working in 1935 than in 1932—an increase of 24 percent—while wages were rising by 70 percent during this same period. The average annual wage of a miner was $677 in 1932 and $1,196 by 1935. In the short term, the NIRA had satisfied the Roosevelt administration's objective of putting people back to work and raising their real income.

Average Hours and Earnings in Bituminous Mines

	Weekly Hours	Hourly Earnings	Weekly Earnings	Annual Earnings
1935	26.5	$.74	$19.58	$ 957
1936	28.8	.79	22.71	1103
1937	27.9	.85	23.84	1170
1938	23.5	.87	20.80	1050
1939	27.1	.88	23.88	1197
1940	28.1	.88	24.71	1235
1941	31.1	.99	24.71	1235[165]

Hourly Wages in Mining Industries

	1939	1950	Increase
Iron Mining	73.8¢	$1.51	103%
Copper Mining	67.9¢	$1.60	136%
Lead and Zinc Mining	68.3¢	$1.56	130%
Bituminous Coal Mining	88.6¢	$2.01	126%[166]

UMWA Struggle with the Steel "Captive" Mines

The five Appalachian Agreements signed by management with the UMWA between 1933 and 1941 were confined to independent coal operators. These contracts negotiated by Lewis restored stability to the bituminous coal industry, but they did not include miners employed in "captive" mines. Coal was being marketed in three ways by the 1930s. Independent mines sold their coal on contract to large consumers, including railroads, factories, electric power plants, and gas plants. Independent mines also sold surplus coal on the open market to wholesale dealers, who sold coal to retail dealers. The retail dealers in turn sold it to small consumers. "Captive mines" had been owned by large railroads, coke and steel companies, and utilities since the 1880s. These companies were principal consumers of coal and coke and had acquired their own mines so that they would not be dependent on fluctuating market conditions. Captive mines controlled about 18 percent of the total coal production from 1913 through 1918 and more than a quarter (26.5 percent) from 1920 to 1929.

Pennsylvania, West Virginia, and Illinois had the largest concentration of captive bituminous coal mines. The steel companies of Pennsylvania produced 31.9 percent of the state's coal in 1924 (75.4 percent of the total captive coal production nationally). The captive mines of Illinois were owned by railroads and public utilities and accounted for 23.8 percent of the state's coal production, while West Virginia's captive mines were owned by steel companies and coke manufacturers and produced 12 percent of the state's coal in 1924.[168] Steel corporations operated 232 mines producing 51.6 percent of the captive-mine coal in the United States in 1924. Bethlehem Steel Corporation (Industrial Collieries Corporation, mining subsidiary company with mines in Cambria, Indiana, Washington, and Westmoreland Counties), U.S. Steel Corporation (H. C. Frick Company with mines in Fayette, Greene, and Westmoreland Counties), Youngstown Sheet and Tube Com-

Captive Coal Mine Production and Percentage Share of National Market

	Production Millions of Net Tons	Percent of Total U.S. Production
1915	83	18.8
1920	113	19.8
1925	121	22.7
1930	117	25.0[167]

pany (Buckeye Coal Company with the Nemacolin mine, Greene County), Crucible Steel Company (Crucible mine, Greene County), Republic Steel Company (Clyde mines in Washington County and the Indianola mine in Allegheny County), and Jones and Laughlin Steel Corporation (Vesta Coal Company with mines in Washington County) were the principal steel companies in western Pennsylvania operating such coal mines.[169]

The steel firms were not interested in establishing industry-wide minimum coal prices since their coal and coke production was consumed internally. The steel industry had been a bastion of the open-shop tradition since the collapse of the Amalgamated Association of Iron and Steel Workers at the Homestead Strike of 1892. Ebert H. Gary of U.S. Steel wrote that "we do not confer, negotiate with, or combat labor unions as such." Gary expressed in these few words the majority attitude held by steel companies toward trade unions since the beginning of the industry. William A. Irwin, president of the U.S. Steel Corporation, clearly voiced the majority position of the steel industry toward union representation of their miners stating, "[A]s long as I live my company will never recognize the United Mine Workers."[170] Steel companies were opposed to unions, charging they were responsible for creating unnecessary antagonism between labor and capital. Charles M. Schwab (1862-1940), former head of the Carnegie Steel Company and later the United States Steel Corporation and president of Bethlehem Steel Corporation since 1904 observed:

> It has always seemed a curious thing to me that people should talk about "conflict" between capital and labor. There is no conflict. It is human nature to want money. Capital wants money, so does labor. Where you see men, either as individuals or in groups, wanting more money, that's not conflict. The interests are identical.[171]

Steel management, like Lewis, saw the UMWA organization of their miners as an entering wedge for steel unionization. The Amalgamated Association of Iron and Steel Workers, affiliated with the AFL, organized some twenty-four thousand workers, or one-fourth of all steel workers in 1891. In 1931, the union had slightly less than five thousand members, or about 1 percent of all steel workers. Steel management approved the wages and hours that commercial coal operators had accepted in the first Appalachian Agreement, but refused to recognize the miners' union or establish dues check-off. Check-off is a provision in a labor agreement whereby the operators are obliged to deduct union dues from workers' pay and forward them to the union headquarters. Officials from the NRA and the UMWA met with steel company representatives in late September 1933 in an attempt to persuade them to recognize the UMWA, but management refused to consider it.

Captive-mine owners had tried to inspire company loyalty in their employees by undertaking a series of paternalistic programs to improve their workers' living conditions. Corporate paternalism was a conscious attempt to attract and maintain a stable and loyal labor force while lessening the appeal of union organizers. W. H. Glasgow, assistant superintendent of the H. C. Frick Coke Company, asserted proudly to a newspaper reporter that "the welfare of the miner and other employees engaged in the mining industry, is the welfare of the industry," and further, "we contend that the employees of the H. C. Frick Coke Company are the best treated and best satisfied coal and coke workers in the world. I know of nothing more conducive to the higher efficiency and uninterrupted service from working men than fair treatment and pleasing surroundings."[172] Many of the "model" industrial coal towns, including Bobtown, Nemacolin, Mather, Muse, Richeyville, Vestaburg, Slickville, and Daisytown, were constructed by steel companies in western Pennsylvania after the 1890s. These communities provided large and clean homes, charged workers fair prices at the company store, and provided a variety of recreational activities for their workers. They sponsored baseball teams and constructed playgrounds for their children, swimming pools, and recreational halls. Their miners were usually paid higher wages than miners employed in independent mines. Some steel companies provided pensions for miners with twenty-five years' service, relief programs in the event of unemployment, and compensation for families of miners killed or hurt in mining accidents. Work at their captive mines was generally steady, and when it was not, a miner could receive relief from the company.[173] The U.S. Steel Corporation provided a pension program for its miners and an opportunity for them to purchase company stock.

Both government and mine workers rejected the steel companies' corporate paternalism for unionization. H. C. Frick Coke Company, Crucible Steel, and Weirton Steel all sponsored the company union, or "employment representation plans" as the corporations preferred to call them. Company unions were formed as a conscious attempt by antiunion steel management to comply with the letter, yet evade the spirit, of the mandate of Section 7a. The company union was sponsored and financed by the companies and these organizations maintained a facade of collective bargaining. They did not carry out real negotiations with their employers, and never went on strike." H. C. Frick Coke Company, a subsidiary of the United States Steel Corporation and the unofficial leader of the captive coal operators, formed the Workmen's Brotherhood, supported by the local chapter of the Ku Klux Klan, and the Miners' Independent Brotherhood, whose leader spent his time writing the commissioner of immigration, in Washington, seeking ways to deport aliens.[174]

The Leisenring baseball team, Fayette County, sponsored by the H. C. Frick Coke Company. Penn State Fayette Campus.

These company strategies to keep the union out of the mines were unacceptable and almost universally rejected by a majority of miners during the turbulent 1930s. Nonunion miners rebelled and demanded union recognition for the UMWA. Miners from Fayette, Westmoreland, and Greene Counties, whose mines were the principal sites of steel-controlled mines, called a wildcat strike and left the pits in September 1933 and did not return until early November. The H. C. Frick Coke Company became the focus of numerous violent confrontations. UMWA sympathizers fought pitched battles for union recognition with company deputies in the Frick company towns of Grindstone and Maxwell, Fayette County, in the summer of 1933. Violence between miners and company deputies also flared at the Star Junction mine and the Colonial Number I mine in Fayette County. Striking miners shut down four H. C. Frick Coke Company and two Jones and Laughlin mines during the year as the strike spread to neighboring counties. Thirty thousand miners were on strike by August 1933 and all Frick mines were idled. The violent confrontations in 1933 were unsuccessful, as the steel barons were intransigent in their virulent antiunion position. Attempts by Lewis, the UMWA, wildcat strikes by nonunion miners, and intervention by President Roosevelt all failed. The 1933 campaign to organize the captive mines was defeated, but the mine leaders and the rank and file were determined to try again.

Steel companies continued to thwart unionization efforts by the UMWA until World War II. The coal and steel industries were operating at full capacity filling war orders in 1941. The UMWA made another attempt in the fall to force the steel corporations to recognize the union. Only Jones & Laughlin Steel Corporation of the twelve largest steel corporations would accept unionization of its mines at this time. The remaining steel corporations, employing about fifty thousand miners in southwestern Pennsylvania, refused a closed shop; instead they offered higher wages and hours concessions. Lewis called three wildcat strikes during the fall of 1941 and they all failed. The National Defense Mediation Board (NDMB) had been established by the Roosevelt Administration in early 1941 to facilitate war production by trying to prevent or settle difficulties between labor and employers. The NDMB panel, on December 7, 1941, granted union representation to all miners employed in the captive mines. Steel companies of southwestern Pennsylvania capitulated and accepted the unionization of their miners.[175] The unionization of "captive mines" permitted the UMWA to negotiate a single contract for the entire coal industry, except for a few small companies. These smaller operations were usually owned by former miners who had saved sufficient money to buy or lease their own property or equipment.

Mechanization of Mining and Its Consequences to the Industry

Coal operators were determined to maintain higher coal prices obtained after 1933, while the UMWA was equally committed to maintaining the economic gains it had attained through the Appalachian Agreements. The average price of a ton of coal increased continually from $1.77 per ton in 1935 to $2.19 in 1941. The higher prices for coal can be credited in part to the consent of most coal operators to maintain collective bargaining agreements.

Coal operators could no longer cut coal prices or cut miners' wages with impunity as had been their practice in the recent past. The agreements equalized wages and it was not possible for one company to undercut competitors with a low-wage policy. The editors of *Coal Age* clearly identified the new economic dilemma confronting coal companies during the 1930s—"How to pay higher wages and yet reduce costs and how to work shorter hours and yet produce the same tonnage?"[177] *Coal Age* was the leading weekly journal of the coal industry and a principal booster of mechanized mining. The rising cost in wages and the shortened workweek forced coal producers to substitute machinery for high-cost labor. Many coal companies had previously regarded machinery and miners as interchangeable and they chose whichever was cheaper.[178] Lewis observed that "the American coal operators never would have mechanized their mines unless they had been compelled to do so."[179]

The principal attraction of machinery for coal operators was that it lowered labor costs by reducing the size of the required labor force and increased productivity per man per hour. The National Recovery Administration issued a report in November 1933 estimating the average cost to mine a ton of bituminous coal at $1.58 in the Appalachian region. This cost was distributed as follows: mine labor at 93 cents; mine supplies at 22 cents; other expenses at 3 cents, royalties and compensation insurance at 15 cents, depreciation, taxes, and insurance at 11 cents and sales and administration at 14 cents.[180] Many factors account for the cost differences among firms and within bituminous coalfields in mining a ton of coal: geological condition, degree of mechanization, wage-rate differential, type and method of mining (surface or underground mine opening), age of the mine, difference in length of work per day and week, difference in amount of coal preparation, age of mining equipment, depth of mine, rate of employee compensation insurance, state and local taxes, and amount of mine acreage being mined.[181] Labor was the principal cost in coal production, varying from 60 to 70 percent of the total expense for each ton of coal mined. Approximately 63 percent of all mine workers in 1929 were "tonnage" workers, who were paid piece-rate based on the number of tons of coal mined daily or on the carload. Their wages constituted about 55 to 60 percent of total labor expenditure in the mine. The average output per man per day rose from 4 tons in 1920 to 5.3 tons in 1931, representing a 32.5 percent increase. This increase in productivity reflected the increased use of undercutting machines, mechanical loading of coal, and general improvement in efficiency of the mining operation.

Coal Price Per Ton/ Production		
	Average Price Per ton	Production (net tons)
1935	$1.77	372,373
1936	1.76	439,088
1937	1.94	445,531
1938	1.95	348,545
1939	1.84	394,855
1940	1.91	460,772
1941	2.19	514,149[176]

"Mechanization of mining" refers to both the method of removing coal from the working face and that of loading the coal in mine cars for their removal by mules, or by motorized storage-battery or electric locomotive to the tipple or cleaning plant.[182] The degree of mechanized coal mining in the nation was not uniform. Coal was still undercut with picks, loaded by hand, and hauled to the surface with draft animals in many Pennsylvania mines as late as the 1930s. Jessie Liotta, a coal loader from Braeburn, Westmoreland County, described the primitive work process still employed to extract coal in the mine at which he worked. His poem was entitled, "I Worked that Mine in '36":

> You'd go straight in, you'd sit in the coal car with your head way down.
> 'Course if you put it up you'd get it kicked off.
> You went in to where you had to work and you crawled out.
> You couldn't stand up; wasn't that high—
> Three foot of coal at the most.
> When you shoveled, you was on your hands, and knees,
> You worked on your own, you worked on tonnage.
> You made
> You got
> So you had to work.
> That was livin'.[183]

The introduction of mechanical loaders made many coal miners superfluous. Hillman Library, University of Pittsburgh.

Some larger coal companies had employed mining engineers to test the feasibility of mechanizing the mining process during the 1920s. The undercutting machine had in fact ushered in the mechanized mining era of the bituminous coal industry during the 1880s. Operators of unionized mines had embraced mechanical undercutting machines as a method to reduce their labor costs and the need for skilled workers, who were often the most militant union miners. Coal, refuse, and miners were hauled to the surface with electric and battery-powered locomotives in the larger mines. However, most coal was still hand loaded with shovels by unskilled loaders until the 1930s. Shoveling coal was the last underground job to yield to machine operation. This was heavy and arduous work requiring many workers to achieve increased production. Hand shoveling of coal was a major impediment in the application of "factory" methods to coal mining.[184] The managing editor of *Coal Age* wrote in 1921: "Mining [of coal] is still in a way a "cottage" industry, only the cottage is a room in the mines."[185] *Coal Age*, a weekly publication, was the leading trade journal of the period and a staunch promoter of mine mechanization. The introduction of a variety of mechanical loading machines effected a revolution in the mining operation. One historian noted the superiority of mechanical loaders over hand loading and how their gradual widespread use fundamentally changed mining methods:

> In loading more than in any other function, mechanization fosters an increased tempo of mine operation. It may indeed be said that the balanced cycle of underground operations is a concomitant of the post-world war mechanization of the loading process. To the extent that loading machines have replaced hand loading, bituminous coal mining [has] become an industry in which many of the old craft traditions have had to be discarded. Each working face does not have its own loading machines; rather loading-machine crews have taken their place . . . as workers performing a specialized function in the larger process of mining. Ideally, a single working face

is attacked in sequence by cutters, drillers and blasters, and loaders, each group working in close coordination with the others . . . The old routine (or lack of routine) has given way to a systematic planning of production with a closely supervised execution of the production process.[186]

A more systematic process to mechanize work in the mines was undertaken by coal companies in the mid-1930s in their frantic attempt to increase daily productivity and reduce higher union labor costs. Mechanical coal loaders were not widely used underground until this decade although mechanical loading equipment had been developed a half-century earlier. The terms "machine loading" and "mechanical loading" are used interchangeably in this study. The first mechanical coal-loading machine used in an American mine was the so-called Stanley Header, designed to break down the coal from the seam as well as to load it.[187] The English-designed machine was brought from England in 1888.[188] The Jones loader or "Coloder" machine was designed as a loading machine between 1893 and 1898 by the Pocahontas Fuel Company in West Virginia. The company received a patent for the machine in 1902. These early experimental coal loaders were unsuccessful and unpopular with coal companies for a variety of reasons. They suffered from a number of technical deficiencies and were subject to continual mechanical breakdowns. They were large and cumbersome machines that were difficult to move from room to room within the mine. Manufacturers improved these machines during the 1920s. Their physical size was reduced and manufacturers made them more mobile by mounting them on caterpillar traction. However, many coal operators remained indifferent to the improved loader. Miners were paid on a piece-rate basis and labor was both abundant and cheaper than investing in this costly machinery.

Mechanical loading equipment is divided into two basic groups: machines that virtually eliminate hand shoveling except for incidental clean-up, and machines that reduce the amount of shoveling. Mobile loaders, scrapers, and duckbills are in the first category, while pit-car loaders and face conveyors fall into the second category.[189] These machines had various horsepower reflecting their daily capacity. The tonnage of bituminous coal loaded by machinery in the United States was 52.3 percent by mobile machine, 23.5 percent by pit-car loaders, 2.4 percent by scraper loaders, and 21.8 percent by face loaders, including those equipped with duckbills in 1935. The mobile loader was the most popular type of loader and 56.1 percent of all mechanically loaded coal was being loaded by one of these machines by 1945.[191] Mobile loaders, which were responsible for the elimination of hand loading, were two different types of machine. One machine consisted of a scoop which after being pushed into the coal pile was lifted mechanically and its contents dumped into a car. Another machine added a pair of claws to the conveyor principle. The two claws were mounted on the lower end of a conveyor, were rotated by motor, and reached into the coal pile on the floor, gathered it up, and drew the coal onto separately motorized conveyors for delivery to the mine car.[192]

Mechanical Loading Equipment			
	Horsepower	Daily Capacity (tons)	Factory Price
Conveyor	5-30	50-300	1,110-2000
Pit-car	1-5	15-25	$ 700-$1,500
Duckbill	15-30	50-300	1,500-2,500
Scraper	7.5-25	50-250	1,500-2,500
Mobile	22.5-50	100-800	6,500-13,500[190]

The scraper consisted of a bucket or scoop attached to an electric hoist. The scoop was dragged past the coal face, dragging the coal to a side entry and into mine cars. The scraper loader tripled the production possible from shoveling coal, but the machine had several disadvantages. In dragging the dislodged coal along the floor it gathered up refuse including rocks, scrap metal, and fire clay. Costly cleaning facilities had to be erected by operators to remove these coal impurities. This was an extraordinary expenditure since less than 8.3 percent of all coal was cleaned mechanically in 1930. The use of scrapers was best applied to long-wall mining.

Pit-car loaders began to be used about 1925 and by 1931 were loading about eighteen million tons of coal annually. These loaders were popular in Illinois, where their numbers increased from fewer than 100 loaders in 1927 to 2,162 in 1931. Pit-car loaders loaded nearly one-half of all the coal loaded by mechanical means in

1931.[193] This loader was a simple conveyor that rose like a ladder from the mine floor to the top of the mine car. The loader scooped coal from the floor to the bottom of the motorized conveyor and it moved the coal forward until it fell into the car. The pit-car loader was difficult to move about in the mine from room to room. The Bureau of Mines reported 2,300 pit-car loaders and 500 mobile loaders in operation in 1928, but by 1943 there were only 300 pit-car and 2,500 mobile loaders in service.[194]

The duckbill loader (self-loading conveyor) is a shovel device with a flared mouth attached to the end of a shaking conveyor. The duckbill is pushed under the piles of broken coal and the differential movement of the conveyor carries the coal backward from the loading head onto the conveyor proper. This loader was used principally by coal operators in Sweetwater County, southern Wyoming, who owned three-quarters of all duckbills in the United States in 1938.

The Coloder Company of Columbus, Ohio, the McKinnley Mining and Loading Company of Fairmont, West Virginia, the Myers-Whaley Company of Knoxville, Tennessee, and the Joy Manufacturing Company of Pittsburgh, Pennsylvania, were the principal manufacturers of these coal-loading machines during the 1920s.[195] Joseph Francis Joy (1883-1977), founder of the Joy Manufacturing Company, was born in Cumberland, Maryland, and attended local school through the fifth grade. Joy was working at a nearby tipple as a slate picker at the age of twelve. Joy held every mine job, from pumper to general superintendent. He was employed at a number of mines in Maryland and western Pennsylvania between 1895 to 1913. He was hired by Pittsburgh Coal Company, as a consulting engineer, to build an experimental coal-loading machine. The loader was tested at Pittsburgh Coal's Somers Number 2 mine at Pricedale, Pennsylvania, where under adverse conditions the machine performed poorly. This unsuccessful experience at Pricedale prompted Pittsburgh Coal to end its business relationship with him to develop a viable loading machine. He subsequently founded the Joy Machine Company of Pittsburgh in 1919. In 1921, the company became a wholly owned subsidiary of Joy Manufacturing Company, which was incorporated in Delaware. The company worked with the Charleroi Iron Works in Charleroi, Washington County, to build different types of loading machines from 1921 to 1924. The firm constructed a variety of mechanical loaders and mining machinery at its new facility at Franklin Plant Number 1, which opened on March 1, 1925. Machinery constructed included coal loaders (models 8BU, 11BU, 12BU, 20BU, and 30BU); shuttle cars (two- and four-wheel drive and steer) models 32E, 42E, 60E-12, 70E-1, 5SC, 6SC, 7SC, 8SC, and 10SC; coal and hard rock continuous miners; mining machine tracks (crawler mounted); timber setters (both track mounted and rubber tired); post pullers; water sprays; and high-speed drifting systems for hard-rock tunnel driving.[196] The company, founded by Joseph Joy, is today the largest manufacturer of mining machinery in the world.

A Joy mechanical loader, circa 1940, dubbed a "Joy Killer" by unhappy miners. Penn State, Fayette Campus.

The name Joy became synonymous with the mechanical loader. Joseph Joy developed a variety of successful coal-loading machines that were used extensively in mines throughout the United States. Miners accustomed to hand loading coal called the Joy mechanical loaders the "man killers." Union miners were able to block or restrict the introduction of mechanical loaders in their mines by refusing to agree to a wage rate for the loading-machine operator. One miner clearly reflected the prevailing opinion that the machines took away jobs and undercut their wages: "The machines have thrown miners out of work. I don't like to work with machines and would rather dig coal with pick and shovel by hand."[197] *Joy Days,* a popular mining camp ballad, clearly expressed the deep-seated fear and apprehension felt by most hand loaders at the increased use of these loading machines by coal companies:

> Here is to Old Joy, a wonderful machine,
> That loads more coal than any we've seen.
> Just picks out the slate and lays up the track,
> Gets plenty of empties and hurries back.
> Ten men cut off with nothing to do,
> Their places needed for another Joy crew.
> Fifty cars a day is the goal they have set,
> With coal and slate together they will get it, I bet.
> The bosses all smile to see Joy at work,
> And keep him well oiled, he never will shirk.
> He uses no shovel and neither a pick,
> But with coal and slate together he loads a car quick . . .[198]

Thereafter, despite miners' protests and resolutions, loading coal by machine continued unchecked. A majority of coal companies were still hand loading their coal in 1930. The introduction of the mechanical loader was gradual, beginning modestly during the 1920s and growing substantially during the 1930s. There were 129 mechanical machines loading some 1,900,000 tons of bituminous coal, or .03 percent of all coal produced, according to the U.S. Geological Survey in 1923. The percentage of coal loaded by machine tripled from 10.5 percent in 1930 to 31 percent in 1939 and grew to 56.1 percent nationwide in 1945.

The United States Bureau of Mines reported this transition from hand loading to mechanical loading in 1938:

Percentage of Coal Undercut by Machinery and Mechanically Loaded				
		PA	WV	IL
Machine Under Cut	1929	70.3	85.8	83.5
	1932	79.8	89.6	85.0
Machine Loaded	1929	1.4	1.5	11.8
	1932	2.4	.6	28.8[199]

> It is well known that the proportion of underground output obtained by mechanical loading has been highest in the coal fields of the northern Rocky Mountains and the Middle West, where [high] wages combined with favorable seam conditions have stimulated the process of mechanization. In the last two years, however, market conditions and the trend of wage rates have tended to stimulate mechanization in the Appalachian region, and a large part of the sales of equipment reported by the eastern and southern fields.[200]

A variety of factors contributed to this expansion of mine mechanization in the industry after 1935—changes in the level and structure of wage rates, rising product demand and a consequent improvement in the earning prospects of the industry, and the availability of improved equipment adaptable to a variety of geological conditions.

UMWA leadership, unlike most miners who were fearful of their future livelihood with the introduction of mechanized mining, wanted coal operators to mechanize their mines by introducing *more* mechanical loaders. The union strategy was to raise wages, shorten hours of work, increase fringe benefits, and improve working conditions for its membership, and this they believed was feasible only if the number of mines and miners was significantly reduced. The official position of the miners' union toward the introduction of machinery in the mine was, "We decided it is better to have a half million men working in the industry at good wages . . . than it is to have a million men working in the industry in poverty."[201] President Lewis, who had been a staunch advocate of mechanized mining since the 1920s and was so as late as 1947, asserted:

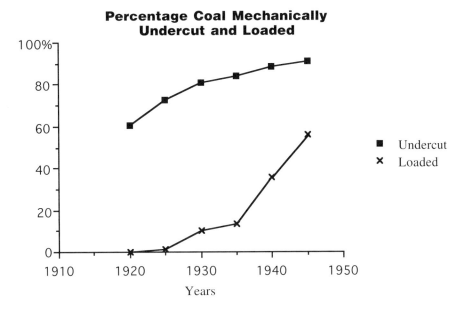

The UMWA takes the position that the only way in which the standard of living could be increased . . . would be by increasing the productivity and lowering the unit costs and utilizing the genius of science and the automatic machine . . . and the usage of power to do the work of human hands.[202]

In 1950 Lewis told an interviewer from *U.S. News & World Report* that "we are not trying to keep men in the mines just to retain jobs. It will be a millennium if men do not have to work underground but can all work in God's sunshine. That'll be a long time in coming, but we would be in favor of it."[203] Lewis viewed the principal purpose of the union as winning higher wages and lower hours in the mines, and the membership dues paid by miners were to help the union leadership gain these goals in direct negotiation with coal operators. Opponents of the mine mechanization program called Lewis "the best salesman the machinery industry ever had." Locals had been able to block the introduction of mechanical loaders in the 1920s by refusing a wage scale for the loading machine operators or by negotiating restrictive work rules. Union miners who opposed the introduction of loaders during the 1930s did not have the same opportunity to express their opposition. Lewis had ignored the traditional processes for ratifying labor contracts with management that had prevailed in the 1920s. He had purged all rank-and-file opponents during the 1930s and centralized the bargaining process and contract negotiations into his own hands. Lewis had used his constitutional authority to revoke the charters of rebellious districts, subdistricts, and locals so that he could replace his opponents with loyal subordinates. Lewis had successfully created a monolithic miners' bureaucracy under his absolute control. A local union spokesman from Fredericktown, Washington County, summarized this usurpation of workers' control by Lewis in 1940: "Whereas, we feel that we should have more say as to the kind of contract that we have to work under; be it Resolved, that the International Convention goes on record in favor of a referendum of the membership before the contract is signed."[204] Job rights of miners were traded by Lewis and his inner circle of advisors in exchange for higher miners' wages, or at least the promise of higher wages for a shrinking membership. Mechanization of mines that began in earnest during the 1930s created a series of new problems within the industry:

(1) Mechanical loaders made work underground easier but the coal loaders required a smaller labor force. Fewer workers increased daily coal production.
(2) The rapid undercutting machines, working with little improvement in ventila-

tion, increased the level of dust in the mechanized mine. High concentrations of dust exposed miners to higher risks of black lung disease and the possibility of mine explosions.

(3) The pace of mining quickened in the mechanized mine. The once independent and mainly unsupervised miners now worked in a more regimented assembly line atmosphere underground. The company foreman was increasingly defining the workday underground.

More and more coal diggers were laid off permanently as the tempo of mechanization increased. By 1945 UMWA membership had steadily dwindled as the unemployment in the coal industry skyrocketed.

The whirlwind union-organization drive directed by John L. Lewis since 1933 had successfully organized more than 90 percent of all miners in less than a decade. The UMWA achieved a guaranteed eight-hour day, a national wage agreement, and the introduction of much-needed safety legislation for its membership. Coal was mined in thirty-one states, with production of 393 million tons of bituminous coal and 50 million tons of anthracite on the eve of World War II. Pennsylvania produced nearly all the anthracite. This nearly 450 million tons represented about four tons for every man, woman, and child in the United States. Pennsylvania, West Virginia, Kentucky, and Illinois produced nearly three-fourths of the nation's bituminous coal output. American involvement in World War II, like the Great War, was an economic boon shared by the coal and coke industry. The average price per ton of coal rose from $2.19 to $3.06 nationally while annual production rose by 63 million net tons between 1941 and 1945.[205] American coal miners were once again enjoying a full workweek during the war years.

Average Coal Price and Production

	Average Value Per Ton	Average Number of Days Worked	Tonnage Mined Per Man Per Day
1936	$1.76	199	4.62
1937	$194	193	4.69
1938	$1.95	162	4.89
1939	$1.84	178	5.25
1940	$1.91	202	5.19
1941	$2.19	216	5.20
1942	$2.36	246	5.12
1943	$2.69	264	5.38
1944	$2.92	278	5.67
1945	No Data	261	5.78

Mechanized loading, cutting, drilling, hoisting, and coal preparation rapidly increased daily tonnage of coal but reduced the number of pick-miners and loaders required. The new machinery increased productivity per worker by reducing the size of the bloated labor force. The number of bituminous coal miners in Pennsylvania declined from 133,703 to 98,764 between 1930 and 1945. The reduction of the work force was a long-term and irreversible process, as the number of miners in the state decreased by almost 84 percent, from 170,000 to about 26,000 between 1950 and 1980. In Pennsylvania, in fact, there were fewer bituminous miners employed in 1970 than in 1880.

The Bituminous Coal Industry in the Post War Decades

Pennsylvania's bituminous coal industry after 1945 was far different from the industry that had developed in the Monongahela Valley during the 1760s. From 1945 to 1960, while the national economy grew by 50 percent, coal production in the Commonwealth fell by 27 percent.[206] By the 1950s the state's production had slumped and bankruptcies were occurring with increasing frequency. Unsuccessful coal companies closed their mines, laid off their miners, and sold off their mining properties including the company-owned houses. Miners who had rented their homes for decades could now own them. Some companies sold the houses directly to individual buyers, with miners having the first chance to buy, while other companies simply sold all their real estate holdings collectively to a real estate investor who then resold the properties. A federal survey of mining families in 1947 found nearly 60 percent were still living in company-owned housing. The government survey also found the average company house was a one-story four-room frame house with a kitchen, a living room, and two bedrooms. Few miners' houses of the period had closets, basements, or indoor bathrooms. The President's Commission on Coal of 1979 found that only 7 percent of miners now rented their houses from the company while 74 percent owned their own houses (most had been purchased from the coal company). The study noted that 25 percent of all miners and their families lived in mobile homes, compared to only 5 percent of the general population.[207] Many former mining communities became incorporated towns or were absorbed by larger neighboring towns. Dozens of the former coal towns and villages dotting the landscape of western Pennsylvania have survived the collapse of the coal industry and today many remain viable communities generations after mining has ceased.

During the 1980s Pennsylvania's soft coal industry hit the skids when statewide production fell by 21 percent while coal production nationally increased by 19 percent. Employment in the industry dropped 61 percent, reflecting lower production and productivity improvement. There were fewer than fourteen thousand miners in the state producing 68.3 million tons in 1989. The bituminous industry of Pennsylvania is now firmly controlled by large corporations which have successfully mechanized the extraction of coal. These few large coal companies have restructured the state's coal industry through mechanization, mergers, and acquisitions. For example, Pittsburgh Coal Company and Consolidation Coal Company had since the 1930s been losing money at a terrific rate before their consolidation on November 23, 1945. George Hutchinson Love, the new president of Pitt-Consol, was a coal company executive with over two decades of practical experience in the industry. He entered the coal industry as the operator of the Union Collieries Company in 1926 and joined Consolidation Coal Company in 1943 when Union Collieries' properties were taken over by Consolidation. The new corporation was created, according to Love, because "we felt that if we could eventually build a company that could afford to close down the poorer properties and concentrate on the better ones, there was a chance to make something out of this industry. If coal was going to be competitive with other fuels, like oil and gas, you had no choice but to mechanize."[208] Pittsburgh Consolidation Coal Company (known as Pitt-Consol) operated thirty-nine underground and nine surface mines in Pennsylvania, West Virginia, and Kentucky in 1945.[209] Pitt-Consol had assets of more than 100 million dollars with a net working capital of approximately 29 million dollars. The three largest coal-producing companies during the 1980s were the Peabody Holding Company (6.4 percent), Pitt-Consol (5.2 percent), and Amax Coal Company (4.4 percent). These three companies alone produced 16 percent of the nation's entire coal production in 1981.

Pitt-Consol, like other successful coal companies that remained solvent in a declining industry and a highly competitive energy market, introduced continuous mining in a desperate attempt to make its industry competitive with oil and natural gas. The post-war mechanized mine made the industry more productive and cost effective by substantially increasing the daily productivity of a smaller labor force. There were fewer than one-half as many underground operations in 1973 (4,744 mines) as in 1950 (9,429 mines). Productivity per miner doubled from 5.78 tons per day to 12.83 tons between 1945 and 1960.[210] The efficiencies of individual underground mines in Pennsylvania differed greatly. Mechanized mines produced more than 20 tons per man per day while less mechanized mines less than 1 ton in 1945. A survey of 622 mines in 1954 noted that 128 mines produced less than 3 tons per man per day, 325 mines produced from 3 to 6 tons per man per day, 125 mines from 6 to 9 tons per man per day, and 44 mines over 9 tons per man per day.[211] Almost 50 percent of Pennsylvania's 91.9 million net tons of coal was mined at fewer than one-half of one percent of the 888 operating mines in 1953. The more efficient mines used large-scale mechanized mining equipment that extracted coal in generally thick seams. The most productive mining operations in Pennsylvania by the mid-1950s were located in the southern parts of Armstrong, Indiana, and Cambria Counties. A typical mechanized coal operation of the period, producing a million tons of coal annually, was at least a $10 million investment.

A bituminous coal miner in 1910 using a pick, shovel, and black-powder explosives could dig about three tons of coal on average per day. By 1988, each American miner produced over nineteen tons of coal, while each surface miner averaged more than forty-six tons per shift. By 1945 90.8 percent of all coal was mechanically undercut and 56.1 percent was mechanically loaded in the United States. Mine mechanization increased the capacity to produce more coal with fewer workers, and furnished greater elasticity in operations by permitting operators to meet peak demands quickly. Mechanization of the mining process linking every operation, from face to the tipple, into a single continuous process was not fully implemented until the 1950s with the introduction of the continuous mining machine. The use of the gathering-type mobile, mechanical loaders of the 1930s had increased productivity, but the undercutting and removing of coal from the face still represented separate mining operations.

The Joy Manufacturing Company of Pittsburgh introduced the 3JCM-2 continuous mining machine in a public demonstration for the news media on December 14, 1948. This machine was designed, according to the manufacturer, to rip two tons of coal per minute out of a solid seam of coal.[212] Mechanical breakdowns and workers' resistance often rendered this production assertion an exaggeration; nevertheless, the machine greatly increased mining productivity. The continuous miner represented the culmination of a gradual process in the technological and social organization of the underground mine as a workplace. Coal is extracted in a sequence of steps in conventional mining. The giant continuous mining machine combined the separate functions of undercutting, drilling, blasting, and loading into a single process. Carbide cutting bits mounted in the rings of rotating cutting discs literally ripped coal from the face, and then giant dual gathering arms swept the broken coal onto a conveyor that passed from the front of the machine to the rear where it was loaded into rubber-tired shuttle cars, or directly onto conveyor belts that transported it out of the mine to the tipple for processing. Water was sprayed on the cutting equipment and the coal face to reduce the extensive airborne dust. Each machine and its collateral equipment cost the coal company more than $250,000 in the 1950s. National data on coal mines employing this machinery were first compiled in 1950 when less than 1 percent (.8) of coal was mined with these large machines (about 3.1 million tons), 91.8 percent was undercut by machine, and 7.4 percent was cut by hand and shot from solid. Some of the earliest continuous mining machines were introduced in Cambria County during the 1940s. Barnes and Tucker Company installed one of the earliest continuous miners at its Lancashire Mine Number 15 near Bakerton in 1949. A second machine with a rotary- (or boring-) type cutting head was installed at Mine Number 31 of Bethlehem Mines Corporation in Cambria County in November 1950. By 1951 there were 157 continuous mining machines in operation throughout the nation.[213] A decade later 27.4 percent of all coal mined nationally was extracted with these machines, while 67.7 percent was undercut by machines and less than 5 percent of all coal was mined by hand. Other machinery introduced in the postwar mechanized mine included the cutting knife, crawler-type loader, and the shuttle car. The cutting knife is a machine on wheels that has a sharp single blade that cuts coal from the seam in either a vertical or horizontal manner. The crawler-type loader picks up coal that is blasted from the coal seam and loads it into cars hauled away by shuttle cars, or

A continuous miner removing coal. Introduced in the 1940s, and widely used during the 1950s.
Penn State, Fayette Campus.

onto a conveyor belt. These machines are equipped with a steel canopy that protects the operators from possible roof fall while working the mine face. These electrically powered machines break up the coal with large, rotating cutting drums studded with carbide teeth at the mine face, while gathering arms located behind the drum scoop the coal directly onto built-in conveyors for loading onto waiting shuttle cars, which carry the coal to the mine conveyor system. The machine advances through the mine face by single cuts limited to the length of the machine, which is approximately twenty feet; as the machine proceeds through the coal seam, miners will install roof supports, usually a combination of cribs, posts, and "roof bolts." After the continuous mining machine removes coal from the working face, miners install temporary supports by installing ten-ton hydraulic roof jacks to provide roof support. The roof bolter and his buddy operate a roof bolting machine that drills holes, positions bolts, and tightens them in the roof. Steel roof bolts of varying sizes are installed using a special torque measuring wrench. The introduction of roof bolts was a major development in the evolution of mine safety. Roof bolts are long steel rods driven in the roof to bind weak, overlaying rock strata into a layer strong enough to support its own weight. The use of roof bolting provided elbowroom underground for both men and machinery.[214] A modern continuous mining machine in 1990 could mine coal at the rate of eight to fifteen tons per minute.

The room-and-pillar system of mining is still, however, the prevalent underground mining method of extraction. Underground coal is mined in rooms separated by walls or pillars left by continuous miners to support the roof. Longwall mining is an alternative method of deep-mining coal. This technique uses a steel plow or rotating cutting drum that passes back and forth across the face of the coal covering an area five hundred to one thousand feet long. The loosened coal falls onto a conveyor belt and is removed from the mine. This method, unlike the room-and-pillar continuous mining method that it challenged following World War II, involves the removal of the entire coal seam, or very large sections of it. American use of longwall mining dates back to the first decades of the twentieth century when posts and cribs were used to support the wall adjacent to the long face. European coal operators had begun using this method extensively in their mines by 1900. A few American coal companies employed longwall mining during this period, although the process was never enthusiastically embraced by a majority of bituminous coal operators until after the Second World War.[215] The mines at Vintondale, Cambria County, and the Maryland Shaft Mine of the Berwind-White Coal Mining Company used this underground mining method during the first decade of the twentieth century. Longwall mining was successfully employed at the Gateway Mine near Clarksville, Greene County, and at Jones and Laughlin's Vesta Mine Number 5, Vestaburg, Washington County, in western Pennsylvania immediately after World War II. Modern mechanized longwall equipment used today in the United States was largely imported from Europe during the 1960s.

Longwall mining is employed today to mine large blocks of coal where the seam is relatively flat and thick, and where the degree of surface subsidence is acceptable. Longwall mining can be accomplished using two methods. The first type uses a cutter called a plough. This mechanical device has teeth positioned along the sides which cut an average six inches of coal per pass as it moves forth across several hundred feet of the seam. The loosened coal is then trans-

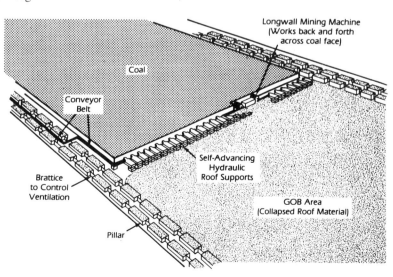

Longwall Mining System. Robert Stefanko, *Coal Mining Technology.*

199

ported from the mine on a series of conveyor belts from the face. The second type of cutting machine is the rotating drum shearer. The metal shearer traverses the coal face cutting thirty inches deep into the coal. As the machine moves along, a chain conveyor is used to support the newly formed roof. Movable steel plates, called props or shields, supported by hydraulic jacks, support the roof over the immediate work area, and as the miners work the machine deeper into the seam, the roof supports are advanced and the roof rock behind the supports are permitted to cave in. Long sections of coal, up to seven hundred feet, can be removed at a time without any pillars to support the mined-out region. The president of the Blue Diamond Company described the longwall extractive method and its superiority to other existing mining methods to his stockholders in the company's 1986 annual report:

> The longwall is a highly mechanized means of mining coal that has been greatly refined in the last five to ten years. It consists of a shearing machine, which hauls itself back and forth on a conveyor frame and cuts 30 feet of coal from a 700 feet face. Each longwall panel will be from 2,500 to 8,100 feet in length. The shearer and conveyor are both located under massive metal roof supports with hydraulic legs and most of the work is performed under these roof supports, greatly enhancing the safety of the work crew. On the longwall face, we are expecting to mine six times as much coal, with a similar size crew, as on a continuous miner section. For this reason, the longwall should make a significant reduction in the mining cost per ton at Scotia.[216]

This method is more efficient than the wasteful room-and-pillar method. Engineers estimated that one-third to one-half of all coal in abandoned mines remains in the form of pillars. Longwall mining offers coal companies the best method for improving daily productivity in underground coal mining by reducing the size of its labor force. A major drawback to longwall mining was the large initial capital outlay for the purchase of the machinery. In 1993 twenty-one mines in Pennsylvania and a total of 112 mines in the United States employed longwall mining.[217] Underground coal production by mining techniques nationwide in 1990 was room-and-pillar with continuous mining machinery, 63 percent; conventional room-and-pillar, 7 percent; shortwall, 1 percent; and longwall, 29 percent.[218]

Oil and gas production had been encroaching on the coal market share since the 1920s, and this trend had greatly accelerated during the post-World War II period. The significant but brief era of "King Coal" was coming to a close during the 1950s and 1960s. Its long-term position as America's primary energy source had been challenged successfully by natural gas and petroleum, with the increasing popularity of these fossil fuels lessening the demand for coal from traditional coal markets. Coal was surpassed, first by petroleum in 1952 and subsequently by natural gas in 1964, as the nation's principal energy source. Automobiles, home heating, and industry were the principal users of oil. The further expansion of natural gas as a viable and major energy source was limited until the construction of an extensive national network of pipelines. This competition made natural gas the choice for home heating and industrial markets. Natural gas exceeded coal as the principal source of energy in the United States in 1960 and four years later surpassed petroleum.[219]

The share of the energy market held by coal was eroded by these fossil fuels and by the continual shrinkage of traditional coal markets. Consumption of steam coal for railroad locomotives ended with a rapid change-over to diesel engines following the war. The steam coal market was once a 130-million-ton-a-year market. The consumption of coal and coke by the steel industry for the production of pig iron and steel stagnated with the decline of these industries. The steel industry also found more-effective ways to utilize coke and, therefore, demand for bituminous coal and coke declined. Coke is still used in iron production, but most steel in the United States is made in electric furnaces. The principal users of Pennsylvania's bituminous coal in 1929 were beehive and by-product coke ovens, 27.5 percent; general manufacturers, 20.0 percent; railroads, 19.5 percent; domestic use, 15 percent; and steelworks, 10 percent. In contrast, the principal coal markets in 1981 were steam coal for electrical utilities, 55 percent; steel companies (metallurgical coal for coke production), 20 percent; industrial and retail consumers (boilers and domestic consumers), 8 percent; and export (both steam and metallurgical coal), 17 percent.[220]

Coal has been extracted from the surface like quarry stone by coal miners since the eighteenth century. The scale of surface or strip mining increased significantly after World War II. The decision to undertake surface mining is dependent upon the average thickness of the overburden and the thickness of the coal seam. Surface mining is employed when the coal seam is located less than fifty to one hundred feet from the surface. "Area" mining and "contour" mining are the two principal methods of surface mining. Area mining is practiced when the coal seam is on relatively flat ground, and it consists of a series of cuts one hundred to two hundred feet wide, with the overburden from one cut being used to fill the mined-out area of the proceeding cut. Contour surface mining follows the coal seam as it outcrops on steep and hilly terrain. Large-scale equipment has enhanced the productivity of surface mining operations. Surface mining equipment includes the auger machine, bucket-wheel excavator, bulldozer, carryall scraper (or pan scraper), continuous surface miner, dragline excavator, walking dragline, front-end loader, and thin-seam miner. Surface mining involves a series of distinct steps from extraction to shipment to market. The first step is the removal of topsoil and surface rocks that are broken up by blasting and then removed by giant power shovels. The second part of the operation is the loosening and removal of coal and the loading of it for shipment. In the final step, coal is transported from the site, usually by truck, to the preparation plant where it is sorted according to a variety of sizes, washed to remove impurities, and then dried. The coal is now ready for transportation to a nearby electric-generating plant or is loaded onto railroad cars or trucks for transport to more-distant local or foreign markets.

Surface mining was not a practical mining method until large-scale excavation equipment became widely available about 1915. Strip mining began in Indiana and Illinois, although less than 1.5 percent of all coal was surface mined during the 1920s. The percentage of coal mined by surface methods increased steadily with each successive decade after the 1920s, from 4.3 percent in 1930, 9.4 percent in 1940, and 23.6 percent in 1950. By 1975 approximately one-half of the 640 million tons of coal produced nationally was mined by surface removal methods. There are numerous advantages of surface mining over conventional underground mining that account for its increasing popularity with coal operators: it is faster, less expensive, and generally a more efficient method of mining. This method requires a smaller labor force and is safer than underground mining. Mechanized surface mining removes from 80 to 100 percent of the coal in the seam while conventional underground mining recovers from 40 to 60 percent. An acre of coal with a three-foot-thick coalbed produces approximately five thousand tons by surface mining, whereas a similar-size area yields thirty-three hundred tons or less in deep mining. The output per man per day in a strip operation was double that of an underground miner in 1960. An average worker in a surface coal operation produces 22.93 tons per day while an underground miner averages 10.64 tons per day. Surface mining is a safer mining method because there is little danger to workers from poison or explosives gases, collapsing roofs, and haulage accidents, which are all major causes of underground injuries and fatalities.

Surface mining has a number of drawbacks including inaccessibility in inclement weather. Furthermore, the process can create severe environmental damage by destroying the natural vegetation of the mined area. The harmful effects of surface mining include disruption of ground water and surface draining regimes, ponding of water forming breeding grounds for insects, and creation of potentially hazardous slopes and cliffs (high walls) and unstable spoils piles. The General Assembly of Pennsylvania passed the Pennsylvania Bituminous Coal Open Mining Conservation Act in 1945, one of the earliest and most comprehensive surface-mine reclamation laws, to deal with the harmful environmental impact of this mining method. The act established standards for surface-mine operators in the restoration of land after mining is completed. The Pennsylvania Assembly enacted in 1963 a more stringent law requiring the restoration of surface-mined land to resemble its prior contour.

Surface or strip mining began in the anthracite region of eastern Pennsylvania during the 1880s, and by the next decade this type of mining had expanded around Eckley and extended east to Sandy Run and west near Jeddo, Ebervale, Japan, Oakdale, and Harleigh. Strip mining accounted for about 1.5 million tons, representing about 2.5 percent of all anthracite mined in 1925. The Pennsylvania Bureau of Mines did not keep precise records of surface mining before 1932, but by 1964, 55 percent of all anthracite production was strip mined.

In contrast, surface or strip mining in the bituminous coalfields of Pennsylvania began only during the late 1930s. The Pennsylvania Department of Mines first reported surface-production statistics in the bituminous coalfields in 1939. Some 2.7 million tons of coal were surface-mined in comparison to 89 million tons mined underground. The Pittsburgh coal seam, located in Allegheny, Fayette, Washington, and Westmoreland Counties, was the first surface-mined in Pennsylvania. After 1940 surface mining increased from 5 percent of the state's coal production in 1940, to 20 percent in 1945, and 25 percent in 1950.

Notes

[1] "Bituminous Coal in 1920," *Coal Age,* vol. 19, No. 3 (January 1921).

[2] Anna Rochester, *Labor and Coal* (New York: International Publishers, 1931), p. 11.

[3] James P. Johnson, *The Politics of Soft Coal* (Urbana: University of Illinois Press, 1979), p. 245.

[4] Sam H. Schurr and Bruce C. Netschert, *Energy in the American Economy, 1850-1975* (Baltimore: Johns Hopkins Press, 1960), p. 63.

[5] *The Decline of the Bituminous Coal Industry in Pennsylvania* (Harrisburg: Greater Pennsylvania Council, 1932), p. 6.

[6] Edward T. Devine, *Coal: Economic Problems of the Mining, Marketing and Consumption of Anthracite and Soft Coal in the United States* (Bloomington, Illinois: American Review Service Press, 1925), p. 21. *Coal Age.* January, 1925. The relative supply of energy by various sources in 1931 were bituminous coal 48.6%, anthracite 7.9 %, natural gas 8.6%, water power 8.5%, imperial oil 1.4%, and domestic oil 25.0%. American coal (anthracite—5.4% and bituminous—45%) supplied less than half of the energy in the United States by 1937. *1950 Bituminous Coal Annual* (Washington, D.C.: Bituminous Coal Institute, 1951), p. 21.

The chart below identifies the principal energy suppliers and percentage of the market share each fuel held between 1920 and 1944:

	Bituminous	Anthracite	Petroleum	Natural Gas	Hydro
1920-24	69.7	11.4	9.3	4.7	4.9
1925-29	67.9	9.3	11.4	7.0	4.4
1930-34	61.6	9.3	14.1	10.2	4.8
1935-39	59.2	7.4	16.3	11.9	5.2
1940-44	59.1	6.0	17.4	12.4	5.1

[7] *Region in Transition: Report of the Economic Study of the Pittsburgh Region* (Pittsburgh: University of Pittsburgh Press, 1963), p. 230.

[8] Barbara Ellen Smith, *Digging Our Own Graves: Coal Miners and the Struggle over Black Lung Disease* (Philadelphia: Temple University Press, 1987), p. 54.

[9] Irving Bernstein, *The Lean Years: A History of the American Worker, 1920-1933* (Boston: Houghton Mifflin Company, 1960), p. 125.

[10] C. L. Packard, "New Methods of Handling Coal Electricity," *Coal Age,* vol. 13, No. 20 (May 1918).

[11] William Fritz and Theodore A. Veenstra, *Regional Shifts in the Bituminous Coal Industry* (Pittsburgh: University of Pittsburgh Press, 1935), p. 34; "How Coal Companies are Cutting Costs," *Coal Age,* vol. 27, No. 3 (January 1925); Devine, *Coal,* p. 190.

[12] Louis Bloch, *The Coal Miners' Insecurity* (New York: Russell Sage Foundation, 1922), p. 13; Elwood S. Moore, *Coal: Its Properties, Analysis, Classification, Geology, Extractions, Uses, and Distribution* (London: John Wiley and Company, 1940), pp. 333-345.

[13] Robert D. Billinger, *Pennsylvania's Coal Industry* (Gettysburg: Pennsylvania Historical Association, 1954), p. 40; Devine, *Coal,* p. 190; *The Bituminous Coal Industry With a Survey of Competing Fuels* (Washington, D.C.: Federal Emergency Relief Administration, 1935), p. 139.

[14] Fritz and Veenstra, *Regional Shifts,* p. 4; *Region in Transition: Report of the Economic Study of the Pittsburgh Region* (Pittsburgh: University of Pittsburgh Press, 1963), p. 231; Douglas Fisher, *Epic of Steel* (New York: Harper and Row, 1963), p. 275.

[15] Fritz and Veenstra, *Regional Shifts*, p. 116.

[16] Howard N. Eavenson, "The Early History of the Pittsburgh Coal Bed," *Western Pennsylvania Historical Magazine* (September 1939).

[17] *The Bulletin Index,* October 1, 1936.

[18] *Pittsburgh: 50th Anniversary of the Engineers' Society of Western Pennsylvania* (Pittsburgh: Cramer Printing Company, 1930), p. 250.

[19] Howard N. Eavenson, *The Pittsburgh Coal Bed: Its Early History and Development* (New York: American Institute of Mining and Metallurgical Engineers, 1938), p. 41; John Boucher, *History of Westmoreland County* (New York: Lewis Publishing Company, 1906), p. 470.

[20] *Pennsylvania's Mineral Heritage* (Harrisburg: The Pennsylvania State College School of Mineral Studies, 1944), p. 28.

[21] Moore, *Coal,* p. 341.

[22] *Region in Transition*, p. 236.

[23] *Department of Mines—Commonwealth of Pennsylvania* (Harrisburg: Department of Mines, 1932), p. 4.

[24] Richard Quin, "Indiana County Inventory," Washington, D.C: National Park Service, 1990, p. 25.

[25] Rochester, *Labor and Coal*, p. 80.

[26] *Annual Report on Mining Activities* (Harrisburg: Department of Enviromental Resources, 1978), pp. 37 and 91.

[27] Ibid., p. 90.

[28] Fritz and Veenstra, *Regional Shifts,* p. 8.

[29] J. V. Thompson, *Coal Fields of Southwestern Pennsylvania* (Copyright John W. Boileau, 1907).

[30] Charles Reitell, *The Shift in Soft Coal Shipments* (Harrisburg: Pennsylvania Industrial Survey, 1927), p. 9.

[31] John E. Gable, *History of Cambria County* (Topeka, Kans.: Historical Publishing Company, 1926), p. 244.

[32] *The Decline of the Bituminous Coal Industry,* p. 4.

[33] Ibid.

[34] *The Bituminous Coal Industry with a Survey of Competing Fuels,* p. 35.

3 million tons or more: 17 producers, 19.9% U.S. Commercial Total;
1 million tons or more: 70 producers, 22.9% U.S. Commercial Total;
500,000 to 1,000,000: 131 producers, 17% U.S. Commercial Total

[35] Ibid.

[36] *Department of Mines,* pp. 66-67.

[37] McAlister Coleman, *Men and Coal* (New York: Farrar & Rinehart, Inc. 1943), p. 128; Edward Eyre, Hunt Tyron, Fred G. Tyron, and J. H. Willits, eds., *What the Coal Commission Found* (Baltimore: Williams and Wilkins Company, 1925), p. 354; Ellis W. Roberts, *The Breaker Whistle Blows* (Scranton: Anthracite Museum Press, 1984), p. 81; Morton S. Baratz, *The Union and the Coal Industry* (New Haven: Yale University Press, 1955), p. 1; Charles E. Beachley, *The History of the Consolidation Coal Company, 1864-1934* (New York: The Consolidation Coal Company, 1934).

[38] Demian Hess, "Berwind-White Coal Mining Company Eureka No. 40 and the Windber Mines, Cambria and Somerset Counties, Pennsylvania," Washington, D.C.: National Park Service, 1988, p. 6.

[39] Mildred Biek, *The Miners of Windber: Class, Ethnicity, and the Labor Movement in a Pennsylvania Coal Town, 1890s-1930s* (Ph.D. diss. Northern Illinois University, 1989), p. 499.

[40] Coleman, *Men and Coal,* p. 105.

[41] Harold M. Watkins, *Coal and Men: An Economic and Social Study of the British and American Coal Fields* (London: George Allen & Unwin Ltd. 1934), p. 226.

[42] Biek, *The Miners of Windber,* p. 80.

[43] Gable, *History of Cambria County,* p. 244.

[44] Rochester, *Labor and Coal,* p. 118.

[45] Keith Dix, *What's a Coal Miner to Do? The Mechanization of Coal Mining* (Pittsburgh: University of Pittsburgh Press, 1988), p. 217; A. F. Brosky, "Interpretation of Progress in Mechanical Loading," *Coal Age,* vol. 33, No. 4 (January 1925). This article gives a full description of the various models of coal loading machines.

[46] Anthony F. C. Wallace, *St. Clair: A Nineteenth Century Town's Experience with a Disaster-Prone Industry* (Ithaca, N.Y.: Cornell University Press, 1985), p. 17.

[47] Eavenson, *The Pittsburgh Coal Bed,* p. 40.

[48] *Mechanization, Employment, and Output per Man in Bituminous-Coal Mining.* (Washington, D.C.: WPA, 1939), p. 40.

[49] Adam T. Shurick, *The Coal Industry* (Boston: Little, Brown and Company, 1924), p. 150.

[50] Hess, "Berwind-White Coal Mining Company," p. 70; Gray Fitzsimons, ed., *Blair County and Cambria County, Pennsylvania: An Inventory of Historic Engineering and Industrial Sites* (Historic American Building Survey/Historic American Engineering Record, National Park Service, 1990), p. 27.

[51] By 1960 some 65 percent of all coal shipped to market was cleaned at the nation's 535 cleaning plants. Contemporary preparation plants separated bituminous coal from heavier impurities mixed with it. A variety of screens were used to separate coal into at least six sizes by the 1960s: carbon or pulverized coal 1/4" or smaller; stoker or pea coal 1/4" to 1"; nut coal 1" to 2"; stove coal 2" to 3"; egg coal 3" to 5"; and lump coal 5" or larger.

[52] *Pennsylvania's Mineral Heritage* (Harrisburg: Pennsylvania State College School of Mineral Studies, 1944), p. 208.

[53] Melvyn Dubosky and Warren Van Tine, *John L. Lewis: A Biography* (New York: Quadrangle Press, 1977); Cecil Carnes, *John L. Lewis: Leader of Labor* (New York: Robert Speller, 1936); David J. McDonald and Edward A. Lynch, *Coal and Unionism: A History of the American Coal Miner's Union* (Silver Spring, MD: Cornelius Printing Company, 1939), p. 161; Roberts, *The Breaker Whistle Blows,* pp. 115-116; Coleman, *Men and Coal,* p. 92.

[54] Thomas Coode, *Bug Dust and Black Damp and Work in the Old Patch Town* (Uniontown: Comart Press, 1986), pp. 3-13; C. E. Lesher, "April 1, 1922," *Coal Age,* vol. 21, No. 13 (March 1922); Heber Blackenhorn, *The Strike for Union: A Study of the Non-Union Question in Coal and the Problems of a Democratic Movement (Based on the Record of the Somerset Strike, 1922-1923)* (New York: H. W. Wilson Company, 1924; reprinted, New York: Arno Press).

[55] Coode, *Bug Dust and Black Damp,* p. 5.

[56] Biek, *The Miners of Windber,* p. 498; Coleman, *Men and Coal,* p. 106.

[57] Devine, *Coal,* p. 179.

[58] Glen Lawhorn Parker, *The Coal Industry: A Study in Social Control* (Washington, D.C.: American Council on Public Affairs, 1940), p. 70.

[59] John Brophy, *A Miner's Life* (Madison: University of Wisconsin Press, 1964), p. 236.

[60] Blackenhorn, *The Strike for Union,* p. 15. This is the best study of the coal strike of 1922 in District 2 with emphasis on strike activities in Somerset County.

[61] Carter Goodrich, *The Miner's Freedom: A Study of the Changing Life in a Changing Industry* (Boston: Marshall Jones Company, 1925), p. 61.

[62] Eileen M. Cooper, "The Magnificent Strike for Unionism: The Somerset County Strike of 1922," *Pennsylvania Heritage,* vol. 3 (Fall 1991).

[63] Margaret Mulrooney, *A Legacy of Coal: The Company Towns of Southwestern Pennsylvania* (Washington, D.C.: National Park Service, 1991), pp. 51-52; John F. Hylan, *Statement of Facts and Summary Committee to Investigate the Labor Conditions at the Berwind-White Company's Coal Mines in Somerset and other Counties, Pennsylvania* (New York, 1922); Lou Athey, *Kaymoor: A New River Community* (Washington, D.C.: National Park Service, 1986), p. 47; Fitzsimons, *Blair and Cambria County,* pp. 11-12; Bruce T. Williams and Michael D. Yates, *Upward Struggle: A Bicentennial Tribute to Labor in Cambria and Somerset Counties* (Johnstown: University of Pittsburgh, 1976); Fred C. Doyle, ed., *50th Anniversary Windber, Pennsylvania* (Dubois: Gary Printing Company, 1947); Frank Paul Alcamo, *The Windber Story: A Twentieth Century Model Pennsylvania Coal Town* (Windber: privately printed, 1983); Hess, "Berwind-White Coal Mining Company," pp. 9-15.

[64] Blackenhorn, *The Strike for Union,* pp. 123-124.

[65] Mulrooney, *A Legacy of Coal,* p. 25.

[66] Muriel Earley Sheppard, *Cloud by Day: The Story of Coal and Coke and People.* (Pittsburgh: University of Pittsburgh Press, 1991), pp. 107-121. Chapter 11, entitled "The Cossack," describes the diverse roles these private police played in the mining communities of Fayette County.

[67] Rochester, *Labor and Coal,* p. 21; Blackenhorn, *The Strike for Union,* p. 101. Blackenhorn had served as field secretary for the Interchurch World Movement's report on the 1919 steel strike. There were 23,900 miners in Fayette County, of whom several thousand were employed at union mines.

[68] Blackenhorn, *The Strike for Union,* p. 125.

[69] Hess, "Berwind-White Coal Mining Company," p. 6.

[70] W. J. Lauck, "The Bituminous Coal Miner and Coke Workers of Western Pennsylvania," *The Survey,* vol. 26 (April 1911).

[71] Blankenhorn, *The Strike for Union*, p. 260. The chart below, from Heber Blankenhorn's study, reflects the ethnic diversity of the miners employed by the Berwind-White Mining Company in their mines in Somerset and Cambria Counties:

	January 1, 1922	January 1, 1923
American	411	507
Slavish	753	275
Polish	495	264
Hungarian	545	358
Italian	415	179
German	37	40
Scotch	25	17
Austrian	35	13
Welsh	6	7
French	3	1
English	23	20
Irish	4	9
Rumanian	47	24
Scandinavian	6	3
Spanish	23	90
Belgian	1	2
Serb	0	3
Croat	17	7
Lithuanian	40	20
Greek	2	17
Mexican	6	15
Porto Rican (sic)	0	1
Macedonian	1	2
Turk	0	1
Bulgarian	1	1
Canadian	0	1
Total	2,894	1,877

[72] *The First Hundred Years, 1890-1990* (Ebensburg: UMWA District Office 2, 1990), p. 17. The returning miners issued the following bitter resolution at the conclusion of their seventeen-month unsuccessful bid for union representation:

Whereas: The long strike of almost seventeen months in the coal fields of Somerset has been terminated and...
Whereas: we recognize the circumstances making necessary the temporary abandonment of our fight against the coal operators for Union recognition, and realize that the failure of the strike to secure our demands was not due to any defects in the principles of unionism, but rather to the brutal tactics and tremendous financial stake of the coal companies, as well as to the weak mindedness, selfishness and un-Americanism of strikebreakers who took our jobs, and reaped the benefit of the wage increases which not they, but we and the union were the means of securing from the coal operators of Somerset County.

[73] Biek, *The Miners of Windber*, p. 516; Hylan, "Statement of Facts."

[74] "Hammond and His Associates End Years's Labors: Completed Report Comprises 800,000 Words," *Coal Age*, vol. 24, no. 13 (September 1923); Eyre, Tyron, Tyron, and Willits, eds., *What the Coal Commission Found*.

[75] Fritz and Veenstra, *Regional Shifts*, p. 23.

[76] *A Medical Survey of the Bituminous-Coal Industry*, commonly referred to as the Boone Report after its director, Rear Admiral Joel T. Boone, was conducted by the Coal Mines Administration. The report, issued in March 1947, was a comprehensive document detailing deficiencies found in housing, sanitation, public health, and hygiene by investigators in many coal towns. *A Report on Community and Living Conditions in the Coal Field* was chaired by John D. Rockefeller IV of West Virginia in 1980.

[77] Eyre, Tyron, and Willits, eds., *What the Coal Commission Found*; Coleman, *Men and Coal*, p. 123.

[78] Devine, *Coal*, pp. 431-437.

[79] Dix, *What's a Coal Miner to Do?* p. 80.

[80] McDonald and Lynch, *Coal and Unionism*, p. 167; Dix, *What's a Coal Miner to Do?* p. 147.

[81] Baratz, *The Union and the Coal Industry*, p. 60.

[82] Workman, *The Fairmont Coal Field*, pp. 131-136.

[83] Rochester, *Labor and Coal*, p. 204.

[84] Shurick, *The Coal Industry*, p. 349.

[85] Ibid., p. 328.

[86] *The First Hundred Years*, p. 2.

[87] Dix, *What's a Coal Miner to Do?* p. 148.

[88] Bernstein, *The Lean Years*, p. 125; Dix, *What's a Coal Miner to Do?* pp. 145-148.

[89] Ibid., p. 147.

[90] Bernstein, *The Lean Years*, p. 129.

[91] George W. Harris, "Montour No. 8 Plant of the Pittsburgh Coal Company," *Coal Age*, vol. 13, no. 13 (March 1918); "Pittsburgh Coal Company Report of 1917," *Coal Age*, vol. 13, no. 3 (April 1918).

[92] Rochester, *Labor and Coal*, p. 63.

[93] Dix, *What's a Coal Miner to Do?* p. 147.

[94] Billinger, *Pennsylvania's Coal Industry*, p. 42.

[95] "Senate Committee Probes Soft-Coal Troubles," *Coal Age*, vol. 33, no. 4 (April 1928).

[96] Rochester, *Labor and Coal*, pp. 86-87.

[97] Coleman, *Men and Coal*.

[98] Harvey O'Connor, *Mellons' Millions: A Biography of a Fortune* (New York: The John Day Company, 1933), p. 219.

[99] Rochester, *Labor and Coal*, p. 53.

[100] Dix, *What's a Coal Miner to Do?* pp. 168-169.

[101] Homer Lawrence Morris, *The Plight of the Bituminous Coal Miner* (Philadelphia: University of Pennsylvania Press, 1934), p. 143.

[102] Hyman Kuritz, "The Labor Injunction in Pennsylvania, 1891-1931," *Pennsylvania History*, vol. 29, no. 3 (July 1962).

[103] Ronald L. Filippelli, "Diary of a Strike: George Medrick and the Coal Strike of 1927 in Western Pennsylvania," *Pennsylvania History*, vol. 43, no. 3 (July 1976).

[104] Carnes, *John L. Lewis*, p. 193.

[105] "Senate Committee Probes Soft-Coal Troubles"; "Fight to Hold Jacksonville Scale Expensive to Organized Labor in 1927," *Coal Age*, vol. 33, no. 1 (January 1928); "How Strong Is the Miners' Union?" *Coal Age*, vol. 28, no. 13 (September 1928).

[106] Ibid., p. 170.

[107] Filippelli, "Diary of a Strike." There exists no in-depth historical study investigating the Strike of 1927. A brief survey of the strike is included in the following sources: Bernstein, *The Lean Years*, pp. 127-136; Rochester, *Labor and Coal*, pp. 204-214.

[108] Glen Lawhorn Park, *The Coal Industry: A Study in Social Control* (Washington, D.C.: American Council on Public Affairs, 1940), p. 71.

[109] Devine, *Coal*, p. 179.

[110] Biek, *The Miners of Windber,* p. 572; Johnson, *The Politics of Soft Coal,* p. 120.

[111] Johnson, *The Politics of Soft Coal,* p. 121.

[112] Coode, *Bug Dust and Black Damp and Work,* p. 28.

[113] Coleman, *Men and Coal,* p. 132.

[114] O'Connor, *Mellons' Millions,* p. 220.

[115] Coleman, *Men and Coal,* p. 132.

[116] Ibid.

[117] Bernstein, *The Lean Years,* p. 130.

[118] "Senate Committee Probes Soft-Coal Troubles," *Coal Age,* vol. 33, no. 4 (April 1928).

[119] Watkins, *Coal and Men,* p. 223.

[120] McDonald and Lynch, *Coal and Unionism,* p. 180; Coleman, *Men and Coal,* p.136; Watkins, *Coal and Men,* p. 191.

[121] Leo Wolman, *Ebb and Flow in Trade Unionism* (New York: National Bureau of Economic Research, 1936), p. 217.

[122] Ibid., p. 231; Rochester, *Labor and Coal,* pp. 224, 225, 226.

[123] Lester V. Chandler, *America's Greatest Depression, 1929-1941* (New York: Harper & Row Publishers, 1970), p. 229.

[124] Bernstein, *The Lean Years,* pp. 131-132; Dix, *What's a Coal Miner to Do?* p. 185.

[125] Harold Barger and Sam H. Schurr, *The Mining Industries, 1899-1939: A Study of Output Employment and Productivity* (New York: National Bureau of Economic Research, Inc., 1944), p. 163.

[126] Johnson, *The Politics of Soft Coal,* p. 192.

[127] Dix, *What's a Coal Miner to Do?* p. 185.

[128] Ibid., p. 102.

[129] Johnson, *The Politics of Soft Coal.*

[130] Bernstein, *The Lean Years,* p. 363.

[131] John F. Bauman and Thomas Coode, "'Old Bill': A New Deal Chronicle of Poverty and Isolation in Southwestern Pennsylvania," *Western Pennsylvania Historical Magazine,* vol. 63 (July 1980).

[132] Thomas Coode, *Bug Dust and Black Damp and Work in the Old Patch Town* (Uniontown: Comart Press, 1986), pp. 103-104; Bauman and Coode, "Old Bill," p. 216.

[133] John F. Bauman, "Orwell's Wigan Pier and Daisytown: The Mine Town as a Stranded Landscape," *Western Pennsylvania Historical Magazine,* vol. 67 (April 1984), p. 102.

[134] Morris, *The Plight of the Bituminous Coal Miner,* p. 183; Bernstein, *The Lean Years,* p. 377; James P. Johnson, "A New Deal for Soft Coal: The Attempted Revitalization of the Bituminous Coal Industry under the New Deal" (Ph.D. diss. Columbia University, 1968), pp. 4-5.

[135] Morris, *The Plight of the Bituminous Coal Miner,* p. 166.

[136] Bernstein, *The Lean Years,* p. 363.

[137] Ibid.

[138] Coleman, *Men and Coal,* p. 144.

[139] George Korson, *Coal Dust on the Fiddle* (Hatboro: Folklore Associates Inc., 1965), pp. 401-402.

[140] Carnes, *John L. Lewis,* p. 239.

[141] Coleman, *Men and Coal,* p. 147; Michael Workman, *The Fairmont Coal Field: Historical Context* (Morgantown, W.Va.: Institute for History of Technology & Industrial Archaelogy, 1992), p. 58. Glen Lawhorn Park, *The Coal Industry: A Study in Social Control* (Washington, D.C.: American Council on Public Affairs, 1940), pp. 85-105; see Chapter 5.

[142] Billinger, *Pennsylvania's Coal Industry,* p. 39.

[143] Johnson, "A New Deal for Soft Coal," p. 20.

[144] James P. Johnson, "Reorganizing the United Mine Workers of America in Pennsylvania During the New Deal," *Pennsylvania History,* vol. 37 (1970), p. 125.

[145] Bernstein, *The Lean Years,* p. 245; Coleman, *Men and Coal,* p. 243. Wechsler was a labor reporter for the newspaper *PM* and authored a Lewis biography entitled *Labor Baron: A Portrait of John L. Lewis,* published in 1944.

[146] Lester V. Chandler, *America's Greatest Depression, 1929-1941* (New York: Harper & Row Publishers, 1970), p. 228.

[147] Johnson, "A New Deal for Soft Coal," p. 51.

[148] Richard O. Boyer and Herbert M. Morais, *Labor's Untold Story* (New York: United Electrical, Radio & Machine Workers of America, 1980), p. 276; Johnson, *The Politics of Soft Coal,* p. 233.

[149] Florence Peterson, *American Labor Unions: What They Are and How They Work* (New York: Harper & Row Publishers, 1945), p. 43.

[150] Ibid., p. 39.

[151] Coleman, *Men and Coal,* p. 155.

[152] Peterson, *American Labor Unions,* p. 29.

[153] *The First Hundred Years,* p. 4.

[154] Associated Press, May 20, 1940; Philip A. Grant Jr., "The Pennsylvania Congressional Delegation and the Bituminous Coal Acts of 1935 and 1937," *Pennsylvania History,* vol. 49, No. 1982.

[155] William Serrin, *Homestead: The Glory and Tragedy of an American Steel Town* (New York: Random House, 1992), p. 193.

[156] Johnson, "A New Deal for Soft Coal," p. 51.

[157] Coleman, *Men and Coal,* p. 149.

[158] Dix, *What's a Coal Miner to Do?* p. 189.

[159] Brophy, *A Miner's Life,* p. 236.

[160] Philip Taft, *Organized Labor in American History* (New York: Harper & Row Publishers, 1964), p. 432.

[161] McDonald and Lynch, *Coal and Unionism,* p. 228.

[162] Dix, *What's a Coal Miner to Do?* p. 200.
First Appalachian Agreement, October 1, 1933 to March 31, 1934
Second Appalachian Agreement, April 1, 1934 to March 31, 1935 later extended to September 30, 1935
Third Appalachian Agreement, October 1, 1935 to March 31, 1937
Fourth Appalachian Agreement, April 1, 1937 to March 31, 1939
Fifth Appalachian Agreement, April 12, 1939 to March 31, 1941

The UMWA signed a sixth Appalachian Joint Wage agreement effective from April 1, 1941 to March 31, 1943.

Coleman, *Men and Coal,* pp. 313-335.

[163] Johnson, "A New Deal for Soft Coal," p. 42.

[164] McDonald and Lynch, *Coal and Unionism,* p. 200.

[165] Baratz, *The Union and the Coal Industry,* p. 91.

[166] Ibid., p. 96.

[167] *The Bituminous Coal Industry With a Survey of Competing Fuels,* p. 67.

[168] Ibid.

[169] Ibid., p. 36.

The Superior Coal Company, owned by the Chicago and Northwestern R.R., and the Union Pacific Coal Company, owned by the Union Pacific R.R., both operated captive mines during the 1930s.

[170] Johnson, "Reorganizing the United Mine Workers of America," p. 127.

[171] William Sisson, Bruce Bomberger, and Diane Reed, "Iron and Steel Resources," p. 98.

[172] Frederic Quivik, "Connellsville Coal and Coke," Washington, D.C.: National Park Service, 1991, p. 43.

[173] Sheppard, *Cloud by Day,* pp. 123-126.

[174] Johnson, "A New Deal for Soft Coal," p. 60.

[175] Coode, *Bug Dust and Black Damp,* p. 122.

[176] Johnson, *The Politics of Soft Coal,* p. 214.

[177] Dix, *What's a Coal Miner to Do?* p. 210; *Coal Age* (October 1933).

[178] Rochester, *Labor and Coal,* p. 242.

[179] *Coal Age,* (June 1986).

[180] *The Bituminous Coal Industry With a Survey of Competing Fuels,* p. 73.

[181] Baratz, *The Union and the Coal Industry,* p. 8.

[182] John N. Hoffman, "Pennsylvania's Bituminous Coal Industry: An Industry Review," *Pennsylvania History,* vol. 5 (1978). Mechanical mining machinery was used in some larger mining operations from the 1890s. Mechanized mining equipment introduced from 1913 to 1941 included the electric drill (1914), pit-car loader (1914), crawler mounted loader (1916), rubbber-belt conveyor (1917), shuttle car (1938), and tungsten carbide tips (1941).

[183] Sue Wrbican, "Portrait of Braeburn," *Pittsburgh History* (Spring 1992).

[184] Schurr and Netschert, *Energy,* pp. 169-180.

[185] Goodrich, *The Miners' Freedom,* p. 19.

[186] Barger and Schurr, *The Mining Industries, 1899-1939,* p. 177.

[187] *Mechanization, Employment, and Output Per Man in Bituminous-Coal Mining,* vol. 1 (Washington: Works Progress Administration, 1939), p. 114.

[188] A brief historical account of the introduction of the mechanical loaders is an article by Dr. L. E. Young, "Coal Mine Mechanization," *Year Book on Coal Mechanization,* 1929, published by The American Mining Congress, pp. 2-6; Morris, *The Plight of the Bituminous Coal Miner,* p. 28.

[189] E. N. Zern, "Underground Coal-Loading Machinery," *Coal Age,* vol. 15, no. 18 (May 1919); Thomas W. Fry, "Development of Coal Cutting Machinery," *Coal Age,* vol. 3, no. 3 (January 1913); A. F. Brosky, "An Approach to Complete Mechanization," *Coal Age,* vol. 33, no. 4 (April 1928); C. E. Warbom, "A New Type of Coal Cutter," *Coal Age,* vol. 4 no. 24 (December 1913); "Machine Loading Reducing Mine Cost 30 Per Cent," *Coal Age,* vol. 24, no. 20 (November 1923); E. N. Zern, "Underground Coal-Loading Machinery," *Proceeding of the Coal Mining Institute* (Crafton: Cramer Printing and Publishing Company, 1918).

[190] *Mineral Year Book* (Washington, D.C.: U.S. Bureau of Mines, 1937), p. 874; *Mechanization, Employment, and Output per Man in Bitumiminous Coal Mining* (Washington, D.C.: Works Projects Administration, 1939), p.116.

[191] Dix, *What's a Coal Miner to Do?* pp. 168-169.

[192] C. L. Christenson, *Economic Redevelopment in Bituminous Coal* (Cambridge: Harvard University Press, 1962), p. 132.

[193] Dix, *What's a Coal Miner to Do?* p. 243.

[194] Christenson, *Economic Redevelopment,* p. 133.

[195] Ibid.

[196] Data on models obtained from material obtained from Joy Manufacturing Company, Pittsburgh.

[197] Morris, *The Plight of the Bituminous Coal Miner,* p. 139.

[198] Korson, *Coal Dust on the Fiddle,* pp. 141-142.

[199] *The Bituminous Coal Industry With a Survey of Competing Fuels,* p. 77.

[200] Baratz, *The Union and the Coal Industry,* p. 107.

[201] Bernstein, *The Lean Years,* p. 125.

[202] Baratz, *The Union and the Coal Industry,* p. 71.

[203] "Continuous Coal Mining," *Fortune Magazine* (June 1950); "Directory of New Technology." *Coal Mining* (June 1986).

[204] Dix, *What's a Coal Miner to Do?* p. 213.

[205] *Mineral Yearbook, 1952* (Washington, D.C., 1953) Volume 11, p. 49; Baratz, *The Union and the Coal Industry,* pp.40-41.

[206] *Coal Age* (June 1986).

[207] Builder Levy, *Images of Appalachian Coalfields* (Philadelphia: Temple University Press, 1989), p. 89.

[208] Smith, *Digging Our Own Graves,* pp. 57-58.

[209] Long, *Where the Sun Never Shines,* pp. 121-122; Beachley, *The History of the Consolidation Coal Company.* Pitt-Consol merged with Continental Oil Company (Conoco Inc) in 1964, which in turn merged with DuPont in 1980. Smith, *Digging Our Own Graves,* p. 57; Workman, *The Fairmont Coal Field,* p. 24; *Consol News 100th Anniversary Edition,* 1964, p. 8.

[210] Robert Stefanko, *Coal Mining Technology: Theory and Practice* (New York: American Institute of Mining, 1983), p. 41.

[211] George F. Deasy and Phyllis R. Griess, *Atlas of Pennsylvania Coal and Coal Mining,* Part 1, *Bituminous Coal.* Pennsylvania College of Mineral Industries, Bulletin of the Mineral Industries Experiment Stations, No. 53, 1959, p. 54.

[212] Smith, *Digging Our Own Graves,* p. 47; *Coal Age* (January 1949).

[213] Jerome White and Samuel Law, *The Coal Industry in Cambria County* (n.p.: Cambria County Historical Society, 1954), p. 84; Billinger, *Pennsylvania's Coal Industry,* p. 47.

[214] "Room-and-Pillar," *Pennsylvania Coal Association Quarterly* (March 1993).

[215] F. C. Cornet, "Proposed Method of Longwall Mining," *Coal Age,* vol. 4, no. 4 (July 1913); Jerome White and Samuel Law, *The Coal Industry in Cambria County* (n.p.: Cambria County Historical Society, 1954), p. 84; William Keyes, ed., *Historic Survey of the Greater Monongahela River Valley* (Harrisburg: National Park Service / Pennsylvania Historical and Museum Commission, 1991), p. 24.

[216] Builder Levy, *Images of Appalachian Coalfields* (Philadelphia: Temple University Press, 1989), p. 9.

[217] *Pennsylvania Coal Quarterly* (Fall 1985).

[218] *Coal Data: A Reference* (Washington, D.C.: U.S. Department of Energy, 1991), p. 10.

[219] Shyamal Majumdar and E. Willard Miller, eds., *Pennsylvania Coal: Resources, Technology, and Utilization* (University Park: Pennsylvania State University Press, 1983), p. 309.

[220] Raymond E. Murphy, *The Mineral Industries of Pennsylvania: Trends of the Mineral Producing and Processing Industries* (Harrisburg: Greater Pennsylvania Council, 1933), p. 32.

Bibliography

Government Documents and Reports

Bureau of the Census. *Tenth Census of the United States. General Analysis of the Bituminous Coal Statistics.* Washington, D.C.: Government Printing Office, 1881.

Bureau of the Census. *Report on Mineral Industries in the United States, 1890.* Washington, D.C.: Government Printing Office, 1892.

Bureau of the Census. *Thirteenth Census of the United States. Mines and Quarries—Statistics for Pennsylvania, (Volume XI).* Washington, D.C.: Government Printing Office, 1909.

Bureau of the Census. *Fourteenth Census of the United States. Mines and Quarries—Statistics for Pennsylvania, (Volume XI).* Washington, D.C.: Government Printing Office, 1919.

Census of Mineral Industries; Coal Mining, 1982. Department of Commerce. Washington, D.C.: Government Printing Office, 1985.

Coal Data: A Reference. Washington, D.C.: U.S. Department of Energy, 1991.

Coal Resources of Pennsylvania. Harrisburg: Commonwealth of Pennsylvania, Department of Environmental Resources, 1980.

Commonwealth of Pennsylvania Department of Mines and Mineral Industries. Bituminous Coal Division, 1931-1944. Harrisburg: Commonwealth of Pennsylvania, 1959.

Deasy, George, and Griess, Phyllis. *Atlas of Pennsylvania Coal and Coal Mining,* Part 1, *Bituminous Coal.* University Park: Pennsylvania State University, College of Mineral Industries: Bulletin of The Mineral Industries Experiment Station, No. 73, 1959.

Dodge, Clifford H., and Glover, Albert. *Coal Resources of Greene County.* Mineral Resource Report 86. Harrisburg: Pennsylvania Geological Survey, 1984.

Edmunds, William, and Koppe, Edwin. *Coal in Pennsylvania.* Harrisburg: Commonwealth of Pennsylvania, Department of Environmental Resources, 1970.

Federal Emergency Relief Administration (Harry Hopkins, Administrator). *The Bituminous Coal Industry With a Survey of Competing Fuels.* Washington, D.C., 1935.

Gardner, James H. *The Broad Top Coal Field of Huntingdon, Bedford, and Fulton Counties, Report # 10.* Harrisburg: William Stanley Ray, State Printer, 1913.

Hickok, W. O., and Moyer, F. T. *Geology and Mineral Resources of Fayette County, Pennsylvania.* Harrisburg: Department of Internal Affairs, 1940.

Industrial Directory of the Commonwealth of Pennsylvania. Harrisburg: Department of Internal Affairs, 1916, 1935, 1941.

Maize, C. H., and Struble, G. S. *History of Pennsylvania Bituminous Coal.* Harrisburg: Pennsylvania Department of Mines, 1955.

Mechanization, Employment, and Output Per Man in Bituminous Coal Mining. Volume 1. Washington, D.C.: Works Projects Administration, Department of the Interior, Bureau of Mines, 1939.

Medical Survey of the Bituminous Coal Industry: Report of the Coal Mines. Washington, D.C.: Department of the Interior, 1947.

Mineral Resources of Johnstown, Pennsylvania, and Vicinity. Washington, D.C.: United States Geological Survey, 1911.

Mineral Resource Reports. Topographic and Geologic Survey Bulletin M6. Commonwealth of Pennsylvania, Department of Forests and Waters.
  Part 1-General Information on Coal—G. H. Ashley, 1928
  Part 2-Detailed Description of Coal Fields—J. D. Sisler, 1926
  Part 3-Coal Resources—J. F. Reese and J. D. Sisler, 1928
  Part 4-Coal Analyses—Prepared U.S. Bureau of Mines, 1928
  Department of Environmental Resources, Pittsburgh Office.

Murphy, Raymond E. *The Mineral Industries of Pennsylvania: Trends of Mineral Producing and Processing Industries.* Harrisburg: Greater Pennsylvania Council, 1933.

Pennsylvania Department of Mines. *Annual Reports of the Department of Mineral Resources.* Report 93. Harrisburg: Office of Resource Management, 1987.

Stevenson, J. J. *Second Geological Survey of Pennsylvania: Greene and Washington District of the Bituminuous Coal-Fields of Western Pennsylvania.* Harrisburg: Board of Commissioners, 1876.

Stone, Ralph W. *Geology and Mineral Resources of Greene County.* Topographic and Geologic Survey Bulletin C 30. Harrisburg, 1932.

The Decline of the Bituminous Coal Industry in Pennsylvania. Harrisburg: Greater Pennsylvania Council, 1932.

The U.S. Coal Industry, 1970-1990: Two Decades of Change. Washington, D.C.: U.S. Department of Energy, 1991.

Wall, J. Sutton. *Second Geological Survey of Pennsylvania: Monongahela River Region.* Harrisburg: Board of Commissioners, 1884.

America's Industrial Heritage Project/HABS/HAER, National Park Service Reports

A Coal Mining Heritage Study: Southern West Virginia. Philadelphia: National Park Service, U.S. Department of the Interior, 1992.
Bennett, Lola M. *The Company Towns of the Rockhill Iron and Coal Company: Robertsdale and Woodvale, Pennsylvania.* Washington, D.C.: National Park Service, 1990.
Brown, Sharon A., Greene, Jerome, and O'Brien, William Patrick. *Historic Resource Survey: Bedford, Blair, Cambria, Fayette, Fulton, Huntingdon, Indiana, Somerset, Westmoreland Counties.* Washington, D.C.: National Park Service, unpublished, 1990.
Fitzsimons, Gray, ed. *Blair County and Cambria County, Pennsylvania: An Inventory of Historic Engineering and Industrial Sites.* Washington, D.C.: Historic American Building Survey/Historic American Engineering Record, National Park Service, 1990.
Heald, Sarah H., ed. *Fayette County, Pennsylvania: An Inventory of Historic Engineering and Industrial Sites.* Washington, D.C.: Historic American Building Survey/Historic American Engineering Record, National Park Service, 1990.
Hess, Demain. *Berwind-White Coal Mining Company Eureka No. 40 and the Windber Mines, Cambria and Somerset Counties.* Washington, D.C.: Historic American Building Survey/Historic American Engineering Record, National Park Service, 1988.
Mulrooney, Margaret M. *A Legacy of Coal: The Company Towns of Southwestern Pennsylvania.* Washington, D.C.: National Park Service, 1989.
Quin, Richard. *Indiana County Inventory.* Washington, D.C.: National Park Service, unpublished, 1990.
Quivik, Frederic L. *Connellsville Coal and Coke.* Washington, D.C.: National Park Service, unpublished, 1991.
Reconnaissance Survey: Brownsville/Monongahela Valley, Pennsylvania /West Virginia. Denver: National Park Service, 1991.
Shedd, Nancy, ed. *Huntingdon County, Pennsylvania: An Inventory of Historic Engineering and Industrial Sites.* Washington, D.C.: National Park Service, 1991.

Pensylvania Historical and Museum Commission Survey and Planning Grant Reports and Miscellaneous Reports

Armstrong County Planning Commission. *Preliminary Researcch for Armstrong County Historic Sites Surveys.* Kittanning, 1980 (unpublished, located at PHMC/BHP).
Cambria County Redevelopment Authority. *Cambria County, Pennsylvania Historic Sites Survey Preliminery Report.* Ebensberg, 1980 (unpublished, located at PHMC/BHP).
DiCiccio, Carmen P. *Survey of Coal Mining Sites in Greene and Washington Counties, Pennsylvania, 1780-1945.* California: California University of PA, 1988 (unpublished, located at PHMC/BHP).
DiCiccio, Carmen P., and Davis, Christine. *Westmoreland County Resource Survey.* Harrisburg, 1993 (unpublished, located at PHMC/BHP).
Keyes, William, ed. *Historic Survey of the Greater Monongahela River Valley.* National Park Service/Pennsylvania Historical and Museum Commission, 1991 (unpublished, located at PHMC/BHP).
Snyder, E. Dennison. *Coal Town Surveys.* Brookville: Jefferson County Planning Commission, 1987 (unpublished, located at PHMC/BHP).
Somerset County Planning Commisson. *Preliminery Report: Somerset County Historical Resource Survey.* Somerset, 1984 (unpublished, located at PHMC/BHP).

General Bibliography

A Town That Grew at the Crossroad. Scottdale: Laurel Group Press, 1978.
Albert, George Dallas. *History of Westmoreland County, Pennsylvania.* Philadelphia: L. H. Evert and Company, 1882.
Alcamo, Frank Paul. *The Windber Story: A Twentieth Century Model Pennsylvania Coal Town.* Windber: Privately printed, 1983.
Aloe, Mark. *Pennsylvania Coal Data, 1986.* Harrisburg: Keystone Bituminous Coal Association, 1987.
Allen, James B. *The Company Town in the American West.* Norman: University of Oklahoma Press, 1966.
Athey, Louis. *Kaymoor: A New River Community.* Washington, D.C.: National Park Service, 1986.
Aurand, Harold W. *From the Molly Maguires to the United Mine Workers: The Social Ecology of an Industrial Union, 1869-1897.* Philadelphia: Temple University Press, 1971.
Baratz, Morton S. *The Union and the Coal Industry.* New Haven: Yale University Press, 1955.

Barger, John Wayman. *Greene County Coal Book and Purchaser's Official Guide.* New Freeport: J. W. Barger Publisher, 1907.

Barger, Harold, and Schurr, Sam H. *The Mining Industries, 1899-1939: A Study of Output, Employment and Productivity.* New York: National Bureau of Economic Research, Inc., 1944.

Baugham, Jon D., and Morgan, Ronald L. *Tales of the Broad Top.* Volume 11 (n.p./n.p.), 1979.

Beachley, Charles E. *The History of the Consolidation Coal Company, 1864-1934.* New York: The Consolidation Coal Company, 1934.

Bernstein, Irving. *The Lean Years: A History of the American Worker, 1920-1933.* Boston: Houghton Mifflin Company, 1960.

Billinger, Robert D. *Pennsylvania's Coal Industry.* Gettysburg: Historical Studies Number 6, Pennsylvania Historical Association, 1954.

Binder, Frederick Moore. *Coal Age Empire: Pennsylvania Coal and Its Utilization to 1860.* Harrisburg: Pennsylvania Historical and Museum Commission, 1974.

Bissell, Richard. *The Monongahela.* New York: Rinehart & Company, 1952.

Bituminous Coal Data, 1960. Washington, D.C.: National Coal Association, 1961.

Bituminous Coal Data, 1950. Washington, D.C.: National Coal Association, 1951.

Black, Herbert, ed. *Black's Directory.* Cincinnati: Coal Analyses and Directory Bureau, 1936, 1938.

Black Diamond's Year Books. 1910-1911.

Blackenhorn, Heber. *The Strike for Union: A Study of the Non-Union Question in Coal and the Problems of a Democratic Movement (Based on the Record of the Somerset Strike 1922-1923).* New York: H.W. Wilson Company, 1924; reprinted, New York: Arno Press, 1969.

Bloch, Louis. *Labor Agreements in Coal Mines.* New York: Russell Sage Foundation, 1931.

Bloch, Louis. *The Coal Miners' Insecurity.* New York: Russell Sage Foundation, 1922.

Blossburg Centennial, 1871-1971. Blossburg, 1971.

Bodnar, John. *Anthracite People: Families, Unions and Work, 1900-1940.* Harrisburg: Pennsylvania Historical and Museum Commission, 1983.

Boileau, John. *Coal Fields of Southwestern Pennsylvania: Washington and Greene Counties.* Privately printed, 1907.

Boucher, John. *History of Westmoreland County.* New York: Lewis Publishing Company, 1906.

Boucher, John. *William Kelly: A True History of the So-Called Bessemer Process.* Greensburg: Published by the Author, 1924.

Brecher, Jeremy. *Strike.* Boston: South End Press, 1972.

Brestinsky, F. Dennis, et al. *Patch/Work Voices: The Culture and Lore of a Mining People.* Pittsburgh: University of Pittsburgh Press, 1991.

Brody, David. *Steelworkers in America: The Nonunion Era.* Harper & Row Publishers, 1960.

Brophy, John. *A Miner's Life.* Madison: University of Wisconsin Press, 1964.

Carawan, Guy and Candy. *Voices from the Mountains.* New York: Alfred A. Knopf, 1975.

Carnes, Cecil. *John L. Lewis: Leader of Labor.* New York: Robert Speller Publishing Company, 1936.

Cassady, John C. *The Somerset County Outline.* Scottdale: Mennonite Press, 1932.

Cassidy, Samuel M., ed. *Elements of Practical Coal Mining.* Baltimore: American Institute of Mining, Metallurgical, and Petroleum Engineers, Inc., 1973.

Chandler, Lester V. *America's Greatest Depression, 1929-1941.* New York: Harper and Row Publishers, 1970.

Christenson, C. L. *Economic Redevelopment in Bituminous Coal: The Special Case of Technological Advance in United States Coal Mines, 1930-1960.* Cambridge: Harvard University Press, 1962.

Coal. Chicago: Field Enterprises Educational Corporation, 1976.

Coal: The Other Energy. Washington, D.C.: Federal Energy Administration, 1987.

Coal Mines, 1910. Cleveland: B. H. Rose Publisher, 1910.

Coal and Coke Operator (devoted to coal mining and coke manufacture).

Coal Field Directory and Mining Catalog. New York: McGraw-Hill Company, 1914, 1928, 1935, 1941, 1947.

Coleman, J. Walter. *The Molly Maguire Riots: Industrial Conflict in the Pennsylvania Coal Region.* New York: Arno Press reprint edition, 1969.

Coleman, McAlister. *Men and Coal.* New York: Farrar & Rinehart, Inc., 1943.

Conti, Philip M. *The Pennsylvania State Police: A History of Service to the Commonwealth, 1905 to the Present.* Harrisburg: Stackpole Books, 1977.

Coode, Thomas. *Bug Dust and Black Damp and Work in the Old Coal Patch.* Uniontown: Comart Press, 1986.

Cordell, Glen R., Lodge, Anne J., Nelson, John H., and Weimer, Barbara A. *McConnellsburg, Pennsylvania—Moments in History.* McConnellsburg: Fulton County Historical Society, 1985.

Corlson, Carl. *Buried Black Treasure: The Story of Pennsylvania Anthracite.* Bethleham: Privately published, n.d.).

Couvares, Francis. *The Remaking of Pittsburgh: Class and Culture in an Industrializing City, 1877-1919.* Albany: State University of New York Press, 1984.

Daddow, Samuel Harries, and Bannan, Benjamin. *Coal, Iron, and Oil; or The Practical American Miner.* Pottsville: Benjamin Bannan Publisher, 1866.

Devine, Edward T. *Coal: Economic Problems of the Mining, Marketing and Consumption of Anthracite and Soft Coal in the United States.* Bloomington, Illinois: American Review Service Press, 1925.

Dix, Keith. *What's a Coal Miner to Do? The Mechanization of Coal Mining.* Pittsburgh: University of Pittsburgh Press, 1988.

Dix, Keith. *Work Relations in the Coal Industry: The Hand-Loading Era, 1880-1930.* Morgantown: West Virginia University Press, 1977.

Dixon, E. C. *Coke and By-Products Manufacture.* London: C. F. Griffin and Company, 1939.

Dodge, Clifford H., and Glover, Albert. *Coal Resources of Greene County.* Mineral Resource Report M 8. Harrisburg: Office of Resource Management, 1984.

Dodrill, Gordon. *20,000 Coal Company Stores in the United States, Mexico and Canada.* Pittsburgh: Duquesne Lithographing Company, 1971.

Doyle Fred C., ed. *50th Anniversary, Windber, Pennsylvania.* Dubois: Gary Printing Company, 1947.

Dubofsky, Melvyn, and Van Tine, Warren. *John L. Lewis: A Biography.* New York: Quadrangle Press, 1977.

Eavenson, Howard N. *Coal Through the Ages.* New York: American Institute of Mining and Metallurgical Engineers, 1942.

Eavenson, Howard N. *The Pittsburgh Coal Bed: Its Early History and Development.* New York: American Institute of Mining and Metallurgical Engineers, 1938.

Eavenson, Howard N. *The First Century and a Quarter of American Coal Industry.* Pittsburgh: Privately printed, 1942.

Edkins, Donald O., comp. *Edkin's Catalogue of Coal Company Scrip.* New Kensington: The Committee of The National Scrip Collectors Association, n.d.

Ellis, Franklin. *History of Fayette County, Pennsylvania, with Biographical Sketches of Many of Its Pioneers and Prominent Men.* Philadelphia: Louis H. Everts and Company, 1882.

Ellsworth Diamond Jubilee: Celebrating 75 Years of Progress, 1900-1975.

Eyre, Edward, Tyron, Hunt, Tyron, Fred G., and Willits, J. H., eds. *What the Coal Commission Found.* Baltimore: Williams and Wilkins Company, 1925.

Fink, Gary M. *Labor Unions.* Westport, Ct.: Greenwood Press, 1977.

Fischer, W., and Bezanson, A. *Wage Rates and the Working Time in the Bituminous Coal Industry, 1920-1932.* Philadephia, 1932.

Fisher, Douglas Allen. *The Epic of Steel.* New York: Harper and Row Publisher, 1963.

Francaviglia, Richard V. *Hard Places: Reading the Landscape of America's Historic Mining Districts.* Iowa City: University of Iowa Press, 1991.

Fritz, Wilbert G., and Veenstra, Theodore A. *Regional Shifts in the Bituminous Coal Industry.* Pittsburgh: University of Pittsburgh Press, 1935.

Gable, John E. *History of Cambria County.* Topeka, Kansas: Historical Publishing Company, 1926.

Gates John K. *The Beehive Coke Years: A Pictorial History of Those Years.* Uniontown: Privately printed, 1990.

Gersha, Charles. *From the Furrows to the Pits.* Van Voorhis: McClain Publishing Company, 1987.

Goodrich, Carter. *The Miner's Freedom: A Study of the Working Life in a Changing Industry.* Boston: Marshall Jones Company, 1925.

Gilfillan, Lauren. *I Went To Pit College.* New York: Viking Press, 1934.

Graebner, William. *Coal-Mining Safety in the Progressive Period.* Lexington: University Press of Kentucky, 1976.

Haley, Karen Anne. *Indiana Township Preliminary History Book.* Indianola: Indiana Township Historical Commission, 1988.

Harvey, George. *Henry Clay Frick: The Man.* New York: Charles Scribner's Sons, 1928.

Hogan, William T. *Economic History of the Iron and Steel Industry in the United States.* Volume 2: Part 3. Lexington, Mass.: D. C. Heath and Company, 1971.

Hughes, Herbert W. *A Text-Book of Coal-Mining for the Use of Colliery Managers and Others.* London: Charles Griffin and Company, 1904.

Hylan, John F., *Statement of Facts and Summary Committee to Investigate the Labor Conditions at the Berwind-White Company's Coal Mines in Somerset and Other Counties, Pennsylvania.* New York, 1922.

Industrial Databook for the Pittsburgh District. Pittsburgh: University of Pittsburgh Press, 1936.

Inzana, Mary Frances. *Bakerton (Elmora), Pennsylvania, 1889-1989.* Indiana: A.G. Halldin Company, Inc., 1989.

Jason, Sonya. *Icon of Spring*. Middletown: Jednota Press, 1988.

Johnson, James P. *The Politics of Soft Coal: The Bituminous Industry from World War One Through the New Deal*. Urbana: University of Illinois Press, 1979.

Kaufman, Jean Troxell. *The Awakening of a Crossroads Village—The Story of Delmont, Pennsylvania (Sesquicentennial Edition, 1833-1983)*. Delmont: Salem Historical Society, 1983.

Keystone Consolidated Publishing Company. *The Coal Catalog Including Directory of Mines*. Pittsburgh: Keystone Consolidated Company, 1914-1987.

Klein, Philip S., and Hogenboom, Ari. *A History of Pennsylvania*. 2nd edition. University Park: Pennsylvania State University Press, 1980.

Knowles, Morris. *Industrial Housing*. New York: McGraw-Hill Book Company, 1920.

Koppers-Beckers Coke Ovens. Pittsburgh: Koppers Company, 1944.

Korcheck, Robert A. *Nemacolin: The Mine—The Community, 1917-1950*. Robert Korcek Copyright, privately printed, 1980.

Korson, George. *Coal Dust on the Fiddle: Songs and Stories of the Bituminous Industry*. Hatboro: Folklore Associates, Inc., 1965.

Kneeland, Frank, ed. *Getting Out the Coal*. New York: McGraw-Hill Publication, 1926.

Lane, John R. *Eliza: Remembering A Pittsburgh Steel Mill*. Charlotteville, Virginia: Howell Press, 1989.

Lewis, John L. *The Miners' Fight for American Standards*. Indianapolis: Bell Publishing Company, 1925.

Long, Priscilla. *Where the Sun Never Shines: A History of America's Bloody Coal Industry*. New York: Paragon House, 1989.

Lytle, Curtis R. *Landrus, Pennsylvania: Pennsylvania Ghost Town and Electric Coal Mine*. Mansfield: The Penny-Saver Printer, 1984.

Majumdar, Shyamal, and Miller, E. Willard, eds. *Pennsylvania Coal: Resources, Technology, and Utilization*. Easton: Pennsylvania Academy of Science, 1983.

Mankin, Guy. *Power Handbook on Fuels*. New York: McGraw-Hill Publication, 1934.

McFarlane, James. *The Coal-Regions of America: Their Topography, Geology, and Development*. New York: D. Appleton and Company, 1873, third edition, 1875.

McDonald, David J., and Lynch, Edward A. *Coal and Unionism: A History of the American Coal Miner's Union*. Silver Spring, Md.: Cornelius Printing Company, 1939.

McLenathan, J. C., et al. *Centennial History of Connellsville, Pennsylvania, 1806-1906*. Columbus, Ohio: Champlin Press, 1906; reprint edition Connellsville Area Historical Society, Inc., 1982).

Miller, Donald, L., and Sharpless, Richard E. *The Kingdom of Coal: Work, Enterprise, and Ethnic Communities*. Philadelphia: University of Pennsylvania Press, 1985.

Miller, E. Willard. *Pennsylvania: Keystone to Progress*. New York: Windsor Publication, 1986.

Miller, George. *A Pennsylvania Album*. University Park: Pennsylvania State University Press, 1986.

Mining Artifact Collector. Redlands, Calif.

Moore, Elwood S. *Coal: Its Properties, Analysis, Classification, Geology, Extraction, Uses, and Distribution*. London: John Wiley and Sons Company, 1940.

Morris, Homer Lawrence. *The Plight of the Bituminous Coal Miner*. Philadelphia: University of Pennsylvania Press, 1934.

Munn, Robert. *The Coal Industry in America*. Morgantown: West Virginia University Library, 1977.

National Coal Association. *Bituminous Coal Facts*, Washington, D.C., 1972.

National Iron and Steel, Coal and Coke Blue Book. 4th Edition. Pittsburgh: R. L. Pold and Company, 1911.

Nicolls, William Jasper. *The Story of American Coals*. Philadelphia: J. B. Lippincott Company, 1897.

Nyden, Paul. *Black Coal Miners in the United States*. New York: The American Institute for Marxist Studies, 1974.

O'Connor, Harvey. *Mellons' Millions: A Biography of a Fortune*. New York: The John Day Company, 1933.

Oliver, John W. *History of American Technology*. New York: The Ronald Press, 1956.

Out of the Dark. Part 1 and 2. Indiana: A.G. Halldin Publishing Company, 1974.

Mine Explosion, 1902, and Its Immigrant Victims. (Johnstown: Johnstown Library, 1977).

Parker, Glen Lawhon. *The Coal Industry: A Study of Social Control*. Washington, D.C.: American Council on Public Affairs, 1940.

Peterson, Florence. *American Labor Unions: What They Are and How They Work*. New York: Harper & Row Publishers, 1945.

Pittsburgh-Buffalo Company. Pittsburgh: Pittsburgh-Buffalo Coal Company, 1911.

Pittsburgh: 50th Anniversary of the Engineers' Society of Western Pennsylvania. Pittsburgh: Cramer Printing Company, 1930.

Pittsburgh and the Pittsburgh Spirit. Pittsburgh: Chamber of Commerce of Pittsburgh, 1927-1928.

Poliniak, Louis. *When Coal Was King.* Lebanon: Applied Arts Publishers, 1970.
Portrait of a Town-Portage, Pennsylvania, 1890-1990. Portage: Portage Area Historical Society, 1990.
Potisek, Lillian, and Singadine, Murchant. *A Bicentennial History of West Bethlehem Township and Marianna Borough: 1776-1976.* Marianna: privately printed, 1976.
Proceedings of the Coal Mining Institute of America. Greensburg: Charles Henry and Company, 1910.
Proceedings of the Coal Mining Institute of America. Crafton: Cramer Printing and Publishing Company, 1920.
Reitell, Charles. *The Shift in Soft Coal Shipments.* Pittsburgh: Pennsylvania Industrial Survey, 1927.
Region in Transition: Report of the Economic Study of the Pittsburgh Region. Pittsburgh: University of Pittsburgh Press, 1963.
Roberts, Ellis. *The Breaker Whistle Blows.* Scranton: Anthracite Museum Press, 1984.
Rochester, Anna. *Labor and Coal.* New York: International Publishers Company, 1931.
Rogers, Henry Darwin. *The Geology of Pennsylvania.* 2 volumes. Philadelphia: J. B. Lippincott and Company, 1858.
Roy, Andrew. *A History of the Coal Miners of the United States.* 1905. Reprint, Westport, Ct: Greenwood Press, 1970.
Saalbach, William F. *United States Bituminous Coal Markets: Trends Since 1920 and Prospects to 1975.* Pittsburgh: Three River Press, 1960.
Salay, David L. ed. *Hard Coal, Hard Times: Ethnicity and Labor in the Anthracite Region.* Scranton: The Anthracite Museum Press, 1984.
Savage, Lon. *Thunder in the Mountains: The West Virginia Mine War, 1920-1921.* Pittsburgh: University of Pittsburgh Press, 1990.
Saward's Annual: A Statistical Review of the Coal Trade. New York: Published annually by Ralph S. Saward, issues 1931, 1945.
Schurr, Sam H. and Netschert, Bruce C. *Energy in the American Economy, 1850-1975: An Economic Study of Its History and Prospects.* Baltimore: Johns Hopkins Press, 1960.
Serrin, William, *Homestead: The Glory and Tragedy of an American Steel Town.* New York: Random House, 1992.
Shedd, Nancy S., and Harshbarger, Jean P., *Second Century—A Huntingdon County Bicentennial Album, 1887-1987.* Huntingdon: Huntingdon County Historical Society, 1987.
Sheppard, Muriel Earley. *Cloud by Day: The Story of Coal and Coke and People.* Pittsburgh: University of Pittsburgh Press, 1991 (reprint).
Shurick, Adam Thomas. *The Coal Industry.* Boston: Little, Brown and Company, 1924.
Sisson, William, and Bomberger, Bruce. *Made in Pennsylvania: An Overview of the Major Historical Industries of the Commonwealth.* Harrisburg: Pennsylvania Historical and Museum Commission, 1991.
Sisson, William, Bomberger, Bruce, and Reed, Diane. *Iron and Steel Resources, 1716-1945.* National Register of Historic Places Multiple Property Documentation Form. Harrisburg: Pennsylvania Historical and Museum Commission, 1991.
Smith, Barbara Ellen. *Digging Our Own Graves: Coal Miners and the Struggle over Black Lung Disease.* Philadelphia: Temple University Press, 1987.
Smith, Helene. *Export: A Patch of Tapestry Out of Coal Country America.* Greensburg: McDonald/Seward Publishing Company, 1986.
Stevens, Sylvester K. *Pennsylvania: Titan of Industry.* New York: Lewis Historical Publishing Company, 1948.
Swank, James M. *Introduction to the History of Ironmaking and Coal Mining in Pennsylvania.* Philadelphia: Published by Author, 1878.
Swank, James M. *History of the Manufacture of Iron in All Ages, and Particularly in the United States from Colonial Times to 1891.* Philadelphia: The American Iron and Steel Association, 1892.
Trachtenberg, Alexander. *The History of Legislation for the Protection of Coal Miners in Pennsylvania.* New York: International Publishers, 1942.
Two Hundred Years of History in New Alexandria, Westmoreland County, Pennsylvania. New Alexandria: Bicentennial Committee, 1976.
The Kernel of Greatness: An Informal Bicentennial History of Bedford County. Bedford: Bedford County Historical Commission, 1971.
Thurston, George H. *Directory of the Monongahela and Youghiogheny Valleys Containing Historical Skeches of the Various Towns Located on Them; with a Statistical Exhibit of the Collieries on the Two Rivers.* Pittsburgh: A. A. Anderson, 1859.
Thurston, George H. *Allegheny County's Hundred Years.* Pittsburgh: A. A. Anderson and Son, 1888.
Thurston, George H. *Pittsburgh's Progress: Industries and Resources.* Pittsburgh: A. A. Anderson and Son, 1886.
Temin, Peter. *Iron and Steel in Nineteenth Century America: An Economic Inquiry.* Cambridge: M.I.T. Press, 1954.

United Mine Workers of America: The First One Hundred Years. Ebensburg: UMWA Publisher, 1991.

Van Kleeck, Mary. *Miners and Management.* New York: Russell Sage Foundation, 1934.

Walkinshaw, Lewis Clark. *Annals of Southwestern Pennsylvania.* New York: Lewis Historical Publishing Company, 1939.

Wallace, Anthony F. C. *St. Clair: A Nineteenth Century Coal Town's Experience With a Disaster Prone Industry.* New York: Alfred A. Knopf, 1987.

Warren, Kenneth. *The American Steel Industry, 1850-1970: A Geographical Interpretation.* Oxford: Clarendon Press, 1973.

Watkins, Harold M. *Coal and Men: An Economic and Social Study of the British and American Coalfields.* London: George Allen & Unwin Ltd., 1934.

Weber, Denise Dusza. *Delano's Domain: A History of Warren Delano's Mining Towns of Vintondale, Wehrum and Claghorn, Volume 1, 1789-1930.* Indiana: A. G. Halldin Publishing Company, Inc., 1991.

Weeks, Joseph D. *Report on the Manufacture of Coke.* New York: David Williams, 1885.

Western Pennsylvanians. Pittsburgh: James O. Jones Company, 1923.

White, Jerome, and Law, Samuel. *The Coal Industry in Cambria County.* N.p.: Cambria County Historical Society, 1954.

Wiley, Richard T. *Monongahela: The River and Its Region.* Butler: The Ziegler Company, 1937.

Williams, Bruce T., and Yates, Michael D. *Upward Struggle: A Bicentennial Tribute to Labor in Cambria and Somerset Counties.* Johnstown: University of Pittsburgh, 1976.

Williams, Bruce T. *Coal Dust in Their Blood: The Work and Lives of Underground Coal Miners.* New York: AMS Press, Inc., 1991.

Wilson, Philip J., and Wells, Joseph H. *Coal, Coke, and Coal Chemicals.* New York: McGraw-Hill Book Company, Inc., 1950.

Wolman, Leo. *Ebb and Flow of Trade Unionism.* New York: National Bureau of Economic Research, 1936.

Workman, Michael E. *The Fairmont Coal Field: Historical Context.* Morgantown, W.Va.: Institute for History of Technology and Industrial Archaelogy, 1992.

Wright, Gwendolyn. *Building the Dream: A Social History of Housing in America.* New York: Pantheon Books, 1981.

Wyer, Samuel S. *The Smithsonian Institution's Study of Natural Resources Applied to Pennsylvania's Resources.* Washington, D.C.: Reprinted May, 1923.

Maps

Atlas of Blair and Huntingdon Counties, Pennsylvania. Philadelphia: A. A. Pomeroy and Company, 1873.

Atlas and Plat Book of Somerset County, Pennsylvania. Rockford, Illinois: Rockford Map Publishers, 1980.

Caldwell, J. A. *Caldwell's Ilustrated Centennial Atlas of Washington County, Pennsylvania.* Conit, Ohio: Record Press, 1876.

Cuff, David, et al. *The Atlas of Pennsylvania.* Philadelphia: Temple University Press, 1991.

General Map of the Bituminous Coal Region of Western Pennsylvania. Harrisburg: Department of Mines, William Stanley State Printer, 1916.

Halberstadt, Baird. *Halberstadt's General Map of the Bituminous Coal Fields of Pennsylvania.* Pottsville: Baird Halbertstadt Publisher, 1901, 1903, 1907.

Hopkins, G. M. *Atlas of the County of Fayette and the State of Pennsylvania.* Philadelphia: G. M. Hopkins, 1872 (reprint 1974).

Map of Blair County, Pennsylvania, from Special Surveys by Geil and Freed. Philadelphia: Geil and Freed, 1859. Reprint by the Blair County Historical Society, Altoona.

Map of Fulton County, Pennsylvania. Philadelphia: A. Pomeroy and Company, 1873.

McConnell, J. L., and Wolfe, G. F. *McConnell's Map of Greene County.* Philadelphia: Tuttle and Company, 1865.

Platt, W. G. *The 1876 County Atlas of Somerset, Pennsylvania.* F.W. Beers and Company; republished by the Somerset County Archaeological Society, 1973.

Rizza, Paul, Hughes, James, and Smith, Allen R. *Pennsylvania Atlas: A Thematic Atlas of the Keystone State.* Berlin, Ct: Atlas Publishing Company, 1975.

Walling, Henry, and Gray, O. W. *Topographical Atlas of the State of Pennsylvania.* Philadelphia: Steadman, Brown and Lyon, 1872.

Dissertations/Master's Theses

Biek, Mildred A. "The Miners of Windber: Class, Ethnicity, and the Labor Movement in a Pennsylvania Coal Town, 1890s-1930s." Ph.D. diss., Northern Illinois University, 1989.

Davis, George Littleton. "Greater Pittsburgh's Commercial and Industrial Development (With Emphasis on the Contributions of Technology), 1850-1900." Ph.D. diss., University of Pittsburgh, 1951.

Enman, John Aubrey. "The Relationship of Coal Mining and Coke Making to the Distribution of Population Agglomoration in the Connellsville (Pennsylvania) Beehive Coke Region." Ph.D. diss., University of Pittsburgh, 1962.

Friscia, August B. "Industrial Retardation and Economic Growth: A Case Study of Secular and Structural Change in the Bituminous Coal Industry of the United States." Ph.D. diss., New York University, 1970.

Gillenwater, Mack H. "Cultural and Historical Geography of Mining Settlements in the Pocahontas Coal Field of Southern West Virginia, 1880-1930." Ph.D. diss., University of Tennessee, 1972.

Johnson, James P. "A New Deal for Soft Coal: The Attempted Revitalization of the Bituminous Coal Industry under the New Deal." Ph.D. diss., Columbia University, 1968.

McCauley, Ray Lemon. "Natural and Cultural Factors That Have Affected Coal Production in Westmoreland County, From 1850-1900." Master's Thesis, University of Pittsburgh, 1950.

Mitchell, Lawrence C. "A Historical Geography of Cambria County, Pennsylvania." Master's Thesis, Michigan State University, 1965.

Smith, Alan McKinley. "The Development of the American Coke Industry to 1875 with Emphasis on the Connellsville Field." Master's Thesis, University of Pittsburgh, 1960.

Sullivan, Charles Kenneth. "Coal Men and Coal Towns: Development of the Smokeless Coalfields of Southern West Virginia, 1873-1923." Ph.D diss., University of Pittsburgh, 1979.

Published Articles

Western Pennsylvania Historical Magazine

Bauman, John. F. "Orwell's Wigan Pier and Daisytown: The Mine Town as a Stranded Landscape." Volume 67 (1984).

Bining, Arthur Cecil. "The Rise of Iron Manufacture in Western Pennsylvania." Volume 16 (1933).

Demarest, David, and Levy, Eugene. "Touring the Coke Region." Volume 74 (1991).

Eavenson, Howard N. "The Early History of the Pittsburgh Coal Bed." Volume 26 (1939).

Garard, Ira D. "Greene County, 1890-1918." Volume 63 (1980).

Rothfus, Robert. "Coal Trains North: The Rochester and Pittsburgh Railroad Company." Volume 54 (1971).

Wardley, C. S. "The Early Development of the H. C. Frick Coke Company." Volume 32 (1949).

Coal Age

"Electricity in Coal Mining." Volume 6 (1914).
"New Methods of Handling Coal Electrically." Volume 13 (1918).
"Storage Batteries for Mine Locomotives." Volume 4 (1913).
"Coal Reserve in Fayette County Contained in Seven Beds." Volume 22 (1922).
"Cost of Mule and Locomotive Haulage." Volume 15 (1919).
"Colliery Dwelling Construction." Volume 1 (1911).
"Mining Methods in the Connellsville Region." Volume 10 (1916).
"Pittsburgh Coal Company Report for 1917." Volume 13 (1918).
"Connellsville and the By-Product Coke Industry in 1915." Volume 9 (1916).
"Coal Mine Ventilating Equipment." Volume 1 (1912).
"Beehive Oven Supremacy Passing—But Not Yet Passed." Volume 15 (1919).
"Connellsville and the By-Product Coke Industry in 1919." Volume 17 (1920).
"Connellsville Coke in 1917." Volume 12 (1918).
"Changes in Beehive Oven Construction Due to Mechanical Operation." Volume 15 (1919).
"Beehive and By-Product Coke." Volume 15 (1919).
"Improvement in Carbide Lamp." Volume 10 (1916).
"Geology and Location of the Coal Fields of Pennsylvania." Volume 6 (1914).
"Obituary for James Jones." Volume 1 (1912).
"The Miner." Volume 13 (1918).

Pennsylvania History

Aurand, Harold W. "Mine Safety and Social Control in the Anthracite Industry." Volume 52 (1985).

Enman, John Aubrey. "Coal Company Store Prices Questioned: A Case Study of the Union Supply Company, 1905-1906." Volume 41 (1974).

Filippelli, Ronald L. "Diary of a Strike: George Medrick and the Coal Strike of 1927 in Western Pennsylvania." Volume 43 (1976).

Grant, Philip A., Jr. "The Pennsylvania Congressional Delegation and the Bituminous Coal Acts of 1935 and 1937." Volume 49 (1982).

Hoffman, John N. "Pennsylvania's Bituminous Coal Industry: An Industry Review." Volume 45 (1978).
Johnson, James P. "Reorganizing the United Mine Workers of America in Pennsylvania During the New Deal." Volume 37 (1970).

Iron Age

"The Coke Industry in the United States." Volume 2 (1902).
"The Connellsville Coke Supply." Volume 28 (1907).
"Connellsville Coke Prices for Thirteen Years." Volume 91 (1913).

Iron Trade Review

"An Innovation in Coke Ovens." Volume 22 (1906).
"Beehive Coke Losing Ground." Volume 66 (1920).
"Beehive Coke Retrogression Is More Marked." Volume 84 (1929).
"By-Product Coke Passes Beehive." Volume 82 (1928).

Engineering and Mining Journal

Fay, C. L. "A Brief Study of the Social Conditions in the Bituminous Coal Region of Pennsylvania." Volume 6 (1907).
Harding, Burcham. "The Largest Collieries in the United States." Volume 69 (1900).
Parsons, F. W. "A Model Coal Mining Town." Volume 3 (1906).
"The Latrobe Coal and Coking Field in Pennsylvania." Volume 8 (1901).

Pennsylvania Heritage

Burbank, Kershaw. "Noble Ambitions: The Founding of the Franklin Institute." Volume 18 (1992).
Cooper, Eileen Mountjoy. "Ernest: Life in a Mining Town." Volume 3 (1977).
Cooper, Eileen Mountjoy. "The Magnificent Strike for Unionism: The Somerset County Strike of 1922." Volume 17 (1991).
Hanney, Joseph M. "Schuylkill County: Built on Coal." Volume 11 (1985).
Hoskins, Donald M. "Celebrating a Century and a Half: The Geologic Survey." Volume 12 (1986).
Kallman, Diane. "Steel on the Susquehanna." Volume 16 (1990).
Parucha, Leonard F. "Bitumen: All Gone With The Wind." Volume 12 (1986).

Miscellaneous Periodicals

Callen, A. C. "Electric Locomotives for Coal Mines." *The Coal Industry* (March 1918).
Ehraber, Tommy. "King Coal." *Pitt Magazine*, vol. 5 (1990).
Gandy, Harry L. "Some Trends in the Bituminous Coal Industry." *American Academy of Political Science*, vol. 147 (1930).
Graebner, William. "Great Expectation: The Search for Order in Bituminous Coal, 1890-1917." *Business History Review*, vol. 48 (1974).
Haldeman, H. L. "The First Furnace Using Coal." *Lancaster County Historical Society*, vol. 1 (1896).
Lauch, W. J. "The Bituminous Coal Mining and Coke Workers of Western Pennsylvania." *The Survey* (1911).
Keighley, Fred C. "The Connellsville Coking Region," *Engineering Magazine*, vol. 20 (1901).
Maclean, Anne Marion. "Life in the Pennsylvania Coal Fields, with Particular Reference to Women." *The American Journal of Sociology*, vol. 14, (1909).
Magnusson, Leifur. "Company Housing in the Bituminous Coalfields." *Monthly Labor Review*, vol. 10 (1920).
Schwieder, Dorothy. "Italian Americans in Iowa's Coal Mining Industry." *Annals of Iowa*, vol. 46 (1982).
Wrbican, Sue. "Portrait of Braeburn." *Pitt Magazine* (spring 1992).
Yates, W. Ross. "Discovery of the Process for Making Anthracite Iron." *Pennsylvania Magazine of History and Biography*, vol. 98 (1974).

Index

Acetylene lamp, 123
Acosta, 164, 166
Adelaide, 70
Adrian, 66
Aladdin Company, 99, 100
Aliquippa, 160
Allegany County, Md., 12, 17, 38
Allegheny City, 21, 33
Allegheny County, 2, 9, 12, 13, 15, 16, 32, 37, 39, 41, 54, 62, 63, 69, 74, 96, 97, 99, 104, 114, 118, 128, 155, 157, 160, 177, 188, 202
Allegheny Portage Railroad, 33, 34
Allegheny River, 17, 22, 33, 67, 99
Allegheny River Mining Co., 64
Allegheny Valley Railroad, 64
Allenport, 92
Allentown, 25, 38
Alliance Iron Furnace, 26
Aluminum, 5
Alverton, 154
Amalgamated Association of Iron, Steel and Tin Workers, 184, 188
Amax Coal Co., 197
Amberson, William, 27, 39
American Coke Co., 78
American Federation of Labor, 132, 133, 138, 184-185, 188
American Institute of Mining Engineering, 31
American Manganese Co., 76
American Miners' Association, 56
Anita, 66; Coal Co., 66
Ankeny, 166
Anschutz, George, 27, 39
Anthracite, 2, 5, 6, 7, 8, 9-11, 13, 14, 24, 25, 27, 31, 33, 38, 39, 47, 55, 56, 57, 61, 62, 68, 74, 89, 105, 114, 116, 122, 126, 136, 137, 155
Anton Brothers, 49. *See also* Sunshine lamps
Appalachian Agreements, 186, 187, 190
Appalachian region, 7, 11, 39, 88, 93, 103, 134, 135, 159, 174, 190
Arbon Coal Co., 36
Ardara, 114
Armstrong County, 2, 12, 22, 27, 38, 39, 41, 63-64, 74, 97, 179, 197
Arnot, 89
Atlantic Crushed Coke Co., 83
Atterbury, Gen. A. A., 175
Atwater, C. M., 76
Aultman, 65
Avella, 160
Avondale, 51

Baer, George F., 64
Baer, William J., 64
Bailey, Berton, 140
Bakerton, 65, 198
Baldwin, William G., 175
Baldwin-Felts Detective Agency, 175
Baltimore and Connellsville Railroad, 64, 164
Baltimore and Ohio Railroad, 34, 35, 37, 41, 63

Banning, 125
Barclay Coal Co., 36
Barnes & Tucker Co., 65, 198
Barnesboro, 65, 74
Barr, 173
Bast, Gideon, 55
Bates, John, 56
Bates Union, 56, 132
Baton & Eliot Co., 99
Beal, John, 26
Bear Creek Furnace, 27
Beatty, Charles, 15
Beaver County, 2, 12, 41, 74, 154
Bedford County, 1, 2, 12, 13, 16, 17, 34, 74, 126
Beechtree, 65
Beehive coke, 28, 61, 69; ovens, 28, 41, 42, 43-46, 64, 69, 70, 73, 74, 75, 76, 78, 79, 80, 81, 111, 113, 153, 154
Beelen, Francis, 27, 39
Belgium oven, 75
Bell, 164, 166
Bellefont and Snow Shoe Railroad, 16
Bennett, James I., 39
Bennington coke plant, 74
Bentleyville, 83, 97
Berks County, 23, 24
Bernice coalfield, 9
Berwind, Charles F., 164
Berwind, E. J., 105, 164
Berwind, Henry, 164
Berwind, John, 164
Berwind-White Mining Co., 64, 65, 96, 102, 105, 131, 157, 161, 163-165, 166, 167-168, 199
Bessemer, Charles H., 66
Bessemer process, 46, 66-68
Bethlehem Iron Co., 38
Bethlehem Mines Corp., 131, 157, 175, 198
Bethlehem Steel Co., 67, 78, 98, 157, 171, 172, 173, 174, 187, 188
Biddle, Nicholas, 24, 25
Biddle, Owen, 14
Birmingham Borough, 18
Bituman, 166
Bituminous Coal Code, 183
Bituminous Coal Conservation Act, 183-184, 202
Black blasting powder, 53
Black lignite. *See* Subbituminous coal
Blacklick coalfields, 65, 163
Blair County, 1, 2, 12, 27, 34, 74, 179
Blair Iron and Coke Co., 74
Blairsville, 65
Blossburg, 16, 36, 140; Coal Co., 36, 37, 89
Blossburg Field, 36, 56
Blossburg seam, 89
Blough, 166
Blue Diamond Co., 200
Bobtown, 98, 188
Boswell, 65
Bouquet, Col. Henry, 14
Boyd, Samuel, 16

Boyd, William, 16
Braddock, 46, 67, 79
Braddock, Gen. Edward, 67
Bradenville, 83
Bradford County, 2, 12, 16, 36, 37, 62, 128
Bradys Bend, 38
Braeburn, 191
Brewster, Thomas, 140
Broad Ford, 46
Broad Top coalfield, 2, 6, 9, 12, 13, 17, 34, 36-37, 63, 74, 89, 158, 163
Broad Top Mountain, 16, 17, 28, 34, 36, 63
Brookings Institution, 150
Brookville seam, 12, 73
Brophy, John, 120, 163, 167, 168, 177, 184
Brown, Walston H., 65
Brown, William H., 18
Brownsville, 14, 16, 18, 21, 27, 28, 32, 83, 92, 176
Buckeye Coal Co., 96, 188
Buffalo and Susquehanna Coal Co., 64
Buffalo, Rochester and Pittsburgh Railroad, 64
Burd, James, 14
Bushnell, Capt. David, 18, 33
Butler County, 2, 6, 9, 12, 13, 41, 74
By-product coke ovens, 75, 76-80, 153, 154

Cairnbrook, 66-67
Calumet, 70
Cambria County, 1, 2, 6, 9, 12, 13, 63, 65, 74, 123, 155, 157, 163, 165, 167, 179, 182, 187, 197, 198, 199
Cambria (district), 170
Cambria Iron Works, 38, 66, 67, 74, 77
Cambria Steel Co., 67, 76, 96
Cameron County, 2, 12, 37, 62, 179
Campbell, James, 43
Cannel coal, 63-64
Canonsburg, 16
Cape Breton Island, 13
"Captive" mines, 80, 96, 102
Carbide lamp, 50, 122-123. *See also* Acetylene lamp
Carbon, 5, 8, 10, 13, 23, 41, 45, 68, 69
Carbon County, 9, 58
Carbon dioxide, 5, 50
Carbon monoxide, 5, 50
Carbondale, 58
Carboniferous Age, 4, 10, 12
Carey, Henry C., 27
Carlisle, 11
Carnegie, Andrew, 46, 67, 70, 73, 114
Carnegie Coal Co., 114
Carnegie Company, 67, 73, 103, 188
Carnegie, Thomas M., 70
Carr, Abner, 16
Carrolltown, 163
Carter Coal Co, 184
Caruther, William F., 35
Cascade Coal and Coke Co., 83

Cassandra, 65
Casselman River, 64
Catasauqua, 25
Catholic Church, 57-58
Center Twp., 65
Central Competitive Field, 135, 140, 158, 159, 162, 169, 170, 171, 173, 175, 176, 186
Central Railroad of New Jersey, 157
Centre County, 2, 9, 12, 16, 34, 163
Charcoal, 22, 23, 24, 25, 26, 27, 28, 38, 39, 42, 53, 68, 69, 74
Charleroi, 193; Iron Works, 193
Chartist movement, 56
Childs, Adelaide Howard, 73
Cincinnati, 17, 18, 33, 43, 55
Cincinnati Mine (Finleyville), 123
Clairton, 78, 79; Coke Works, 79; Steel Co., 78, 79
Clarion County, 2, 6, 12, 74
Clarion seam, 12
Clarksville, 199
Clay Furnace, 47
Clearfield, 170; Bituminous Coal Corp., 65, 173, 175; Coke and Iron Co., 27
Clearfield County, 9, 12, 13, 16, 27, 37, 62, 65, 74, 83, 155, 157, 163, 164, 179
Clemons, David, 16
Cleveland Agreement, 167, 169
Cleveland Rolling Mill, 81
Clinton County, 2, 12, 13, 27
Clinton Furnace, 39
Clinton, Gov. DeWitt, 33
Clinton Iron Works, 39
Clyde Iron Works, 25
Clyde mines, 188
Clymer, 65, 182
Coal and Iron Police, 128, 163, 166, 172, 175, 178
Coal beds, 5, 9, 12, 13, 18, 21, 49
Coal Center, 92
Coal Hill Mine, 15, 16, 19, 22
Coal loaders, mechanical, 192-193, 198-200
Coal Mine Health and Safety Act (1969), 121
Coal Mining Act (1952), 127
Coal scrip, 56, 129
Coal washing, 114-115
Coalification, 5
Cochran, James, 43; and Co., 70
Cochran, Mordecai, 43
Cochran, Sample, 43, 70
Code of Fair Competition, 183
Cokeburg, 97
Colebrookdale, England, 23
Coleman and Weaver Co., 65
Colonial Iron Co., 74
Columbia, 16, 33, 38
Columbia County, 9, 58
Colver, 65
Commodore, 65, 174
Conemaugh, 166
Conemaugh River, 22, 33
Congress of Industrial Organizations (CIO), 163, 184-185
Connell Run, 28

Connellsville, 16, 21, 28, 41, 43, 46, 54, 76, 154, 170
Connellsville and West Newton Navigation Co., 32
Connellsville Central Coke Co., 83
Connellsville Coke and Iron Co., 111
Connellsville Coke District, 12, 32, 39-41, 43, 46, 57, 61, 69-70, 73, 75, 78, 79, 80, 81, 83, 88, 113, 127, 129, 153, 155, 158, 163, 182, 185
Consolidation Coal Co., 64, 96, 101, 102, 156, 164, 171, 172, 174, 176, 182, 197. *See also* Pittsburgh Consolidation Coal Co.
Continental Coke Co., 78
Continuous mining, 198-199
Corning & Blossburg Railroad, 36
Covington Machine Co., 84-85
Covode, John, 35
Cowanashannock Coal and Coke Co., 64
Craig, Col. Isaac, 22
Cramer, Zadok, 19, 21
Crane, George, 25
Cresson, 74, 163
Crucible, 98
Crucible Fuel Co., 92
Crucible Steel Co., 78, 155, 188, 189
Cumberland and Elk Lick Co., 64
Curtisville, 99

Daisytown, 92, 188
Danville, 38
Darby, Abraham, 23
Darr Mine, 123, 125
Dauphin County, 9
Davis, Samuel, 14
Davy lamp, 50
Davy, Sir Humphrey, 50
Delano, Warren, 182
Delaware and Hudson Railroad, 157
Delaware, Lackawanna and Western Railroad, 51, 157
Delaware River, 11, 23
Denbo, 92
Denys, Nicolas, 13
Devine, Edward T., 101
Dillingham Commission, 88
Dix, Keith, 47
Drake, Edwin Laurentine, 151
Dravo, Michael, 54
Douglas, William O., 184
Dubinsky, David, 184
DuBois, 66
Dufield, Reverend, 15
Dunbar Furnace Co., 28, 76
Dunlo, 65
Duquesne steel works, 79
Durham, England, 28

East Broad Top Railroad, 37
Eastern Middle Field, 9
Eavenson, Howard N., 15, 31
Ebervale, 202
Eckley, 202
Edgar Thomson Steel Works, 46, 67, 79
Edison battery lamp, 123
Eleanor, 66
Elk County, 2, 12, 62, 74

Ellsworth, 97
Ellsworth, James W., 97
Enzian, Charles, 161
Erie and Beaver Canal, 47
Erie and Pittsburgh Railroad, 47
Erie Canal, 33
Erie Railroad, 36, 157
Ernest, 65
Eureka Mines Nos. 30-42, 165, 168
Evans, Eastwick, 19
Evans, Lewis, 14
Evans, Oliver, 11, 21
Everson, 38

Fairchance Furnace, 24, 28
Fairmont Coal Co., 164
Fairmont field, 158, 170
Fairmont, W.Va., 83, 125
Fairmount Pennsylvania Nail & Wire Works, 11
Fall Creek Coal Co., 37
Fallbrook, 89; Coal Co., 37, 89
Farrandsville, 27
Fawcett & Brothers, 18
Fayette County, 1, 2, 9, 12, 13, 14, 16, 22, 24, 26, 27, 28, 32, 37, 39, 41, 46, 62, 63, 69, 70, 73, 81, 83, 85, 87, 96, 104, 111, 114, 131, 153, 154, 157, 166, 183, 189, 202
Fear, Thomas G., 99
Fell, Judge Jesse, 11
Felts, Thomas L., 99
Findley, J. B., 103
Firmstone, William, 27, 28
First Geological Survey of Pennsylvania, 17, 24
Fisher, Gov. John S., 175
Food and Fuel Control Act, 138
Ford Collieries Co., 99
Fort Crevecoeur, 14
Fort Pitt, 14, 15, 16, 22
Fort Redstone, 14
Foster, William Z., 139
Franklin, 76, 77
Franklin Institute, Philadelphia, 24
Fredericktown, 92, 195
French Creek, 17, 32
Frick, Elizabeth, 46
Frick, Henry Clay, 46, 70, 103, 104; and Company, 70, 73, 81; Coke Co., 43, 73, 78, 81, 83, 85, 97, 102, 111, 113, 126, 129, 131, 153, 154, 157, 175, 188, 189
Frick, John W., 46
Frostburg, Md., 28
Frostburg Mining Co., 164
Frozen Run, 27
Fuel Administration, 138
Fulton County, 1, 2, 12, 16, 34, 74
Fulton, John, 74
Fulton, Robert, 33

Gallatin, Albert, 22
Garfield, Dr. H. A., 138, 139
Gary, Judge Elbert, 188
Geisenhainer, Dr. Frederick W., 25
Georges Creek Coal Co., 28
Georges Creek Field, 2, 8, 12, 17, 34, 164, 170

Georges Creek, Md., 22, 164
Georges Twp., 83
German Twp., 83
Gist, Christopher, 14
Gill, John, 18
Glasgow Gas Works, 25
Glasgow, W. H., 188
Glass, Frederick, 11
Glen Campbell, 65
Glenwood Coal & Coke Co., 65
Goodman Manufacturing Company, 110
Goodtown, 64
Gompers, Samuel, 132, 133, 138, 162, 184
Gore, Obodiah, 11
Gowen, Franklin B., 58
Graceton, 65
Graff, Bennett & Co., 39
Graff, John, 39
Gray, 164, 166
Great Allegheny Coalfield, 34
Great Depression, 151, 154, 178-186
Great Steel Strike of 1919, 140
Great Western Iron Co., 38
Green, Ralph, 56
Green, William, 140, 184
Greenback-Labor Party, 58
Greene County, 2, 12, 27, 63, 84, 92, 96, 97, 98, 118, 126, 153, 155, 157, 166, 188, 199
Greene Furnace, 27
Greensburg, 83, 158
Greenwich, 163
Greenwood, Miles, 43
Grindstone, 97, 189
Guffey, Joseph F., 183
Guffey-Snyder Act. *See* Bituminous Coal Conservation Act
Guibal fan, 51

Hammond, John Hays, 169
Harding, Warren G., 167; Administration, 93, 167; Commission, 93-94, 168-169
Harleigh, 202
Harris, Mary, 131, 138
Harrisburg, 35
Harrison undercutting machine, 115-116
Hastings, 74
Haupt, Herman, 35
Haywood, William D. (Big Bill), 138, 139
Hazard, Erskine, 11, 25
Hazard's Register of Pennsylvania, 24
Hazleton, 136
Hecla, 113
Heisley Coal Co., 65
Helvetia Coal Mining Co., 64
Hennepin, Louis, 13, 14
Herriman, Frank E., 173
Herron & Peterson, 18
Hickman Run, 43
High-volatile coal, 5, 6, 8, 12, 13
Hillman Coal and Coke Co., 157
Himrod and Vincent Co., 47
Hirshfield, David, 167
Hocking Valley (Ohio), 46, 56, 116, 128
Holley, Alexander, 67
Hollidaysburg, 33, 34
Hollsopple, 65, 166
Holmes, Joseph A., 126

Homer City, 154
Homestead steel works, 79
Hoover, Herbert, 169, 174, 178, 181, 185
Hooversville, 65
Hopewell Village Iron Plantation, 24
Horatio, 66
Horton, Nathan Port, 17
Houtzdale, 164
Hulme, Thomas, 19
Humphries, E. A., 83
Huntingdon and Broadtop Railroad and Coal Co., 36, 37
Huntingdon County, 1, 2, 12, 16, 17, 28, 34, 37, 74
Hurst, Fannie, 179
Husband, 166
Hutchins, Capt. Thomas, 15
Hutchinson, A. A. and Brothers, 70, 111
Hydrogen, 5, 6, 37
Hylan, John F., 167

Imperial, 160
Indiana, Pa., 170
Indiana County, 1, 2, 9, 12, 14, 22, 63, 65, 154, 155, 157, 166, 182, 187, 197
Indiana Twp., 99
Indianola, 96, 99, 188; Coal Co., 99
Industrial Workers of the World, 138, 139
Inland Collieries Corp., 96, 99, 173, 187
Inland Steel Corp., 96, 99, 155
Interborough Rapid Transit Co., 165, 167
Irwin Coal Basin, 35
Irwin Gas Basin, 36, 158
Irwin, 35, 114
Irwin, William A., 188
Iselin, 65
Islin, Adrian, 65

Jacksonville Agreement, 169-171, 173, 174
Jacobs Creek, 26, 123, 125
Jamison Coal & Coke Co., 83, 129, 131
Japan, Pa., 202
Jeddo, 202
Jefferson Coal Co., 66
Jefferson County, 2, 12, 13, 63, 65-66, 74
Jeffrey Manufacturing Co., 110
Jenkins, John Sr., 11
Jenners, 65, 96, 164
Jerome, 64, 65, 166
Jevons, William Stanley, 61
John Bates' Union, 56
Johnson, Gen. Hugh S., 181
Johnston, William, 22
Johnstown, 33, 34, 38, 66-67, 76, 96, 122, 123
Jones & Laughlin Co., 38, 83, 92, 154, 157, 160, 188, 189, 199
Jones, John H., 97
Joy, Joseph Francis, 193
Joy Manufacturing Co., 110, 171-172, 193-194, 198
Juniata River, 17, 33, 34

Kanawaha Field, 85, 104, 158, 170, 174
Karthaus Furnace, 27
Karthaus, 16
Keeny, Frank, 140

Kelly, William, 66-67
Kemble Coal & Iron Co., 74
Keystone, 64
Keystone Coal & Coke Co., 83, 102, 129, 182
Keystone Coal and Manufacturing Co., 64
Kiel Run, 166
Kiskiminetas River, 14, 22, 33, 99
Kittanning seam, 12, 99
Klondike Coke District, 12, 78, 81, 83
Knights of Labor, 57, 58, 131, 132, 133, 138
Koppers Co., 77
Koppers, Heinrich, 77
Ku Klux Klan, 189

Labor's Non-Partisan League, 185
Lackawanna County, 9
Lackawanna Steel Co., 97
Lackawanna Valley, 9, 26
Lancaster County, 16
Landon, Alf, 185
Larimer Coke Plant, 114
Larimer, Gen. William, 35
La Salle, Robert de, 13
Latrobe, 38, 39, 158, 170
Lauch, W. J., 91, 168
Lawrence County, 2, 6, 12, 13, 46, 47
Lebanon, 25
Lebanon County, 9
Lechner undercutting machine, 116
Ledlie, George, 18
Lehigh and New England Railroad, 157
Lehigh Canal, 11
Lehigh Coal and Navigation Co., 11, 33
Lehigh Crane Iron Co., 25
Lehigh River, 11; Valley, 11, 25, 26
Lehigh Valley Railroad, 157
Leisenring, 70
Leisenring, John, mines, 111
Leith, 70
Lesher, C. E., 157
Lesley, J. Peter Jr., 17, 48, 63
Lewis, John L., 139, 140-141, 151-152, 162, 163, 167, 170, 171, 173, 176-177, 180-181, 182, 183, 184-185, 189-190, 194-196
Lewis Manufacturing Co., 99
Lick Run, 36
Licking Creek, 14
Lignite, 5, 6, 7, 8
Ligonier, 27, 38, 158
Ligonier Valley Railroad, 38
Limpus, Lowell, 176
Listie, 64
Little Moshannon Creek, 16
Lloyd, Henry Demarest, 102
Lloyd, Thomas, 56
Lock Haven and Tyrone Railroad, 16
Lonaconing Furnace, 28
Long-wall Mining, 199-200
Love, George Hutchinson, 197
Low-volatile coal, 5, 6, 8, 13, 47
Lower Connellsville Coke District. *See* Klondike Coke District
Lower Kittanning seam, 12, 65, 73, 165
Lower Freeport seam, 12, 65, 73, 83
Lucerne Mines, 65, 154

Lucy Furnace, 25, 67
Luzerne County, 9, 51, 57
Lycoming County, 2, 12, 16, 27, 36, 37
Lycoming Creek, 36
Lyman, William, 25

Macdonaldton, 166
Magee, James, 35
Mahoning Creek, 22
Mahoning Iron Works, 47
Mahoning Valley, 47, 57, 67
Main Bituminous Field, 2
Mallory, William, 36
Mammoth Mine, 123
Manatawny Creek, 23
Maple Creek Mine, 118
Marianna, 97, 98, 123; Mine, 123
Martin, Pierre, 67
Mary Ann Furnace, 28
Maryland No. 1 and 2 Shaft Mine, 165, 168, 199
Mather, 96, 97, 98, 99, 188; Collieries, 99
Mauch Chunk, 9, 11, 58; Railroad, 33
Maxwell, 189
McBride, John, 132, 135, 141
McBryde, Patrick, 133
McClure Coke Co., 70
McCormick, Provance, 43
McKean County, 2, 12, 37
McKeesport, 28, 41, 54
McIntyre, 65
Mcintyre Coal Co., 37
McParland, James, 58
Meason, Col. Isaac, 27, 28
Medium-volatile coal, 5, 6, 8, 13
Mellon, Andrew W., 104, 171, 174, 175
Mellon, Judge Thomas, 46
Mellon, Richard, 171, 175
Menallen Coke Co., 154
Menallen Twp., 83
Menzies, Michael, 115
Mercer County, 2, 12, 13, 46, 47, 62
Mercer seam, 12
Methane, 5, 50, 106, 122
Meyersdale, 64, 158, 170
Middle Kittanning seam, 12
Midvale Steel and Ordinance Co., 67
Miller, William L., 18
Milnor, Col. Robert W., 32
Miners' and Laborers' Benevolent Association, 57, 139
Miners' Independent Brotherhood, 189
Miners' National Association, 57
Miners' National Progressive Union, 132
Mingo Coal Works, 55
Mining machines, 84-85, 115-119, 190-196, 198-202
Mississippi River, 14, 32, 33, 35
Mitchell, John, 125, 132, 133, 134, 135-136, 141, 170, 182
Mitchell, T. J., 81
Molly Maguires, 58, 166
Monarch, 111
Monongah Mine Disaster, 125
Monongahela City, 39, 49, 97, 103
Monongahela Navigation Co., 32
Monongahela River, 2, 14, 15, 16, 17, 18, 19, 22, 26, 28, 32, 33, 34, 37, 41, 46, 63, 67, 78, 79, 80, 92, 104, 118, 155

Monongahela River Consolidated Coal & Coke Co., 103, 104
Monongahela Valley, 17, 18, 21, 32, 33, 34, 35, 49, 55, 67, 103, 196
Montour Mine 4,
Montour Rolling Mill, 38
Morgan, J. Pierpont, 102, 103
Morrel, 70
Morrell, David, 66
Morris Run, 89; Coal Mining Co., 37, 89
Morrow, James D. A., 171
Mother Jones. *See* Harris, Mary
Mount Pleasant, 38, 41, 70, 111, 123
Mount Savage Iron Co., 164
Mount Savage, Md., 64
Mount Savage Rolling Mill (Maryland), 38
Mount Union, 37
Mount Washington, 39. *See also* Coal Hill Mine
Muddy Creek, 155
Mulrooney, Margaret, 95
Murray, Philip, 182, 185
Muse, 188

Nanticoke Coal Co., 51
Nanty Glo, 65, 163
National Defense Mediation Board, 190
National Federation of Miners and Mine Laborers. *See* Miners' National Progressive Union
National Industrial Recovery Act, 178, 181, 182, 183, 186
National Miners' Union, 177-178
National Recovery Administration, 181, 183, 188, 190
The Navigator, 19
Neilson, James B., 25
Nemacolin, 96, 98, 188
New Deal, 158, 168, 169, 180, 181, 183
New Eagle, 118
New Geneva, 22, 32
New Orleans, 33
New River and Pocahontas Consolidated Coal and Coke Co., 104, 156, 165
New River Field, 8, 17, 85, 104, 158, 174
New Salem, 83
New York, Ontario and Western Railroad, 157
Nichols, Mr., 28
Nicholson Twp., 83
Nitrogen, 5, 6, 50
Norris, Sen. George, 177
Norris-LaGuardia Act, 177-178
Norristown, 38
North Central Fields, 2, 12, 16, 36, 37, 63, 89
Northern Field, 9
Northumberland County, 9, 58
Norton, Lester Leroy, 28
NuMine, Pa., 64

Oak Grove Mine, 35
Oakdale, 202
Ocean Steam Co., 164
O'Hara, Gen. James, 22
Ohio River, 14, 17, 32, 33, 34, 35, 43, 55
Old Ben Coal Co., 156

Oldtown, 16
Oliphant, F. H., 24, 28
Oliver, Henry W., 104
Osceola Mills, 86
Otto-Hoffman (manufacturer), 76, 77
Overholt, Abraham, and Co., 46
Oxygen, 5, 6, 28, 50, 68

Paisley Coal Co., 173
Patch towns, 88-101, 119
Patterson, Burd, 27
Pattin, John, map, 14
Patton, 65
Paul, 70
Peabody Coal Co., 156
Peabody Holding Co., 197
Peat, 4, 5, 8
Penn family, 15, 16
Pennsylvania Bituminous Coal Open Mining Conservation Act, 202
Pennsylvania Bureau of Industrial Statistics, 121, 128
Pennsylvania Canal, 33, 34, 35
Pennsylvania Coal & Coke Co., 63, 65, 74, 102
Pennsylvania Department of Environmental Resources, 114, 154
Pennsylvania Department of Mines, 87, 106, 123, 154, 202
Pennsylvania Gas Coal Co., 101, 114
Philadelphia, 11, 14, 17, 21, 25, 33, 34, 35, 36, 64, 102, 111, 165
Philadelphia and Reading Coal and Iron Co., 58
Phoenixville, 38
Phosphorus, 28, 80
Picklands, Mather and Co., 96, 98, 155
Pinchot, Gov. Gifford, 179
Pinkerton Detective Agency, 58
Pioneer Furnace, 25
Pittsburgh, 12, 13, 14, 15, 16, 17, 18, 19, 21, 22, 31, 32, 33, 34, 35, 38, 39, 46, 64, 67, 77, 81, 99, 104, 107, 113, 122
Pittsburgh and Lake Erie Railroad, 38, 39
Pittsburgh Coal Co., 102, 104, 125, 131, 135, 156-157, 159, 160
Pittsburgh Consolidation Coal Co., 197
Pittsburgh district, 6, 11, 13, 37, 56, 58, 61, 63, 70, 78, 79, 80, 88, 104, 158, 160, 174, 175
Pittsburgh seam, 6, 12, 13, 16, 17, 18, 28, 39, 41, 45, 63, 65, 73, 76, 83, 84, 98, 111, 118, 171-172, 173, 174, 175, 197, 202
Pittsburgh Terminal Railroad Company, 131, 157, 172, 175
Pittsburgh, Westmoreland and Somerset Railroad, 38
Pittsburgh-Buffalo Co., 97
Pittsburgh-Westmoreland Co., 83
Plummer Mine, 28
Plymouth, 51
Pocahontas Field, 8, 17, 85, 119
Pottstown, 23, 38, 58
Pottsville, 25, 58
Powderly, Terence V., 57, 58, 131, 132
President's Commission on Coal, 197

Pretoria, 166
Progressive Miners of America, 177

Quemahoning Coal Co., 64

Rachel and Agnes Mines, 98
Rae, John B., 133, 141
Rainey, W. J., 81; Coal Co., 70, 81, 102, 131, 156-157
Ralphton, 64, 166
Rea, Henry R., 104
Reading, 38
Reading Railroad, 35, 157
Rectangular coke oven, 81-85
Redi-Cut Homes, 100
Redstone, 14
Redstone Creek, 27
Redstone seam, 12, 73
Renton Mine, 118
Republic Steel, 188
Revloc, 65, 166
Richeyville, 92, 188
Riddle, Samuel, 17
Riddlesburg, 74
Rist, Joseph, 46
Ritner, Gov. Joseph, 24
Ritter, Peter, 27
River Coal Co., 83
Robbins, F. L., 135
Roberts & Schaeffer Co., 161
Robertsdale, 89
Robinson, Moncure, 33
Roche, Josephine, 176
Rochester and Pittsburgh Coal and Iron Co., 65, 66, 102
Rochester & Pittsburgh Coal Co., 63, 65, 131, 154, 157
Rockefeller, John D., 102, 104, 175, 182
Rockefeller, John D. Jr., 164, 174, 182
Rockhill Furnace, 74
Rockhill Iron & Coal Co., 74, 89
Rockwood, 64
Rodgers, Henry D., 24
Rolling Mill Mine, 77, 122, 123
Room and pillar system, 199
Roosevelt, Franklin, 169, 179, 181, 185, 189, 190
Roosevelt, James, 182
Roosevelt, Nicholas, 33
Roosevelt, Theodore, 136
Roscoe, 92
Rossiter, 65, 173
Rostraver Twp., 38, 125
Ruff Creek, 155
Rust, H. B., 77

St. Bernard Coal Co, 115
St. Clair, 57
St. Clair Steel Co., 78
St. Michael, 165
Saleeby, Dr. C. W., 77
Salisbury and Baltimore Railroad, 64
Saltsburg, 14
Sample Run, 65
Say, John, 27
Schuylkill County, 9, 23, 25, 56, 57, 58, 132
Schuylkill Field, 9, 11, 58

Schuylkill River, 11, 23; Falls, 11; Valley, 26
Schwab, Charles M., 174, 188
Schwaider, Dorothy, 91
Scott, John, 35
Scranton, 38, 58, 87, 136
Second Geological Survey of Pennsylvania, 48, 63, 74
Semet-Solvay Co., 76, 77
Semianthracite, 5, 6, 7, 8, 10
Semibituminous, 7, 8, 13, 16, 34, 36, 37, 63, 74, 89, 163
Sewickley seam, 12, 73
Shadyside Furnace, 27, 39
Shannopin Mine, 118
Sharon seam, 12
Shawmut Mining Co., 102
Shaws Mine, 64
Shenango and Allegheny Railroad, 47
Shenango Valley, 47, 57, 67
Shephard, Muriel Earley, 127
Sheridan, Peter B., 136
Sherman Anti-Trust Act, 104
Shreeves Run, 17
Siemens-Martin open-hearth process, 66
Siney, John, 57
Slack coal, 114
Slickville, 96, 188
Smallwood, 92
Snow Shoe, 16, 37
Snyder, John Buell, 183
Somerset, 64, 158, 170
Somerset and Mineral Point Railroad, 64
Somerset Coal Co., 164
Somerset County, 1, 2, 9, 12, 38, 62, 63, 64-65, 74, 96, 126, 155, 157, 158, 163, 165, 166, 167, 168, 179, 182
South Fork, 65, 163
South West Connellsville Coke Co., 78
South Union Twp., 83
Southwestern Railroad, 41
Spangler, 65
Standard Oil Co., 102, 103, 104
Standard Shaft Mines, 111, 113
Star Junction, 96
Stauffer, Joseph R. & Co., 70
Steelton, 38, 67, 78
Stephen, Capt. Adam, 14
Stephens, Uriah Smith, 57
Stiles, Henry A., 64
Subbituminous Coal, 5, 6, 7
Sukes Run, Pa., 33
Sulfur, 5, 6, 8, 10, 23, 28, 41, 53, 69, 80
Sullivan County, 9, 10
Summit Twp., 64
Sunshine lamp, 50, 122
Superanthracite, 6
Surface mining, 201-202
Susquehanna County, 9
Susquehanna River, 11, 16, 23, 25, 26, 27, 33
Sykesville Mine, 83

Taft, Philip, 185
Taylor, John, 28, 43
Taylor, Richard Cowling, 13, 16
Terry, Parshall, 11
Thick Freeport seam, 12, 99

Thomas, David, 25
Thurston, George, 55
Tinstman, A. O., 46
Tioga, 16
Tioga County, 2, 12, 13, 16, 36, 56, 62, 74, 89, 128, 139
Tioga Navigation Co., 36
Tioga River, 36
Titusville, 151
Towanda Mountain, 16
Towanda, 36; Coal Co., 37
Trotter, 70
Trough Creek Valley, 28
Tuit, T. J., 84
Turner, William Sr., 43; Jr., 43, 133
Tuscarawas Valley (Ohio), 57
Twin Creek, 166
Two Mile Creek, 27, 39

Union Collieries Co., 197
Union Furnace, 28
Union Supply Co., 131
United Coal and Coke Co., 78
United Mine Workers of America, 61, 120, 121, 122, 123, 125, 126, 132-141, 150-151, 162-190
United Otto (manufacturer), 77
United States Coal and Coke Co., 157
United States Steel Corp., 77, 78, 102, 104, 118, 126, 139, 155, 157, 188, 189
Upper Freeport seam, 12, 65, 73, 76
Upper Kittanning seam, 12, 65, 73, 165
Upper Middletown, 27
Upper Potomac Field, 8, 170
U.S. Bituminous Commission, 150
U.S. Bureau of Mines, 8, 126-127, 162, 194
U.S. Coal Commission. *See* Harding Commission
U.S. Geological Report, 125
U.S. Geological Survey, 194
U.S. Immigration Commission, 168
U.S. Mine Safety and Health Administration, 127

Van Meter, 125
Vanderbilt, 87
Venango County, 2, 12, 37
Vesta Coal Co., 92, 157, 188
Vestaburg, 92, 188, 199
Vintondale, 65, 199
Volatile matter, 5, 6, 8, 10, 41, 63

Walston, 66
Walter Forward, 33
War Labor Board, 138
War of 1812, 11
Ward, Maj. Edward, 15
Warden, William G., 172
Warren County, 37
Washington Coal & Coke Co., 96
Washington County, 2, 12, 16, 32, 37, 41, 49, 55, 62, 63, 74, 83, 92, 97, 114, 118, 123, 153, 155, 157, 160, 177, 179, 187, 188, 193, 195, 199, 202
Washington, Pa., 21
Washington seam, 12
Watchom, Robert, 133

Watson, James, 18
Watson, Robert, 18
Wayne County, 9
Waynesburg seam, 12
Wealth Against Commonwealth, 102
Weaver, Daniel, 56
Wechsler, James A., 182
The Weekly Miner, 56
Weeks, Joseph, 39
Wehrun, Henry, 183
Weirton Steel Co., 189
Weld, Henry T., 64
West Bethlehem Twp., 97
West Leisenring, 111
West Overton, 46
Western Middle Field, 9
Western Pennsylvania Railroad, 65
Westinghouse Co., 110
Westmoreland Coal Co., 35, 101, 102, 114, 157
Westmoreland County, 1, 2, 9, 12, 13, 14, 16, 27, 32, 35, 36, 37, 38, 39, 41, 46, 62, 63, 69, 70, 73, 81, 83, 85, 87, 96, 101, 104, 111, 113, 114, 123, 125, 131, 153, 154, 155, 157, 158, 165, 166, 179, 182, 187, 191, 202

Westmoreland (district), 170
Westmoreland Furnace, 27
Westmoreland Mining Co., 131
Wharton, Thomas Jr., 14
White, John, 27
White, John P., 141, 162
White, Josiah, 11, 25
White, Judge Allison, 164
Wickhaven, 125
Wilkinson, 47
Williamsport, 16
Williamsport and Elmira Railroad, 36
Willkie, Wendell, 185
Wilmore, PA, 165
Wilpen, 38
Wilputte (manufacturer), 77
Wilson Creek, 64
Wilson, William B., 139, 140
Wilson, Woodrow, 104, 137, 138, 139
Windber, 65, 96, 161, 163, 164, 165, 166, 167
Wolf and Koehler, 51
Wolf, Charles, 107
Wolf Creek Colliery, 55

Wolman, Leo,
Wood, Morrell and Co., 66
Woodruff, Cornelius, 16
Wynn, 70
Wyoming County, 10
Wyoming Valley, 9, 11, 26

Yatesboro, 64
Yellow-Dog contract, 57, 173, 175
Youghiogheny & Ohio Coal Co., 173
Youghiogheny (district), 170
Youghiogheny Navigation Co., 32
Youghiogheny River, 17, 18, 27, 28, 32, 37, 38, 41, 43, 46
The Young Millwright and Miller's Guide, 21
Youngstown Sheet and Tube Co., 96, 187-188

Zimmerman, 64
Zimmermann, Daniel B., 64
Zirconium, 5

```
622.33
DiC
DiCiccio, Carmen
COAL & COKE IN PENNSYLVANIA
37214000258685
```

MAR 2 4 2008

BLUE RIDGE SUMMIT FREE LIBRARY
BLUE RIDGE SUMMIT, PA. 17214